Intonation

Text, Speech and Language Technology

VOLUME 15

The titles published in this series are listed at the end of this volume.

Intonation

Analysis, Modelling and Technology

Edited by

Antonis Botinis

University of Skövde, Sweden and
University of Athens, Greece

KLUWER ACADEMIC PUBLISHERS

DORDRECHT / BOSTON / LONDON

Library of Congress Cataloging-in-Publication Data

Intonation : analysis, modelling and technology / edited by Antonis Botinis.
p. cm. -- (Text, speech, and language technology ; v. 15)
Includes index.
ISBN 0-7923-6605-0 (alk. paper)
1. Intonation (Phonetics) 2. Linguistic models. I. Botinis, Antonis. II. Series.

P222 .I454 2000
414'.6--dc21

00-062193

ISBN 0-7923-6605-0 (HB)
ISBN 0-7923-6723-5 (PB)

Published by Kluwer Academic Publishers,
P.O. Box 17, 3300 AA Dordrecht, The Netherlands.

Sold and distributed in North, Central and South America
by Kluwer Academic Publishers,
101 Philip Drive, Norwell, MA 02061, U.S.A.

In all other countries, sold and distributed
by Kluwer Academic Publishers,
P.O. Box 322, 3300 AH Dordrecht, The Netherlands.

Printed on acid-free paper

Printed in the Netherlands

Contents

Section IV. Intonation Modelling

Section V. Intonation Technology

Contributors

AVESANI, CINZIA
Inst of Phonetics & Dialectology, University of Padova, Italy

BANNERT, ROBERT
Dept of Phonetics, University of Umeå, Sweden

BARTKOVA, KATARINA
France Telecom, CNET, France

BOTINIS, ANTONIS
Dept of Languages, University of Skoevde, Sweden

BRUCE, GÖSTA
Dept of Linguistics & Phonetics, University of Lund, Sweden

CAMPIONE, ESTELLE
Équipe DELIC, Université de Provence, France

DI CRISTO, ALBERT
Laboratoire Parole et Langage, Université de Provence, France

DI CRISTO, PHILIPE
Laboratoire Parole et Langage, Université de Provence, France

FANT, GUNNAR
Dept of Speech, Music & Hearing, KTH, Sweden

FILIPSSON, MARCUS
Dept of Linguistics & Phonetics, University of Lund, Sweden

FOUGERON, CÉCILE
Laboratoire de Psycholinguistique, Université de Genève, Suisse

FRID, JOHAN
Dept of Linguistics & Phonetics, University of Lund, Sweden

GRANSTRÖM, BJÖRN
Dept of Speech, Music & Hearing, KTH, Sweden

GUSTAFSON, KJELL
Dept of Speech, Music & Hearing, KTH, Sweden

HAAN, JUDITH
Dept of Linguistics and Dialectology, University of Nijmegen, The Netherlands

HEUVEN, VAN J. VINCENT
Phonetics Laboratory, Dept of Linguistics, University of Leiden, The Netherlands

HIRSCHBERG, JULIA
Bell Labs, Lucent Technologies, USA

HIRST, DANIEL
Laboratoire Parole et Langage, Université de Provence, France

HORNE, MERLE
Dept of Linguistics & Phonetics, University of Lund, Sweden

HOUSE, DAVID
Dept of Speech, Music & Hearing, KTH, Sweden

HOUSE, JILL
Dept of Phonetics & Linguistics, University College of London, UK

JUN, SUN-AH
Dept of Linguistics, University of California at Los Angeles, USA

KOOPMANS-VAN BEINUM, FLORIEN
Inst of Phonetic Sciences, University of Amsterdam, The Netherlands

KRUCKENBERG, ANITA
Dept of Speech, Music & Hearing, KTH, Sweden

LILJENCRANTS, JOHAN
Dept of Speech, Music & Hearing, KTH, Sweden

MÖBIUS, BERND
Inst of Natural Language Processing, University of Stuttgart, Germany

RIETVELD, TONI
Dept of Language & Speech, University of Nijmegen, The Netherlands

ROSSI, MARIO
Laboratoire Parole et Langage, Université de Provence, France

SHIH, CHILIN
Bell Labs, Lucent Technologies, USA

TATHAM, MARK
Dept of Language & Linguistics, University of Essex, UK

VAN DONZEL, MONIQUE
Inst of Phonetic Sciences, University of Amsterdam, The Netherlands

VAN SANTEN, P.H. JAN
Center for Spoken Language Understanding, Oregon Graduate Institute of Science and Technology, USA

VÉRONIS, JEAN
Équipe DELIC, Université de Provence, France

WICHMANN, ANNE
Dept of Cultural Studies, University of Central Lancashire, UK

Acknowledgements

The present volume is an integrated selection of papers stemming from the European Speech Communication Association (ESCA) Workshop on Intonation in Athens, in September of 1997. The Workshop took place under the auspices of Athens University and was financially supported by the Hellenic Ministry of Education and the Hellenic Ministry of Development - General Secretariat of Research and Technology.

Cinzia Avesani, Robert Bannert, Mary Beckman, Gösta Bruce, Albert Di Cristo, Grzegorz Dogil, Björn Gränstrom, Julia Hirschberg, Jill House, Dieter Huber, George Kokkinakis, Mario Rossi, Christel Sorin and Marc Swerts have been members of the International Scientific Committee on the Workshop. Most of them were also involved in the review process and contributed to this volume as authors.

The initial set up of this volume took place at the Department of Linguistics, University of Athens, and the main body of editorship was carried out at the Department of Languages, University of Skövde, Sweden. I believe that the joint effort of all who worked on this volume has made a significant contribution to the field of intonation and considerable new knowledge has been produced for the benefit of the educational and international scientific communities.

In addition to authors, reviewers, colleagues and organisations referred to above as well as the University of Skövde for economic support in the production of the volume, many more have contributed in various respects, too numerous to mention individually.

My sincere thanks to all.

Antonis Botinis

Acknowledgements

1

Introduction

ANTONIS BOTINIS

1.1 Background

This introduction provides essential information about the structure and the objects of study of this volume. Following the introduction, fourteen papers which represent current research on intonation are organised into five thematic sections: (I) Overview of Intonation, (II) Prominence and Focus, (III) Boundaries and Discourse, (IV) Intonation Modelling, and (V) Intonation Technology. Within the sections the papers are arranged thematically, although several papers which deal with various aspects of intonation and prosody are basically intersectional.

As the title indicates, "Intonation: Analysis, Modelling and Technology" is a contribution to the study of prosody, with major emphasis on intonation. Intonation and tonal themes are thus the central object of the volume, although temporal and dynamic aspects are also taken into consideration by a good number of papers. Although tonal and prosodic distinctions have been dealt with throughout man's literate history with reference to the study of language, for example by classical philosophers such as Plato and Aristotle, it is in recent decades that we have witnessed the most fertile growth in intonation studies, as with experimental phonetics and speech technology in general. As Rossi (this volume) points out, intonation research really began to blossom in the sixties with a multi-fold increase in prosodic studies, reflected in contributions to the International Congress of Phonetic Sciences (ICPhS), and in the international literature.

In the present decade, intonation has become established as a major research area in speech and language-related subjects and disciplines, as

A. Botinis (ed.), Intonation, 1-10.

evidenced by contributions to the International Congress of Phonetic Sciences and to European Speech Communication Association (ESCA) Workshops, such as the Workshop on Prosody in Lund, Sweden (1993), the Workshop on Intonation in Athens, Greece (1997), and the Workshop on Dialogue and Prosody in Eindhoven, The Netherlands (1999). Significant progress has been made on relations between intonation form and function, demonstrating that a wide range of functions in lexicon and morphology, syntax and semantics, as well as discourse and dialogue, are conditioned by intonation. Intonation is also the most characteristic vocal means for communicating paralinguistic and indexical information. The significance of intonation is furthermore widely acknowledged in areas of speech technology, particularly in speech synthesis and speech recognition systems. A high-quality implementation of intonation e.g., contributes to a highly natural as well as highly intelligible phonetic output at both segmental and prosodic levels. On the other hand, synthetic speech with a mediocre intonation implementation is typically judged to be of low quality. Thus, intonation is not only related to tonal and prosodic distinctions *per se* but affects substantially the overall performance of a speech synthesis system.

1.2 Overview of Intonation

Section I, Overview of Intonation, consists of one paper, "Intonation: Past, Present, Future", by Mario Rossi. The major turning points and research paradigms in intonation are outlined from a diachronic perspective with reference to numerous languages (e.g. Karcevskij, 1931; De Groot, 1945; Bolinger, 1951; Trager & Smith, 1951; Abe, 1955; Faure, 1962; Öhman, 1967; Fry, 1967; Halliday, 1967). Theoretic and analytic aspects of intonation, in accordance with the European and the American traditions, are evaluated, with a penetrating review of the development and state-of-the-art of intonation models (e.g. Bruce, 1977; Pierrehumbert, 1980; Rossi, Di Cristo, Hirst, Martin & Nishinuma, 1981; Gårding, Botinis & Touati, 1982; Fujisaki, 1983; 't Hart, Collier & Cohen, 1990; Grønnum, 1992; Ladd, 1996; Kohler, 1997). The criteria for a Phonetics Science with reference to intonation are considered and current approaches to its study are discussed in depth. The paper concludes with future perspectives and the significance of comparative studies of intonation with reference to languages with different prosodic and phonetic systems (e.g. Gussenhoven, 1990; Cruttenden, 1997; Rossi, 1999; Hirst & Di Cristo, 1998b).

1.3 Prominence and Focus

Section II, Prominence and Focus, is concerned with intonation analysis and consists of three papers in which phonetic and phonological aspects of prosodic distinctions and prominence relations are considered in a number of languages. The first paper, "Acoustic-phonetic Analysis of Prominence in Swedish", by Gunnar Fant, Anita Kruckenberg and Johan Liljencrants, covers the main prosodic categories in Swedish, including stress, accent and focus. Thorough acoustic analyses are reported, and physiological, perceptual and phonological aspects are all taken into consideration. The prosodic parameters of F_0, duration and intensity for prominence distinctions are examined and their perceptual as well as voice source interrelations are discussed. The study reported is related to a series of earlier studies by Fant and his associates (e.g. Fant, Kruckenberg & Nord, 1991; Fant, 1997) and is mostly associated with Swedish research on prosody (e.g. Bruce, 1977; Bruce & Granström, 1993; Gårding, 1998).

The second paper, "Prosodic Disambiguation in English and Italian", by Julia Hirschberg and Cinzia Avesani, examines the relationship between choice of prosodic pattern and syntactic and semantic specifications, and the role of prosody in disambiguation in English and Italian. The authors report an experiment using read speech in which they investigate accentuation and phrasing. The study continuous earlier research on this area and is related to autosegmental analysis of intonation and prosody (e.g. Pierrehumbert, 1980; Avesani, Hirschberg & Prieto, 1995).

The third paper, "Contrastive Tonal Analysis of Focus Perception in Greek and Swedish", by Antonis Botinis, Robert Bannert and Mark Tatham, deals with the perception of focus in two experiments, one for each of the investigated languages. It studies the identification of sentence productions with different focus placements as well as perceptual effects of tonal manipulations with different focus associated tonal targets. The tonal manipulations include local tonal range, global tonal structure as well as tonal alignment shifts and tonal neutralisations. The study is based on earlier work on Greek and on contrastive intonation (Botinis, 1989, 1998; Botinis & Bannert 1997; Fourakis, Botinis & Katsaiti, 1999) and is mostly related to prosodic research on English and Swedish (e.g. Bruce, 1977; Bannert, 1986; Beckman, 1986).

1.4 Boundaries and Discourse

Section III, Boundaries and Discourse, is also concerned with intonation analysis. It consists of three papers which deal mainly with acoustic analyses from different languages relating tonal boundaries and segmentation to issues of intonation form and discourse structure. The first paper, "Phonetic Correlates of Statement versus Question intonation in Dutch", by Vincent J. van Heuven and Judith Haan, considers sentence types and intonation form of the question vs. statement paradigm in Dutch. Three experiments are reported, one acoustic and two perceptual, with reference to four sentence types, i.e. one statement and three questions with different grammatical markings. This study is associated with intonation analysis and modelling in Dutch (e.g. 't Hart, Collier & Cohen, 1990; Haan, van Heuven, Pacilly & van Bezooijen, 1997).

The second paper, "Pitch Movements and Information Structure in Spontaneous Dutch Discourse", by Monique van Donzel and Florien Koopmans-van Beinum, investigates tonal patterns in relation to discourse structure in Dutch. One experiment using spontaneous speech reports on the acoustics of tonal gestures and tonal boundaries in relation to perceptual effects in accentuation and grouping. The study is related to discourse analysis with reference to intonation and information structure of spoken discourse (e.g. Brown, Currie & Kenworthy, 1980; van Donzel, 1997; Hearst, 1997).

The third paper "Discourse constraints on F_0 peak timing in English", by Anne Wichmann, Jill House and Toni Rietveld, examines the effect of discourse structure on the temporal alignment of tonal contours associated with accented syllables. Two studies are reported: the first is an analysis of a naturally-occurring scripted monologue, and the second is a study of read speech in an experimentally controlled environment. The paper develops earlier work on segmental constraints on the timing of tonal contours and also contributes to existing research on the effect of discourse structure on intonation (e.g. Brown *et al.*, 1980, Wichmann 1991, Swerts, 1994).

1.5 Intonation Modelling

Section IV, Intonation Modelling, consists of four papers which deal with model prototypes and the acoustic modelling of intonation in different languages. The first paper, "Automatic Stylisation and Modelling of French and Italian Intonation", by Estelle Campione, Daniel Hirst and Jean Véronis, deals with the stylisation of raw tonal contours, and the modelling of intonation with reference to discrete prosodic categories in French and

Italian. A language-independent system of intonation transcription is presented, "INTSINT", and applied to selected passages in French and Italian. This study builds on work on stylisation, modelling and transcription of intonation (see Hirst & Di Cristo, 1998a; Véronis, Di Cristo, Courtois & Chaumette, 1998) and is related to international work on these areas (e.g. 't Hart, Collier & Cohen, 1990; Silverman, Beckman, Pitrelli, Ostendorf, Wightman, Price, Pierrehumbert & Hirschberg, 1992; D'Alessandro & Mertens, 1995).

The second paper, "A Phonological Model of French Intonation", by Sun-Ah Jun and Cécile Fougeron, deals with the phonetics and phonology interface, i.e. the relation between underlying representations of tonal patterns and surface realisations of prosodic units, in French. The authors report on an acoustic investigation, using read speech, of focus realisation in different types of carrier sentences (i.e. declaratives vs. interrogatives, etc.); a theoretical model of French intonation with phonological specifications is presented adopting the framework proposed by Pierrehumbert and her colleagues (e.g., Pierrehumbert, 1980; Pierrehumbert & Beckman, 1988). This work is based on recent research (see Jun & Fougeron, 1995; Fougeron & Jun, 1998) and is related to models of intonational phonology (e.g., Gussenhoven & Rietveld, 1992; Féry, 1993; Jun, 1998).

The third paper, "A Declination Model of Mandarin Chinese", by Chilin Shih, takes up aspects of intonation modelling in Mandarin Chinese with emphasis on tonal realisation and declination effects. Acoustic experiments with read speech are reported, in which prosodic factors associated with declination, including prominence and focus, sentence length and finality, are investigated. This type of work is related to analysis and modelling of sentence intonation in a good number of languages (cf. Bruce 1977; Grønnum, 1992; Prieto, Shih, & Nibert, 1996).

The fourth paper, "A Quantitative Model of F_0 Generation and Alignment", by Jan van Santen and Bernd Möbius, focuses on an analysis of the time course of F_0 contours using a corpus of American English read speech. A quantitative model is presented, describing how F_0 contours are aligned with segment and syllable boundaries, and how segmental perturbations of the F_0 contour due to vowel height and obstruency are incorporated. The model builds on earlier work by Möbius, Pätzold and Hess (1993), and is most closely associated with the superpositional model proposed by Fujisaki (1983), of which it can be considered to be a generalisation. Recent applications of the model to other languages have been reported in van Santen, Möbius, Venditti and Shih (1998).

1.6 Intonation Technology

Section V, Intonation Technology, consists of three papers, which deal with theoretical and modelling issues of intonation and prosody on the one hand and with applications in speech synthesis and speech recognition systems on the other. The first paper, "Modelling of Swedish Discourse Intonation in a Speech Synthesis Framework", by Gösta Bruce, Marcus Filipsson, Johan Frid, Björn Granstöm, Kjell Gustafson, Merle Horne and David House, deals with intonation modelling in Swedish and its application in speech synthesis. The prosodic categories of Swedish are outlined and realisation prototypes are exemplified with reference to a series of read and spontaneous dialogues. A comprehensive analysis of tonal structures including tonal range, tonal register, down-stepping, and declination is presented with reference primarily to prominence and grouping. The paper also refers to tonal alignment with the segmental representation and its implementation in speech synthesis. This work is related to intonation modelling and the development of speech synthesis systems in Swedish as well as other languages (Bruce, 1977; Carlson, Granström & Hunnicutt, 1990; House, 1990; Grønnum, 1992).

The second paper, "A Prosodic Model for Text-to-speech Synthesis in French", by Albert Di Cristo, Philippe Di Cristo, Estelle Campione and Jean Véronis, is an application of French prosody in a speech synthesis system. The prosodic structure and distribution of major prosodic categories of French are outlined, together with the interactions between rhythm and intonation and other components of the grammar. The theoretical framework presented is the basis for a model of French prosody, which is implemented in a speech synthesis system and evaluated with reference to a sample of extracts from unrestricted speech material. This work is related to prosodic analysis, modelling and speech synthesis application and evaluation in French as well as other languages (see Rossi, 1985; Di Cristo, 1998; Hirst & Di Cristo, 1998).

The third paper, "Prosodic Parameters of French in a Speech Recognition System", by Katarina Bartkova, is an implementation of French prosody in a speech recognition system. Following acoustic analysis of F_0, duration and intensity of major categories and units, a prosodic model is outlined. The prosodic model and the performance of the system are evaluated with two types of speech material, read and spontaneous. This type of work is related to prosodic analysis, modelling and speech recognition implementation in French as well as other languages (e.g. Ljolje & Fallside, 1987; Bartkova & Jouvet, 1995).

1.7 Perspectives and Prospects

In summary, this volume includes current paradigms of intonation research from a variety of scientific and discipline environments, with reference to different languages, both European and non-European. This introduction outlines the structure of the volume and the objects of study, as stated earlier, along with the scientific context of the research, but no comprehensive review of the state-of-the-art, or assessment of the scientific significance and research impact of the papers, has been attempted. The former is largely covered by Mario Rossi's paper "Intonation: past, present, feature"; the latter is left to the individual reader for consideration and ultimate evaluation.

A significant amount of knowledge about intonation and prosody has been accumulated and a good number of languages have been studied extensively. The majority of the studies have concentrated on acoustics and the relation between the underlying and surface representations, i.e. the phonetics and phonology interface. These types of studies constitute a solid basis for further development of prosodic research, which, for a deeper understanding of prosody, should include perceptual investigations on a large scale. Future work must include physiological studies building on increased understanding of production theory. Comparative studies and prosodic typology of variable prosodic structures should bring to light language-general and language-specific prosodic features, which mostly define the linguistic characterisation of each language. Deeper analysis and knowledge of prosody should make a significant contribution to linguistic theory and lead to more sophisticated models and thus higher-quality technology and diverse applications. Some of the aspects outlined above are central to current research and some are just emerging or initiated in the present volume.

In conclusion, this volume is expected to make a real contribution to current research, whether in intonation itself or in the study of individual languages. Further, it is hoped that the volume will serve the academic and scientific communities in the development of advanced courses in a variety of University subjects and disciplines, such as Phonetics and Linguistics, Speech and Language technology, Speech and Hearing Sciences and subjects with a language-specific orientation, as well as being of value to the individual researcher and professional with interests in intonation and linguistic research in general.

References

Abe, I. 1955. Intonational patterns in English and Japanese. *Word* 11, 386-398.

Ainsworth W.A. (ed.). 1990. *Advances in Speech, Hearing and Language Processing*. London: JAI Press.

Avesani, C., J. Hirschberg and P. Prieto. 1995 The intonational disambiguation of potentially ambiguous utterances in English, Italian, and Spanish. *Proc. 13th ICPhS* (Stockholm, Sweden), vol. 1, 174-177.

Bannert, R. 1986. Independence and interdependence of prosodic features. *Working Papers* 29, 31-60. Dept of Linguistics and Phonetics, Lund University.

Bartkova, K. and D. Jouvet. 1995. Using segmental duration prediction for rescoring the N-best solution in speech recognition. *Proc. 13th ICPhS* (Stockholm, Sweden), vol. 4, 248-251.

Beckman, M.E. 1986. *Stress and Non-Stress Accent*. Dordrecht: Foris.

Bolinger, D.L. 1951. Intonation: levels versus configurations. *Word* 7, 199-210.

Botinis, A. 1989. *Stress and Prosodic Structure in Greek*. Lund University Press.

Botinis, A. 1998. Intonation in Greek. In Hirst and Di Cristo (eds.), 288-310.

Botinis, A. and R. Bannert. 1997. Tonal perception of focus in Greek and Swedish. *Proc. ESCA Workshop on Intonation* (Athens, Greece), 47-50.

Brown, G., K. Currie and J. Kenworthy. 1980. *Questions of Intonation*. London: Croom Helm.

Bruce, G. 1977. *Swedish Word Accents in Sentence Perspective*. Lund: Gleerup.

Bruce, G. and B. Granström. 1993. Prosodic modelling in Swedish speech synthesis. *Speech Communication* 13, 63-74.

Carlson, R., B. Granström and S. Hunnicutt, 1990. Multilingual text-to-speech development and applications. In Ainsworth (ed.), 269-296.

Cruttenden, A. 1997 (2nd edition). *Intonation*. Cambridge University Press.

D'Alessandro, C., and P. Mertens. 1995. Automatic pitch contour stylisation using a model of tonal perception. *Computer Speech and Language* 9, 257-288.

De Groot, A.W. 1945. L'intonation de la phrase néerlandaise et allemande considérée du point de vue de la linguistique structurale. *Cahiers Ferdinand de Saussure* 5, 17-31.

de Hoop and M. den Dikken (eds.), 1997. *Linguistics in the Netherlands* 1997. Amsterdam: John Benjamins.

Di Cristo, A. 1998. Intonation in French. In Hirst and Di Cristo (eds.), 195-218.

Fant, G. 1997. The voice source in connected speech. *Speech communication* 22, 125-139.

Fant, G., A. Kruckenberg and L. Nord, 1991. Durational correlates of stress in Swedish, French and English. *Journal of Phonetics* 19, 351-365.

Faure, G. 1962. *Recherches sur les Caractères et le Rôle des Éléments Musicaux dans la Prononciation Anglaise*. Paris: Didier.

Féry, C. 1993. *German Intonational Patterns*. Tübingen: Niemeyer.

Fougeron, C. and S.-A. Jun. 1998. Rate effects on French intonation: phonetic realisation and prosodic organisation. *Journal of Phonetics* 26, 45-70.

Fourakis, M., A. Botinis and M. Katsaiti. 1999. Acoustic characteristics of Greek vowels. *Phonetica* 56, 28-43.

Fry, D.B. 1967. The present-day tasks of the phonetic sciences. *Proc. 6th ICPhS* (Prague, Czechoslovakia), 87-89.

Fujisaki, H. 1983. Dynamic characteristics of voice fundamental frequency in speech and singing. In MacNeilage (ed.), 35-55.

Gårding, E. 1998. Intonation in Swedish. In Hirst and Di Cristo (eds.), 112-130.

Gårding, E., A. Botinis and P. Touati. 1982. A comparative study of Swedish, Greek and French intonation. *Working Papers* 22, 137-152. Dept of Linguistics and Phonetics, Lund University.

Grønnum (Thorsen), N. 1992. *The Groundworks of Danish Intonation.* Copenhagen: Museum Tusculanum Press.

Gussenhoven, C. 1990. Tonal association domains and the prosodic hierarchy in English. In Ramsaran (ed.), 27-37.

Gussenhoven, C. and T. Rietveld. 1992. A target-interpolation model for the intonation of Dutch. *Proc. ICSLP* '92 (Banff, Canada), vol. 2, 1235-1238.

Haan, J., V.J. van Heuven, J.J.A. Pacilly and R. van Bezooijen. 1997. An Anatomy of Dutch question intonation. In de Hoop and den Dikken (eds.), 97-108.

Halliday, M.A.K. 1967. *Intonation and Grammar in British English.* The Hague: Mouton.

Hart, J., t', R. Collier and A. Cohen. 1990. *A Perceptual Study of Intonation.* Cambridge University Press.

Hearst, M.A. 1997. TextTiling: segmenting text into multi-paragraph subtopic passages. *Computational Linguistics* 23, 33-64.

Hirst, D.J. and A. Di Cristo. 1998a. A survey of intonation systems. In Hirst and Di Cristo (eds.), 1-44.

Hirst, D.J. and A. Di Cristo (eds.). 1998b. *Intonation Systems.* Cambridge University Press.

House, D. 1990. *Tonal Perception in Speech.* Lund University Press.

Jun, S.-A. 1998. The Accentual Phrase in the Korean prosodic hierarchy. *Phonology* 15, 189-226.

Jun, S.-A. and C. Fougeron. 1995. The Accentual Phrase and the Prosodic structure of French. *Proc. 13th ICPhS* (Stockholm, Sweden), vol 2, 722-725.

Karcevskij, S. 1931. Sur la phonologie de la phrase. *Travaux du Cercle Linguistique de Prague* 4, 188-227.

Kohler, K. 1997. Modelling prosody in spontaneous speech. In Sagisaka *et al.* (eds.), 187-210.

Ladd, D.R. 1996. *Intonational Phonology.* Cambridge University Press.

Ljolje, A. and F. Fallside. 1987. Modelling of speech using primarily prosodic parameters. *Computer Speech and Language* 2, 185-204.

MacNeilage P.F. (ed.). 1983. *The production of Speech*. New York: Springer-Verlag.

Möbius, B., M. Pätzold and W. Hess. 1993. Analysis and synthesis of German F_0 contours by means of Fujisaki's model. *Speech Communication* 13, 53-61.

Öhman, S.E.G. 1967. Word and sentence intonation: a quantitative model. *STL-QPSR* 2-3, 20-54.

Pierrehumbert, J.B. 1980. *The Phonology and Phonetics of English Intonation*. PhD dissertation, MIT (published 1988 by IULC).

Pierrehumbert, J.B. and M.A. Beckman. 1988. *Japanese Tone Structure*. Cambridge, Mass.: MIT Press.

Prieto, P., C. Shih, and H. Nibert. 1996. Pitch downtrend in Spanish. *Journal of Phonetics* 24, 445-473.

Ramsaran, S. (ed.). 1990. *Studies in the Pronunciation of English*. London: Routledge.

Rossi, M. 1985. L'intonation et l'organisation de l'énoncé. *Phonetica* 42, 35-153.

Rossi, M. 1999. *L'Intonation, le Système du Français: Description et Modelisation*. Paris: Ophrys.

Rossi, M. This volume. Intonation: past, present, feature.

Rossi, M., A. Di Cristo, D.J. Hirst, Ph. Martin, and Y. Nishinuma. 1981. *L'Intonation, de l'Acoustique à la Sémantique*. Paris: Klincksieck.

Sagisaka, Y, N. Campbell and N. Higuchi (eds.), 1997. *Computing Prosody*. New York: Springer-Verlag.

Silverman, K., M.E. Beckman, J. Pitrelli, M.F. Ostendorf, C.W. Wightman, P.J. Price, J.P. Pierrehumbert and J. Hirschberg. 1992. ToBI: a standard for labelling English prosody. *Proc. ICSLP* '92 (Banff, Canada), vol. 2, 867-870.

Swerts, M. 1994. *Prosodic Features of Discourse Units*. PhD dissertation, Eindhoven University of Technology.

Touati, P. 1987. *Structures Prosodiques du Suédois et du Français*. Lund University Press.

Trager, G.L. and H.L. Smith. 1951. *An Outline of English Structure*. Norman, Oklahoma: Battenburg Press.

van Donzel, M. 1997. Perception of discourse boundaries and prominence in spontaneous Dutch speech. *Working Papers* 46, 5-23. Dept of Linguistics and Phonetics, Lund University.

van Santen, J.P.H., B. Möbius, J. Venditti, and C. Shih. 1998. Description of the Bell Labs intonation System. *Proc. ESCA Workshop on Speech Synthesis* (Jenolan Caves, Australia), 93-98.

Véronis, J., Ph. Di Cristo, F. Courtois and C. Chaumette. 1998. A stochastic model of intonation for text-to-speech synthesis. *Speech Communication* 26, 233-244.

Wichmann, A. 1991. Beginnings, Middles and Ends: Intonation in Text and Discourse. PhD dissertation, University of Lancaster.

Section I

Overview of Intonation

2

Intonation: Past, Present, Future

MARIO ROSSI

2.1 Introduction

I do not intend to draw up an inventory of the past and present publications on intonation. Partial or exhaustive references can be found in Léon and Martin (1970), Di Cristo (1975), Gibbon (1976), Bertinetto (1979), Rossi, Di Dristo, Hirst, Martin and Nishunuma (1981), Selkirk (1984), Cruttenden (1986), Ladd (1996), Cutler, Dahan and van Danselaar (1997), Hirst and Di Cristo (1998), Lacheret and Beaugendre (1999). It would require a huge amount of effort to go back and look for past studies on prosody or intonation, which have been increasing exponentially in number over the years. Such a historical overview is beyond the scope of this paper. But it would be worth reporting the milestones which contain the seeds of the concepts underlying current theories on intonation and explain their development.

In this paper, after a brief presentation of the historical and methodological preliminaries to modern intonational studies, I intend (i) to put forth the criteria defining intonational phonology as a science and (ii) to tackle the problem of the background of current models on intonation. The definitions of intonation will be given as a conclusion of this presentation: as a matter of fact the definitions of intonation generally do not appear as primes, but as a posteriori statements of a methodology: I shall follow this logical approach. I shall conclude with a programmatic glance at the future of intonation.

A. Botinis (ed.), Intonation, 13-52.

2.2 Preliminaries

If we take international conferences and congresses as a fairly good indicator of the scientists' interest in a given topic, it is interesting to note that between the third international congress of phonetic sciences held in Ghent in 1938, to the fourteen in San Francisco, the number of papers on prosody increases forty fold. In Ghent, after the publication by Willem Pée of his *Beitrag zum Studium der Niederländischen Intonation*, only three papers where devoted to prosody, all on the intonation of Dutch: the first, on the definition of the foot as a stress group in its "optimal form": the second on intonation as a cue for dialectal differentiation: and the third in which Ledeboer van Westerhoven defined intonation and accent by their domain – sentence and word respectively – and where accent was viewed as the substratum of Intonation.

Apart from the pivotal report by Pike on *The grammar of intonation*, only four papers on prosody were presented, twenty six years later, in 1964, in Münster (Ten contributions from Soviet Union had been announced but not received. From a long time, Soviet linguists, motivated by didactic purposes, have shown a great interest for intonation). Seven papers can be found in the Proceedings of the 6th ICPhS in Prague in 1967, so that in his keynote speech on "*The present-day tasks of the phonetic sciences*", Dennis Fry was compelled to complain about the situation, claiming that the study of prosodic features in speech had to be "*one of our present-day tasks*". His exact words were: "*We do not have enough direct measurements and observations in the area of prosodic features and a good proportion of the data we have are not particularly well organised. To take as an example tone and intonation and their relation to fundamental frequency, we need a more systematic approach to observations in this area and in particular a much sharper awareness of the different functions of affective and grammatical intonation.*"

Fry's programmatic alarm was heard by researchers. Four years later in Montreal, 42 papers were presented by prosodists, including the report by Pierre Léon, *Où en sont les études sur l'intonation*. However the state of publications in conference proceedings is only an indication of the growth of prosodic studies, not the actual state of the art. In his exhaustive bibliography covering the years from 1900 to 1973, Albert Di Cristo collected more than four thousands studies on prosody, one half of which were published during the sixty five years before 1967, and the other half (about two thousands) during the five years following the Congress of Prague! These facts clearly demonstrate that the end of the sixties was a "turning point" for prosodic studies, especially for studies on intonation.

This "intonational boom" was prepared by pioneers in three areas: 1- by those who tried to outline the content of intonation, at the pragmatic level, like Karcevsky (1931), Mathesius (1937), De Groot (1945) and Daneš (1960), at the modal and emotional levels, like Wodarz (1960), or at the syntactic level, like Bierwish (1966): 2- by the descriptivists of (i) the so-called tone languages, as Chao (1933), Abe (1955), Haugen and Joos (1953), and Nien Chuang Chang (1958), (ii) English intonation, like Palmer (1922), Bolinger (1951, 1958), Jassem (1952), Schubiger (1958), Kingdon (1958), O'Connor and Arnold (1961), Faure (1962), Crystal and Quirk (1964), Wode (1966), Halliday (1967), (iii) and other languages, like Fónagy (1952, 1967), Rossi (1965), and so forth, who attempted to account for prosodic and intonational facts within particular linguistic frameworks where priority was given to contours and configurations: 3- by the proponents of American structuralism, like Wells (1945), Trager and Bloch (1941), Pike (1941), and Trager and Smith (1951), who were the first to propose a phonological model of intonation.

It is worth emphasising the great impact of language pedagogy on the development of intonation studies, which has been a strong and constant motivation, particularly of the descriptivists.

In the European context the claim that intonation played a linguistic and phonological role was overridden by a reductionist view of structuralism: for Martinet (1961, 29): "*Anything may be said to be prosodic that does not fit in the monematic and phonematic segmentation: so that the American 'suprasegmental' is not a bad substitute. I just think nothing is gained by speaking of 'suprasegmental phonemes*."

Nevertheless the first in France, maybe in Europe, to argue for an intonational phonology was Georges Faure (1962, 34), who said: "... *les tons, comme les phonèmes, ne sont que des éléments de signes et demeurent, comme eux, polyvalents lorsqu'ils jouent un rôle distinctif...un même ton pouvant contribuer, en collaboration étroite avec les autres éléments phoniques, à la discrimination de "signifiés" extrêmement variés*."

Dennis Fry was well aware of this situation, however. He emphasised the need for intonational studies based on spontaneous speech and experimental data. Experimental data became available more specifically in the mid-sixties when phoneticians had the opportunity to use the appropriate devices, like intensity and pitch meters and intended to apply their results to speech synthesis. Dennis Fry argued for an experimental method "*to find out how the features which appear in the data are used by the people who employ the particular language*". If we relate this contribution at the Prague Congress to his 1960 contribution on *Linguistic theory and experimental research*, we must acknowledge that he viewed linguistics as a science. Well before Prague, the work by Denes (1959), Uldall (1960), Hadding-Koch (1961), Delattre (1962), Lieberman and Michaels (1962), Isacenko and Shädlich (1963), Mettas (1963), Cohen and 't Hart (1965), and Öhman (1967), based on the method required by Fry's programmatic contributions, must be considered as milestones in modern intonational research, and a testimony to the fundamental role of speech technology, particularly of speech synthesis, in prosodic knowledge improvement.

From the thirties to the mid-sixties, this environment of fundamental contributions, devoted to intonation and the relationships between intonation and tones and/or stress, constitutes the theoretical background of present intonation research. It contains the seeds that explain the conception of phonology, and specifically of intonational phonology, as a science, seeds that will inform the recent hierarchical, superpositional and linear views of intonation. We shall examine these two aspects of intonation research at the present time: phonology as a science and the background of current approaches to intonation.

2.3 Intonational Phonology as a Science

2.3.1 Criteria for a Science

Against the destructive divorce between an exclusively descriptivist phonetics and a reductionist glossematic view of phonology, a significant event was the publication in 1973 of the book by W.E. Jones and J. Laver, *Phonetics in Linguistics*. As the title indicates, the authors claim that phonetics, including its prosodic domain, is an integral part of linguistics. Fry, whose 1960 contribution was appropriately reproduced in this book, claimed that if a discipline which has language for its object wants to be a science, it has to face the consequences. As already said, this requires a theory tested by experimentation which in turn relies on the behaviour of listeners providing a necessary means of validation of the theory. In his programmatic claim, Fry was probably thinking about the fundamental research carried out at the Haskins. Fry, and most of his colleagues in English-speaking countries, being educated in a different cultural and scientific climate from that of continental Europe, under the empiricist influence of Bertrand Russell and the realism of William James, assumed the phonological theory to be a model whose validity should be tested by experimentation, that is by assessment at different levels of sound transmission. As such, we can better understand the modern views of Ohala (1991) on phonetics and the premises and objectives of the Laboratory

Phonology movement which has given rise to regular symposiums and has generated the impulse for fruitful research with specific applications in the domain of speech synthesis. The much admired Bertil Malmberg significantly contributed to the development of this concept in Europe.

Today all phoneticians identify with this modern concept of phonetics, which I shall hereafter refer to as *speech linguistics* following Ferdinand de Saussure. Yet later in 1983 in their introduction to *Prosody: models and measurements*, Cutler and Ladd, noted that, at that time in the field of prosody, there was a dichotomy between measurers and model-builders, the former being mere instrumentalists and descriptivists, and the latter, being experimentalists and theorists. So Vaissière, for example, was presented as a measurer against the model-builder Pierrehumbert. Cutler and Ladd explained the difference between the two as having to do with the function and representation of prosody more than with methodology, i.e., with the "directness" of the link between the function and representation of prosody, and with "abstractness". The concrete approach defines prosody in physical terms and conceives of this link as a relatively direct mapping between functions and acoustic shapes, while model-builders assume the existence of a "well defined abstract level of representation that mediate between specific prosodic function ... and specific acoustic traits". I go along with this interpretation of the dichotomy, and I am sure that Vaissière would agree with it. But I think that the dichotomy between a concrete approach and a more abstract one is not a difference between measurers and model-builders. Let us take an example: is Fujisaki a mere measurer or a model-builder? It is obvious that even though he advocates a concrete approach, he is still a model-builder. So it would be appropriate to question the epistemological status of speech linguistics for a better understanding of the problem under discussion. Going a bit further in Fry's assumption, two essential criteria are necessary for a branch of learning to be called a science:

(i): A science must deal with data that obstruct thought: these data appear in the form of empirical facts.

(ii): A science does not speak of observables in terms of objects, but rather in terms of relations existing between those objects: science seeks to uncover a structure, a Reality, beneath empirical facts, using a model whose results must feed the theory, which need revision in order to send back new questions to the model.

Thus two components are essential in order to transform a discipline into a science: the object and the tool. The object is represented by the empirical facts which are outside the observer and can be structured: the tool is the theory which serves to describe, explain, and predict, and which must be evaluated by experimentation. Empirical facts for the speech linguistics are what is exchanged between the sender and the receiver, a noise that manifests mental representations and is produced by articulators from the biological world, the latter having its own logic, different from that of the symbolic representation in the minds of speakers. Hence speech is a cross-roads for mutually independent elements coming from different systems, each of which possesses its own logic. This independence is the source of variability and the lack of a one-to-one correlation between levels. The aim of speech linguistics consists of looking, beneath acoustical and biological observations, to discover the regularities of symbolic representation and to identify the structuring system underlying pre-systematic phenomena.

2.3.2 Theories: Measurers or Model-builders?

All theorists on intonation agree with the view that intonation is a level in the cognitive phonology device, whatever the place of the latter may be in the hierarchy of cognitive level modules. This assumption, under the influence of a long American tradition that began with Wells and Pike, has given rise to a thorough interpretation of intonational phonology as a tonal phoneme system, reinforced by Pierrehumbert's influential work. Not all intonation theorists agree with this latter view. If we consider intonation as pertaining to *speech linguistics*, which together with the 'Langue' system was conceived of by de Saussure (Bouquet 1997, 264) as a necessary level of representation, comprised of its own abstraction levels above the acoustic signal, there is a place for another interpretation of intonational phonology, which might also be viewed as an abstract representation in terms of relational 'holistic gestalts' (Aubergé, Grønnum), of distinctive features (Kohler, Martin), or of prosodic morphemes as axioms of the theory (Rossi).

Given this clarification, current theories on intonation are all phonological theories, and, if we disregard overlapping aspects, they may be roughly classified into three main categories: hierarchical, superpositional and linear. In my view, hierarchical approaches are those where intonation is conceived of as a sequence of prosodic morphemes representing the outcome of a hierarchy of linguistic modules (Gibbon, Couper-Kuhlen, Kohler, Lehiste, Martin, Möbius, Rossi, Stock & Zacharias, and so on). In accordance with this view, the sequence of morphemes is predicted and layered by ranked linguistic levels. In superpositional approaches, or 'contour interaction models', prosody is regarded as the superposition of contours from different prosodic levels: these approaches are implemented by several groups: Fujisaki's, Gårding's, 't Hart and Collier's, Möbius, Mixdorff's, O'Shaughnessy's, Thorsen-Grönum's, and Vaissière's. Among the proponents of this trend some, like Möbius, Bruce and so forth, may conceive of accent and phrase contours as a phonological sequence of tones, others, like Grønnum, deny that intonation could be taken inside phonology. In linear views, where axioms are phonemes embodied in tones, the hierarchy is reversed with respect to superpositional theories, the linear sequence of tones and accent shapes forms the sentence intonation contour, or in Ladd's terms pitch accents are the "building blocks of an intonation contour" (1996, 286). The main variants of this trend are represented by Pierrehumbert and Beckman, Hirst and Di Cristo, Ladd, and Mertens. Models that take tone phonemes as the axioms of the theory will hereafter be identified as pertaining to the 'narrow phonological trend'.

All these intonational approaches try to process data by achieving empirical pertinence outside the observer with a theory aimed at discovering relational regularities beneath facts, even though some scholars have still not developed a clear-stated or elaborate model. In other words, the different individual approaches can be ranked along a continuum ranging from pre-scientific to scientific. The dichotomy between measurers and model-builders is no longer acceptable, however. In a recent review entitled *Prosody in the comprehension of spoken language*, Cutler *et al.* (1997, 143) recognise that "such a division no longer accurately characterises the field". The "*directness*" of link criterion describing the link between the function and representation of prosody may be an aspect of the theory, it's not a criterion of scientificity. At the two ends of the scale of abstractness, we have, for example, Pierrehumbert's model with "*well defined abstract levels of representation*" and Fujisaki's with a "*direct mapping between functions and the acoustic shapes*". Both are model-builders.

As I said above, the aim of speech linguistics is to try to discover the regularities of symbolic representation in the minds of speakers, or in Fry's words (1960), "*to find out how the features which appear in the data are used by the people who employ the particular language*". Umberto Eco (1988), already suggested that the human reception of speech is the methodological guarantee of its significance. This goes a step further in intonational research: for now, the symbolic representation of most models, with the likely exception of perception-based models like 't Hart and

Collier's, are operational phonological representations, not competence models nor performance models. We all consider the representation of intonation as a *cognitive* representation, but truly symbolic representations constructed in all models are intermediate, falling between the empirical facts and the cognitive level. Collier (1992, 205) addressed the true issue, as an echo to Fry's 1960 program: "*...can the physical data on duration and fundamental frequency be taken as a faithful reflection of the prosodic 'targets' the reader wants to attain?*". The expected answer: "*There are indications to the contrary*", is an incentive, as argued by Collier, to derive abstract intonational representation not from the acoustic facts in a direct way, but after a perceptual reduction process has been applied to them, in order to incorporate into the system "*those principles that guide the phonetic behaviour of a human speaker*".

2.4 The Background of Intonational Models

If for a moment we disregard the European phonological models derived from Pierrehumbert's 1980 work, we can see an initial dichotomy between the European and American trends.

2.4.1 The European Trend

2.4.1.1 Morphological Approaches

The European trend was largely influenced by both Saussurian semiology and the Prague School. The later definition of de Saussure's sign (Bouquet, 1997) as the close union of a "signifiant" and a "signifié" prompted phonologists to consider intonation, which apparently lacked the so-called second articulation, as a sign that directly links an acoustic substance to a meaning. Hence the concrete approach and the conception of intonation contours by Delattre and other prosodists, and the morphological view of intonation by most European scholars and some American linguists (such as Pike, Bolinger, Lehiste, among others). This trend was strengthened by the fact that since prosodic substance is difficult to grasp, prosodic contributions focused more particularly on content. This trend includes the ex-soviet school which was conditioned by a somewhat different background, the Russian School, particularly the Leningrad School, being influenced by the psychological views of Baudouin de Courtenay. For Sčerba, for instance, the phoneme is that subset of sounds of which listeners and speakers are aware. Hence a twofold consequence: importance given to meaning and to a substance-based approach in the phonological analysis. This approach was reinforced by the influence of Reformatskiy's

Moscow school; Reformatskiy believed the phonemic level to be superfluous and the phonetic level to be the only necessary level below morphonology.

The study of content was mainly conducted by the advocates of the Prague School: conceiving of intonation as a syntagmatic organisation coterminous with the sentence and its parts, Mathesius, Karcevsky, De Groot, Daneš and others, developed a theory of "sentence phonology" regarded as an organisation of semantic theme and rheme units. This is the origin of the trend identified by Ladd as the "*Linguist's theory of intonational meaning*" (Ladd, 1987, 1996). This trend finds a modern justification in the Selkirk's view of the cognitive levels hierarchy: because phonological representation is a second level below the semantic and syntactic ones, it receives from these top levels all the necessary information for its syntactic and semantic interpretation. This mechanism implies a strong constraint defined by Hirst, Di Cristo and Espesser (forthcoming) as the ***Interpretability condition*** *"Representations at all intermediate levels must be interpretable at both adjacent levels: the more abstract and the more concrete"*. Hence "*functional representations which encode the information necessary for the syntactic and semantic interpretation of the prosody*" in the morphological models are justified by the above condition.

This background fed the European work on intonation from before the sixties until now. European studies on intonation are characterised by the emphasis they place on the functions of intonation, especially the grammatical and semantic functions: the descriptivists of British intonation from Jassem to Crystal, the modern descriptivists of most languages (Cruttenden, Fónagy, Kratochvil, Potapova, Svetozarova, and so forth: see Hirst & Di Cristo, 1998a, 1998b, for an exhaustive account for intonation modern descriptivists), and the modern partisans of a morphological approach to intonation (Brazil, Gussenhoven, Couper-Kuhlen, Kohler, Martin, Rossi, among others, belong to this trend).

Thus, in the hierarchical approach where intonation is viewed as a sequence of prosodic morphemes, we ought to distinguish, between those who conceive of such a sequence as the output of a hierarchy of linguistic modules (for example: Kohler, Martin, Rossi and, in some sense, Selkirk), and those who conceive of it as the result of a pragmatic meaning (Bolinger and, in some sense, Brazil, Gussenhoven, for example). Concerning the particular relationship between intonation and syntax, three main approaches may be distinguished: intonation is associated to the syntactic structure in a (i) strict, (ii) independent, or (iii) conditional way. The first approach, which implies homomorphy between syntax and intonation, is no longer considered by anyone (for more information on the topic, see Rossi, 1997b, 1999). The second approach is espoused by those who consider prosodic morphemes as reflecting a pragmatic meaning only: so Cruttenden and Gussenhoven are reluctant to extract systematic differences by comparing forms and grammatical functions: "*...three facts argue against attributing grammatical function to tone-unit boundary placement: scarcity, exceptions, optionality*" (Cruttenden, 1970), and "*intonational data should be seen as autonomous in the sense that dependence on segmental linguistic structure...should have no place in the description*" (Gussenhoven, 1983). The third approach is well represented by Martin, Kohler and Rossi. Martin (1987) states that specific rules and constraints, such as stress and syntactic clash conditions, maximal number of syllables of a prosodic word, and eurythmicity, govern the possible association mechanism that links one or more prosodic structures to a given syntactic structure. For Kohler (1997) the abstract phonological categories specified by prosodic distinctive features are integrated into a syntactic, semantic and pragmatic environment with the corresponding markers: phrasal accentuation is derived from the syntactic component preceding the prosodic module. Rossi (1993, 1997b, 1999) hypothesises that the prosodic structure (pragmatic and boundary morphemes, surfaced stress, rhythmical accent) is given by pragmatic, syntactic and phonotactic-rhythmical serially connected modules. In short, rhythmic and phonotactic conditions on interrelatedness between syntax and intonation may come first (Martin), second (Rossi), or first and second (Kohler).

Advocates of the narrow phonological trend claim that intonation constitutes an independent level of analysis whose form must be described by an appropriate model in order to account for facts (substance), and later for content: some Europeans have nevertheless remained influenced by the strength of content. Among them the proponents of the superpositional model, Gårding, Botinis (1989) and Bannert (1995). However, Grønnum, although not an advocate of the phonological trend, expressed her reluctance to account for prosodic functions, deeming this quest a 'futile attempt' (Grønnum, 1992, 78). This assumption is in accordance with Hjelmslev's conception which viewed modulations, that is intonation, as pertaining to the plane of expression only. The IPO exponent case is different: 't Hart and Collier (1990, 67) for instance refuse any kind of linguistic categorisation for intonation patterns, because "*such a strategy would be contrary to the IPO approach*"; the IPO strategy is indeed psychoacoustic and psycholinguistic in nature: evidence for the categorisation of intonation patterns has been found in listeners responses during appropriate experimentation. It is worth emphasising that, for IPO exponents, the conception of meaning, i.e. "*listener's internal representation of the intonational system*", reminds us of Bloomfield's behavioural definition (see below).

Among the intonational phonologists who were influenced by or are followers of Pierrehumert's model, some have a practice that differs from the narrow phonologist view. In his debate with Bolinger on the notion of focus, Gussenhoven (1983, 1985) gave an interesting example of a contrasting approach. On the other hand, in his criticism of the work by Pierrehumbert and Hirschberg (1990) on the meaning of intonational contours, Ladd's main argument (1996, 101) is that similarity of meaning should be reflected by similarity in phonological representation. With such an assumption he looks like an advocate of the contrasting approach whose aim is "*to construct a phonology which gives the same underlying representation to contours with the same meaning*" and vice versa (Pierrehumbert, 1980, 59). This position is reminiscent of his early theory for interpreting stylised intonations whose study clearly demonstrated "*the inadequacy of any approach to English intonation which treats contours as sequences of significant pitch levels*" and the fundamental connection between form and function (1978, 517).

2.4.1.2 The superpositional Approaches

Another aspect of some European models is the construction of sentence (= utterance) prosody conceived of as the superposition of accents on phrase intonation contour or of stress contours on the sentence intonation contour. As emphasised by Ladd (1996, 25), the mathematical implementation of such a model was first attempted by Öhman (1967) and by Fujisaki (1981). Fujisaki's model is implemented in many systems, while Öhman's directly inspired Gårding and Grønnum approaches (Figure 2.1).

Öhman hypothesised that sentence and word 'intonations' must be separate commands to larynx and may be added in different ways to account for different intonation systems. But the idea of the superposition of two components in the prosody of utterance derived (i) from the Prague School conception of an intonation component as a sign, which has scope over stress and whose domain is the sentence (Karcevskij, Mathesius), and (ii) from studies on African and Oriental languages, with the identification of the downdrift phenomenon and the recurring problem of tones and intonation relationships. In the latter studies tones and intonation are viewed as different functions of pitch, the problem being knowing how the two are "superimposed" so a compromise is reached. Abe (1955), stated, for instance, that: "*Japanese tone and intonation, both being pitch phenomena, overlap or clash in some instances*". More precisely, for the Chinese dialect of Chengdu, Chang Nien Chuang (1958) argued that:

"*Intonation does exist in the Chengdu dialect. It is superimposed on the sentence as a whole. And it is this superimposed intonation that modifies the individual tones*".

Precisely this view, although with different words, is designed in the superpositional models: that is the view of a global slope of intonation contours to which stresses and/or accents are subordinated. In the Fujisaki's model, an F_0 value representing the baseline, a phrase component and an accent component are additively superposed: functions by which control mechanisms are realised are generated by two different sets of parameters, so that local F_0 movements associated with accents are superposed on the phrase contour, following its course.

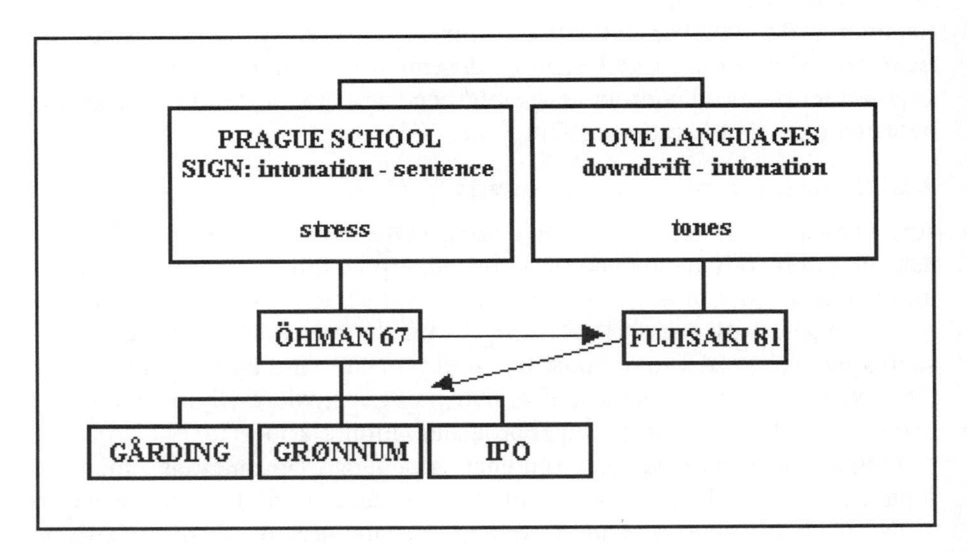

Figure 2.1. Background of the superpositional approaches.

In the Grønnum's approach (1992, 1995), the intonation contour (its course, its slope, its length) is given by a line connecting stressed syllables. Intonation contours are categorised as intonation patterns. Stress groups follow the course of a given intonation pattern without interfering with it: the features of the intonation pattern are decisive for the manifestation of stress groups. In the IPO view (ibid., 67), "*it is the intonation pattern which dictates the local choice among the various pitch movements, and the order in which they may appear*". In the Lund model sentence intonation is represented by parallel lines, *the tonal grid,* into which accents and boundaries are inserted: word accents are subordinated to sentence intonation that "*constraints the pitch values of the target points of accents*" (Gårding, 1993, 25). The general idea underlying the different models is that accents are local prominences subordinated to the global trend of sentence or phrase F_0 called intonation contour or sentence intonation. Depending on models, accents are more (IPO) or less (Grønnum) constrained by the intonation contour. Intonation patterns are empirically (Gårding, Grønnum), mathematically (Fujisaki) or psycholinguistically ('t Hart & Collier) defined.

The superpositional hypothesis gave way to thorough and interesting discussions at the 13th Stockholm ICPhS (Beckman, 1995; Grønnum, 1995; Ladd, 1995; Möbius, 1995) and recently in Grønnum (1998) and Ladd (1998). Conclusion was that all models, included the linear models, are in some way superpositional to the extent they have to deal with the interactions between accents or stress groups and downdrift: as emphasised by Möbius (1995) the difference lies on how models represent this mechanism of interactions. Yet it goes without saying that linear and superpositional implementations of these mechanisms are motivated by very different principles and induce important different consequences, as emphasised by Beckman (1995).

2.4.2 The American Trend

Whitney (1867, 101), together with the comparativists (and de Saussure in his first period), considered solely the phonological plane to be a sign: words are signs for ideas. This assumption supported the relative independence view of the phonological plane against the content plane. So Bloomfield devised a behavioural model of speech communication wherein words are represented as concrete stimuli separated from meaning defined as the hearer's reaction and the situation: "*In principle, the student of language is concerned only with the actual speech*" (1935, 74) which is studied without reference to its meaning. This view was taken at face value by Harris (1951) who claimed that the phonological system could be defined with distributional criteria without using meaning. Clearly, as shown by the Bloomfield conception of speech communication, the American structuralist view is a concrete, bottom up approach, contrasting with that of European structuralists: for instance, one of the criteria defined by Trager and Smith, and others, for classifying sounds as allophones of the same phoneme is phonetic similarity. This criterion was dismissed by

European structuralists, except by British School phonologists for the reasons stated above. On the other hand, given that the utterance speech-sound contains prosodic features that trigger a hearer's response, the language student has to count them as linguistic features. Bloomfield and the other American linguists (such as Wells, Pike, Bloch & Trager, Trager & Smith) were then led to process accents and intonation on a par with phonemes.

In this roughly sketched background, we find the basic ideas underlying Pierrehumbert's intonational model: (i) the requirement of treating intonational data as an autonomous level of analysis: (ii) the attempt "*to deduce a system of phonological representation for intonation from observed features of* F_0 *contours*" (Pierrehumbert, 1980, 59); as a consequence, (iii) a relatively direct link between F_0 contours and the abstract phonological level, but an indirect link between the acoustic signal and functions: (iv) a method of discovery akin to the distributional model: and (v) the compositional conception of contour meaning similar to Harris's definition of meaning (Harris, 1951, 190).

Pierrehumbert's model took advantage of other linguistic trends, specifically Metrical Phonology (Liberman, Liberman & Prince, and Prosodic Phonology (Selkirk), akin to the chomskyan criticism of structuralism. Chomsky (1962) questioned the linearity principle, arguing that not only phonetic, but also phonological segments can be superimposed at the same point in the speech sequence. Such a criticism of a basic tenet of strict structuralism, which is related to the suprasegmental view of prosodic features proposed by Wells (1945), Trager and Smith (1951) and others, together with the Scripture's early assumption (1902) of the sentence as being made up of two parallel lines, the syllable line and the melody or intonation line, could be considered as the roots of non-linear or autosegmental phonology (Leben, 1973; Goldsmith, 1976). On this basis, Liberman (1975) developed his Metrical Theory whereby the relationships between accents, tunes and phrasing are constructed through the F_0 feature. Also based on the framework of generative phonology, Selkirk's contribution deals more specifically with rhythmic structures, syntax-phonology mapping, and phrasing as the domain of phonological rules.

In these structuralist and generative backgrounds we can found the basic ideas and the main tools that embody Pierrehumbert's theory and more generally the trend (Ladd, 1996) refers to as the autosegmental-metrical, or AM theory (Table 2.1).

Table 2.1. Linearity substituted by Non-Linearity.

NON	
- SCRIPTURE	1-line of syllables 2-line of melody
- WELLS 1945, TRAGER	Suprasegmental phonemes
- CHOMSKY	Criticism of linearity
- LEBEN 73, GOLDSMITH	Autosegmental theory
- LIBERMAN 75 … & PRINCE	Metrical theory
- SELKIRK 78	Prosodic theory

These ideas, especially intonational level autonomy, and acoustic features as a necessary level of analysis in phonology, were very influential both within the United States (Beckman, Hirschberg, Ostendorf, Price, Wightman, and so forth), and outside the States where they resounded as a renewal and rehabilitation of a field largely ignored or scorned by European structuralists. They impinged on already existing models such as the morphological and the superpositional models, and were adopted as a standard by many scholars with or without major variants: Arvaniti, Avesani, Graber, Grice, Gussenhoven, Hayes and Lahiri, Ladd, Sosa, Uhmann, Wunderlich, and so forth (see Table 2.5 for more details). They also gave rise to related models, like those proposed by Mertens, and Hirst and Di Cristo.

A glance at the relationships between some of the most different variants of the phonological approach is necessary to gain insight into the interferences between the cultural background and the theories. With this aim in view, the models of Pierrehumbert, of Mertens and of Hirst and Di Cristo will be compared.

Pierrehumbert (1980, 59) alluded to two main approaches in intonational phonology, (i) one approach attempting to deduce the phonological representation for intonation from observed features of F_0 curves: (ii) the other, the so-called contrastive approach, trying to construct an intonational phonology by identification of meaning. She clearly rejected the latter I have already presented as the morphological view. As expected she selected the first phonological approach. Two are the axioms: the phonemes (H/L tones), and the domain of selection of categories, that is the intonational phrase of the prosodic hierarchy proposed by Selkirk (1984), and by Nespor and Vogel (1986). Pierrehumbert's method of identification of tonal phoneme sequence, on the stressed syllables given by the metrical grid, is a kind of bottom up coding from signal, liable to predict F_0 contours through appropriate rules. Abstract categories are constructed from the sequences of tones or complexes of tones. Some of these categories are meaningful: Pierrehumbert shares with Bolinger's the view that pitch accents are morphemes, and, following Trager and Smith, that boundary tones have meaning. The intonational phrase is defined as a well formed sequence of these abstract categories, specifically: {x (≥0 pitch accents + 1 nuclear accent + 1 phrase tone + 1 boundary tone}. Implemented rules evaluate the target values of tones locally with regard to the preceding tone without look-ahead procedure. The output of rules, i.e. the F_0 contour, is considered as the phonetic representation; the model does not contain any level of systematic phonetic representation, because there is no well defined level of representation below the derived phonological representation of intonation which is the more concrete before the implemented F_0 contours (1980, 10).

In Mertens model the primitives are (i) the H/L relative and absolute tone phonemes, (ii) the tone paradigms: unstressed (NA), initial (AI) and stressed syllables (AF), and (iii) the intonation group (IG). Tones are identified by a bottom up procedure that starts with the F_0 signal, and uses psychoacoustic intervals in some cases: hence, tone identification is local and does not include pre-head processing. As in the case of Pierrehumbert's procedure, the Mertens method of tone identification is also a kind of direct coding from signal but with intervening perceptual criteria which are reminiscent of psychoacoustical approaches like that of 't Hart and Collier or of Rossi *et al.* (1981).The construction of the intonation group depends on the syntactic government principle and semantic cohesion.

Hirst and Di Cristo's model consists of three levels of abstraction. The first (MOMEL), which locates a string of target points, is considered as the phonetic representation of F_0 curves. The second (INTSINT) a system for intonational transcription, is said to be a surface phonological representation characterising the target points of the phonetic representation: its primitives are tones that are absolute (Top, Mid, Bottom) or relative (Higher, Lower, Same, Upstepped, Downstepped). The third level is the deep phonological representation that interacts with the abstract prosodic parameters given by the previous module (1997). The primitives of this level are very abstract: H/L tones and three prosodic units: the Tonal Unit, the Rhythmic Unit and the Intonation Unit. Tones are assigned by an up down procedure to the output of the metrical grid. Intonation units are defined a priori by syntactic criteria, tonal and rhythmical units by a given tone sequence.

The main differences between the models concern abstractness. For a proper assessment of the abstractness criterion, we must refer to the abstractness levels defined by Shaumjan (1968, 1971).

Table 2.2. Abstraction Levels by Shaumian (1971).

		CONCRETE	ABSTRACT
SEMIOTIC LEVEL	IDEALIZED PHONOLOGICAL ELEMENTS	PHONEMES DIFFERENTORS	PHONEMES DIFFERENTORS
	RELATIONAL PHYSICAL ELEMENTS	PHONEMOIDS DIFFERENTOIDS	PHONEMOIDS DIFFERENTOIDS
PHYSICAL LEVEL	PURELY PHYSICAL ELEMENTS	CONCRETE SOUNDS	SOUND TYPES

On the basis of the statement that sounds are physical elements, that phonemic substrata are relational physical elements, and that phonemes are purely relational elements or constructs, three levels are distinguished: (i) constructs or phonemes, at the semiotic level, which are 'free from any physical substance', (ii) observation of sounds at the physical level, and (iii) in between, a third level of relational physical elements, also called phonemoids and differentoids, which come into play as an interface: phonemoids and differentoids are defined by physical elements from the level of observation and 'stand in the relation of differentiation to signs'. At each level, phonemes, phonemoids and sounds are differentiated as concrete or abstract, i.e. by whether they are individual items or a class of items (Table 2.2).

Given this framework, how can we interpret the above approaches? Mertens method is pretty inductive: the procedure used may be viewed as a kind of transcription or description based on direct observation: the H and L 'phonemes' are classes of relational physical elements, hence abstracts phonemoids of the interface level, while the other categories, such as the absolute tones, NA, AI, and AF are sound types or abstract categories of the observation level. In the same way, the elements described under INTSINT for the prosodic transcription, in the Hirst and Di Cristo model, don't constitute a surface phonological representation, in the sense of a transformation of an underlying level by specific rules, but more precisely are sound types as in Mertens model. In contrast, the elements, tones and units, of the very deep level are abstractly designed hypotheses, hence phonemes of the level of constructs meant for interacting later with the abstract elements of the observation level. Interestingly we can recognise in

Hirst and Di Cristo's phoneme conception the mere discovery methodology of the European structuralists. Pierrehumbert's looks like an inductive method. It is in fact more complex: Pierrehumbert first considers a sample of data, and then sets forth hypotheses to explain the data constructing very abstract operators whose action accounts for the facts adduced for verification: examples of such operators are the unstarred spreading tones or the phrase tone. The categories identified by Pierrehumbert are either constructs of the highest level (e.g. abstract operators) or relational physical elements, also said phonemoids (tones and pitch accents). Hence, hers is a hypothetic-deductive method allowing the statement of hypotheses on the basis of direct physical observation. Her method is well in line with the American structuralist tradition, specifically the post-bloomfieldian linguists who, rejecting the level of systematic phonetic representation as the lowest level of representation, were led to consider the phonemic level as the lowest level of representation, close to actual sounds (Table 2.3).

Table 2.3. Position of the 'strict' phonological models with regard to the Shaumian's abstraction levels.

		CONCRETE	ABSTRACT
SEMIOTIC LEVEL	IDEALIZED PHONOLOGICAL ELEMENTS		Hirst/DiC's **H,L** Pier's **Spreading tones**
	RELATIONAL PHYSICAL ELEMENTS		Mertens's **H,L** Pier's **starred H, L**
PHYSICAL LEVEL	PURELY PHYSICAL ELEMENTS	Hirst/DiC.'s **Momel**	Mertens **NA, AI...** Hirst/DiC's **Intsint**

The three levels of success required for a theory to be a scientific model are achieved, although by different means, in the above models: observational adequacy, descriptive adequacy and explanatory adequacy (Chomsky, 1962).

Given its degree of abstractness, each model has a profile which explains the presence vs. absence of an intermediate phonetic level of representation. In the Mertens approach indeed the proposed symbolic representation coincides with such a level: in the Hirst and Di Cristo model, the module (INTSINT) I have defined as the abstract level of observation is a necessary phonetic intermediate stage induced by the assumed discovery procedure; in the Pierrehumbert model, there is no place for a systematic phonetic level of representation because the method is based on direct physical observation and because, as explained by Pierrehumbert herself, "*the target values calculated by rules are like acoustic physical values as those calculated at the output of a phonetic level of representation*" (1980, 53).

Given that intonation is conceived of as an autonomous cognitive level in the phonological device, then theoretically, functions are not to be taken into account in the formal analysis. But theory and practice differ to variable extent. In the Hirst and Di Cristo model, the proposed abstract phonemes are formal elements without content or functions, but intonation unit is defined on a syntactic basis (Hirst & Di Cristo, 1984). In Mertens device, some of the primitives have meaning, for example elements with pragmatic content. In Pierrehumbert's model, the results of the tone identification process are pitch accents and boundary tones considered to be intonational morphemes. Intrusion of such intonational morphemes at the formal analysis stage led her away from attempting "*to deduce a system of phonological representation for intonation from observed features of F_0 contours*". For instance, the long discussion concerning the L*+H- pitch accent (1980, 74) is highly indicative of the mismatch between intention and practice.

In fact functions are introduced, in all models, through the metrical grid. As explained by Pierrehumbert (1980, 34), the syntax-based metrical grid is modified by pragmatic constraints whose strength can reverse the *ws* basic order at the phrase level: for example, in

we are looking / for an apartment to rent (from Pierrehumbert, 1980)
w s w s
w s w

where the basic S→ *ws* → *ws /ws*, is reversed to give S → *ws* → *w / sw*, because of the presence of a contrastive prominence on *apartment*. Following Gussenhoven (1983, 1984) the accent on *apartment* is an example of a phrase accent marking the whole phrase as focused: this accent is determined by SAAR (Sentence Accent Assignment Rule) based on semantic-syntactic criteria (see below under 5).

This is not a value judgement of mine; it is a mere statement which leads me to wonder if we can properly account for intonation without relying on functions or meaning. The sole tangible mark of meaning indeed is the sign proper. The sole tangible mark of the sign semantic face is the sign phonological face. Hence meaning is both represented by the 'Signifiant', the phonological face, and embodied as a form by its representation in the phonological face, which, by virtue of its status as a discrete sequence, partitions the universe of meaning substance (Bouquet 1997, 358). In this basic interrelatedness between phonological and meaning forms, I find evidence in support of a model where the construction of the phonological level takes place side by side with the elaboration of meaning form. It is not overstating the case to say that, if phonology and meaning tasks are considered as separate stages, there is a chance of reaching a deadlock: (i) the quest for intonational meaning could end in an amorphic paralinguistic substance, as would that of segmental meaning if conceived of apart from phonological form; (ii) in an a posteriori theory of intonational meaning, semantic contents could be constructed that do not fit the previous phonological analysis or are devised for a peculiar phonological construct, with the expected distortions (regarding this last point, see Ladd criticism (1996, 100) of Pierrehumbert and Hirschberg's paper (1990) on intonational meaning).

2.4.3 Definitions of Intonation

Added to the first dichotomy between the Prague School and the American concrete structuralism is a second dichotomy concerning the conceptions of intonation, largely governed by the first. Because they are epistemological definitions, i.e., not a priori programmatic definitions, but a posteriori statements of a practice and methodology, I did not present the definitions of intonation at the beginning of this paper.

As a rule of thumb, in the American trend, based on a tradition where speech sounds are signs for ideas, as exemplified in the behavioural model, intonation is equated with the main prosodic parameter, F_0, which generally plays the major role in prosody. For Pike "sentence melodies" are synonymous to "intonation contours": yet this conception is widespread in countries around the world. To take some examples, for Abe (1955), Faure (1962), Collier and 't Hart (1975) and Mixdorff and Fujisaki (1995), intonation is confounded with "*speech melody*" or F_0; Hadding and Studdert-Kennedy (1964) unambiguously stated that: "*intonation was varied by means of the intonator...*".

Based on the assumption that prosody is a phonological structure, the definition of intonation proposed by Trager and Smith (1951), wherein intonation is a sequence of pitches with a terminal juncture, implies the previous construction of an abstract representation of discontinuous items. In this line, intonation is regarded by Pierrehumbert (1980, 11) as a phonological structure: "*The complete phonological representation for intonation is thus a metrical representation of the text with tones lined up in accordance with rules*".

In Europe two main trends conditioned the concept of intonation: (i) radical structuralists, particularly Hjelmslev, who considered intonation as a mere modulation of substance: (ii) the Prague School scholars, who defined intonation by its domain, the sentence and its parts, and by its functions, intonation being conceived of either as a two-sided sign at the morphological level or as a system of prosodic features, generally F_0, endowed with systemic functions (Daneš, 1967). It should be borne in mind that, following an erroneous interpretation of De Saussure, structuralists considered sentence and utterance, and more generally any 'syntagmation', as mere substance-based realisations of the linguistic system. In this way intonation was seen as the speech modulation whose aim was to implement the system in discourse. Hence substance-based conceptions of intonation can be found in Europe as well as in the States, but for different reasons. Generally, as seen above, intonation is restricted to pitch movement alone. Some prosodists define intonation by pitch movement and its domain, so for Gårding (1994, 36): "*Intonation is the global melody of a phrase or sentence to which the local pitch movements are subordinated*".

Other prosodists however, like Crystal (1969, 196), proposed a parametric definition: "Intonation refers to a phenomenon which has a very clear centre of pitch contrast, and a periphery of reinforcing and occasionally contradicting contrasts of different order".

Among functionalists, Lehiste is in line with Daneš: intonation will be used when pitch is functioning at the sentence level. Rossi referring to the Saussure view of the sign defines intonation as a linguistic form differing from stress by its domain (utterance and its parts) and by its threefold function (modality, hierarchy, expression or emphasis). Kohler's view (1997) is also a functional one, but emphasis is placed on pitch in its analogic (peaks and valleys) and abstract forms (pitch categories, sentence stress) in their domain (utterance and its parts).

The definition proposed by Ladd (1996: "Intonation...refers to the use of **suprasegmental** phonetic features to convey 'postlexical' or **sentence-level** pragmatic meanings in a **linguistically structured way**", sounds like an interesting conflation of two trends: it refers to an abstract phonological representation (linguistically structured) and integrates the main overarching function of intonation (pragmatic meanings over its domain (sentence-level). In line with Crystal and the more general European trend it refers implicitly to pluri-parametric features (suprasegmental). I would agree with this definition except for the restricted view of functions. Ladd's definition contains the two components of Hirst's and Di Cristo's (1998) definition: that is to say "intonation proper" or "non-lexical prosodic system", and "prosodic parameters", with the difference that Hirst and Di Cristo define the link between these two systems as a phonetic interface also called intonation.

One conclusion we could draw from the conceptions of intonation would be whether or not they influence the practice of scholars: as a posteriori statements of a practice they do not, but they provide some interesting indications about the introspective capabilities of researchers, and the scientific background for a potential classification I've presented.

2.5 The future of Intonation

In the days to come, I think the first task should be to significantly improve the identification of similarities and differences among languages promoting multi-language descriptions, in order to become wise enough, as hoped by Hirst (1991), to design "*a third generation model*" defined by "*a number of independent levels of representation determined by more general principles*". All present models contain an extensive number of language-dependent features or principles that it would be necessary to clarify by testing them with other languages, possibly with unrelated languages, as Pierrehumbert and Beckman (1988), among others, did with Japanese, and the Gårding's team is doing with French (Touati), German (Bannert), Greek (Botinis), Japanese, Eskimo and Yoruba (Nagano-Madsen), and Swedish (Bruce), etc., to mention only some of the fruitful examples. We must emphasise however that a great deal of work has been done in this direction. I am developing a database of all languages for which today we have intonation studies based on at least one of the approaches alluded above. The for now very provisional results are included in the Tables 2.4 and 2.5.

Table 2.4. The languages analysed by concurrent models.

LANGUAGES		**MODELS**		
Languages	Number of languages	Hierarchical	Superpositional	Linear
Indo-European	27	8	18	23
Asiatic	7	3	5	3
Finno-Ugric	5	1	2	2
African	5	3	3	5
Arabic	4	1	1	2
Creoles Pidgins	2	2	-	1
	50	18	29	36

Table 2.5. Authors involved in the description of the languages mentioned in Table 2.4.

Hierarchical: Aubergé, Bailly, Bolinger, Caelen, Couper-Kuhlen, Cruttenden, Crystal, Dascalu, Diara, Fougeron, Gibbon, Hazael-Massieux, Kohler, Kratochvil, Martin, Möbius, Nishinuma, Potapova, Rafitoson, Rossi, Stock and Zacharias, Svetozarova, Tortorelli, etc.

Superpositional: Adriaens, Bannert, Beaugendre, Botinis, Bruce, Donzel, Ebing, Fujisaki, Gårding, Göskens, Grønnum, Horne, Iivonnen, Koopmans-van Beinum, Mixdorff, Möbius, Mountford, Nagano-Madsen, Odé, Sanderman, Swerts, 't Hart & Collier, Terken, Touati, van Heuven, Zhang, etc.

Linear: Alcoba, Andreeva, Arvaniti, Avesani, Beckman, Connell, Féry, Fretheim, Frota, Gibbon, Grabe, Grice, Gumperz, Gussenhoven, Hayes & Lahiri, Horvath, Inkelas and Zec, King, Ladd, Laniran, Lindau, Mennen & den Os, Hirschberg, Hirst & Di Cristo, Jun, Mertens, Nicolas, Pierrehumbert, Poser, Post, Prieto, Price, Reis, Reyelt, Sosa, Mora, Murillo, Kubozono, Uhmann, Varga, Vallduvì, Vella, Verluyten, Wunderlich, etc.

Interestingly some languages gave rise to analyses with several concurrent models (for instance, French with seven, English and German with six, and Japanese with four) and some models, namely the linear ones, spread their web over many unrelated languages. With the above aim in mind, it would be desirable to organise (i) cross-model meetings devoted to the study of one and the same language analysed by concurrent models, (ii) cross-language meetings devoted to the study of several languages analysed by the same model, as that organised by ToBI at the 13th ICPhS of Stockholm. In this way, as wished by Gibbon twenty years ago, we could "*give an explicit account of intonation universals or outline a theory of explanatory adequacy for intonation within which they would have their place.*" (1976, 290.

Ladd (1996, 102) deplored that "*for the present, proposals about the meaning of intonation are not a reliable source of evidence on intonational phonology*". I think the main task, in this perspective, would be to (i) compare results obtained on intonational meaning by the different trends involved in intonation studies; (ii) construct a typology of the pragmatic means (intonational, morphological, syntactic), used in the languages of the world; (iii) for each language identify intonational words or morphemes on which a proper phonology could be constructed, following the very influential methodology used by Bruce (1977) in analysing Swedish accents; (iv) define a theory of emotion and attitude, as proposed by Couper-Kuhlen (1986, 130), clearly distinguished from a theory of the informational function of intonation, or from a theory of prominence; and (v) look for the factors that determine intonational phrasing, in order to find out whether the correlation between intonation units and syntactic constituents is only probabilistic (Cruttenden 1986, 79) or whether a more fundamental correlation is blurred by pragmatic and/or phonotactic factors and, if so, to what extent.

It would be a stimulating enterprise "*to investigate how prosodic patterns function in human communication*" (Terken, 1993). This was the main task that Fry (1960, 1967) assigned to phoneticians more than thirty years ago: at the present time, it is still a topical question "*to find out how the features which appear in the data are used by the people who employ the particular language*". This task amounts to matching intonational patterns with the symbolic representation in the minds of speakers and listeners, to discover a model of the competence of the latter. Cutler *et al.*'s review (1997) of the psycholinguistic studies on recognition by prosodic means provides interesting insight for this venture.

All of the work I quoted and other studies as well have brought out undefined topics to investigate (in particular see Ladd, 1996); it would be tedious here to present a mere list of them. I would like nevertheless to mention some of them.

IP or intonational phrase, also called, depending on models, intonation unit, intonation group, tone group, and so forth, is defined by a boundary tone and an internal structure. Ladd (1996, 235) argued that IP boundaries are difficult to identify and that the assumption of an internal prosodic structure creates a potential for *theoretically incompatible observations.* Following the issue on the finite-state grammar structure, addressed by Pierrehumbert (1980) in the case of tag constituents, and by the Beckman and Pierrehumbert work on Japanese (1986), it was assumed that the internal structure of IP is constituted of intermediate phrases. Intermediate phrase notion was criticised, by several, particularly by Gussenhoven and Rietveld (1992) who proposed, as a response to the unresolved problem, to introduce an AD or Association Domain constituent (Gussenhoven, 1990), defined by the structure of the "*conventional prosodic constituents*", but being not part of the prosodic hierarchy (Gussenhoven & Rietveld, 1992, 287). This solution, like others, is far from uncontroversial. As argued by Selkirk (1995, 567) "*at present, principles governing intonational phrasing are not well understood*".

Gussenhoven's FTA, Focus-To-Accent theory, together with the related language-dependent SAAR, Sentence Accent Assignment Rule, and MFR, Minimal Focus Rule, states that "*a given [-focus]-[+focus] structure may well require the nucleus to fall outside the material marked [+focus]*"

Gussenhoven (1984, 15). In other words, multi-word constituents could be marked as focused by a sentence accent on one word only. According to Ladd (1996, 163) indeed, "*it is difficult to assume a straightforward bi-directional correspondence between focus and accent*": it is difficult then to accept a radical FTA view like Bolinger's and Pierrehumbert's. Hence the following questions need to be resolved: "*Where do we place accent, given focus?*", and "*How to determine the breadth of focus, given accent?*" in the different languages: are there universal or general principles underlying the language-dependent SAAR?

A clear-cut case which casts doubt on the reliability of a radical FTA view is the following French example:

> *[Ils laisseront]* $\mathbf{CT^{-2}}$ *[la place]* $\mathbf{CT^{-1}}$ *[pour deux agricultures]* **CT** *[sur les marchés mondiaux]*
> ([They will leave] [the place] [for two crops]
> [on the world-wide market])

where CT, CT^{-1} and CT^{-2} are continuative boundary morphemes whose hierarchy (resp. from the strongest to the weakest) is explained by semantic-syntactic criteria (Rossi 1993, 1997b, 1999); CT^{-2} is erased by constraints in the phonotactic module, but the remaining syntax-dependent hierarchy is not modified by any constraints in the phonology device. Yet if this hierarchy is upset:

> *[Ils laisseront la place]* CT *[pour deux agricultures]* CT *[sur les marchés mondiaux]*

the constituent *[pour deux agricultures]*, isolated between the two high CT boundaries, is focalised without a supplement of any accent: moreover the numeral *deux* constitutes a narrow focus because a quantifier is always informative, even without any focal accent. Hence the need for a pragmatic theory fitting intonation, as proposed before.

The above example raises the issue of the intonational unit hierarchy, whose knowledge is crucial in speech technology, especially in speech synthesis and automatic speech recognition (ASR). Generally, intonational phonologists deal with linear phrasing by disregarding the hierarchy among the intonational phrases (IP). They are mostly concerned with the hierarchy of the categories in the Strict Layer Hypothesis and that of the prominence of words in intonational phrases. Yet recent studies by Beckman (1990) and Price (1991) acknowledge that the distribution of the pitch and lengthening cues in English plays a role in ranking prosodic units, confirming previous or contemporary results for other languages (Klatt, 1975; Rossi *et al.*, 1981; Bartkova & Sorin, 1987; Fant, Kruckenberg & Nord, 1991). On the other hand, studies on discourse structure (Hirschberg & Pierrehumbert, 1986; Botinis 1992; Swerts, 1997) and work on database labelling have emphasised the role of intonation in the hierarchy of discourse units.

For instance, in an analysis of discourse boundaries of different strengths, Swerts (1997) found six significant pause duration categories, two pitch resetting groups, and two boundary tone classes signalling the degree of embeddedness of phrases in the discourse.

For database labelling, four significant hierarchy levels (break indices) were identified by Price, Ostendorf, Shattuck-Hufnagel and Fong (1991) and by Wightman, Shattuck-Hufnagel, Ostendorf and Price (1992), which correspond to the four levels found by Ladd and Campbell (1991), although with a different conception of the depth of structure. The levels identified by the former are not related solely to IPs, but are also related to the different strings in the prosodic hierarchy (utterance, IP, intermediate phrase, and so forth); those defined by the latter refer to domains, such as IP, and to superdomains, which are compounds of domains of a given type (e.g. IP). Interestingly, this conception of the depth of structure is tantamount to the recognition of a hierarchy among IPs.

And in a study aimed at implementing phrase boundaries in synthetic speech essentially on the basis of pitch contours, Sanderman and Collier (1996, 3396) adduced empirical evidence that listeners are sensitive to a five-level hierarchy.

A glance at certain corpora in the above studies shows that in most cases, boundary levels are determined by the syntactic hierarchy. So studies on boundary strength, as well as research on prosodic phrasing, argue for syntactic constraints on prosody and for matching rules between prosody and syntax. Yet can we be satisfied with the informal observation that there is a rather close relationship between semantic and syntactic structures, and that prosody is sensitive to major syntactic boundaries like clauses and major phrases?

Cutler *et al.* (1997, 169) noted that "*the studies reviewed have not always been explicit about the exact role of prosodic information in the parsing process*", and that there has been regrettably little attention paid "*to establishing the range of possible parsing effects of prosody*". We need a suitable model to handle linguistic structures that constrain prosodic boundaries (see Rossi, 1997b).

These are all very interesting points to discuss, and, a last but not least important point concerns the application of prosody to speech technologies.

There is a dialectic relation between application domains and phonetic research. Synthesis and ASR constitute two means of experimental validation of the theory which are outside the observer and consequently invaluable to the foundation of 'speech linguistics' as a science. Today, all models are oriented towards applications to text-to-speech synthesis in order to improve its quality; for the future the main task will be achieving credible applications to ASR. Interesting results have already been obtained in recent work (Campbell, 1993; Hunt, 1997; Ostendorf, 1998; Bartkova, this volume). But, as emphasised by Granström (1997) in his overview of applications of intonation, the general difficulty encountered in applications, especially in ASR, is "*in handling and integrating suprasegmental phenomena in situations where a focus on segmental effects has been prevalent*". A glance at the architecture of these systems demonstrates that models achieving optimal fusion between syntax and other linguistic levels need to be developed in order to improve the results of prosody processing in ASR systems.

2.6 Conclusion

Three main trends came to light, with a considerable amount of overlapping. I am ill-informed about the cultural background of countries other than those in Europe and America, but it would be interesting to examine the extent to which local cultural backgrounds have modified the profile of the trends derived from structuralism, which has had such a strong world-wide influence. I think perhaps that the main effect on theories and models is due to the strength of the particular languages. Theories and models are governed by cultural background, by the specificity of the languages spoken, and by original principles: there is room, then, for all possible theories and models, each of which is entitled to its dignity. Mine is not an eclectic position. An essential and potentially fruitful task for every scholar, in the future assessment of theories, is to try to drive her or his model to its final consequences, instead of blurring it with borrowed features which generally do not fit the architecture of the model. Discussion and controversy are highly desirable, but no value judgements should be made. Bolinger, of whom I spoke little because it is hard to classify Giants, was presented by Crystal (1969, 54) "*as debunker of superficial intonation generalisations*". We can and must all be debunkers of shortcomings, inconsistencies and superficial generalisations, all the while being aware of the fact that the life of our own model, like every model, is limited in time.

The future is open to novelties we can not imagine now. Concerning the future of intonation, I have discussed a few issues to be addressed. Among the current problems to resolve, some will look as a mere trifle in the near future, while others that will crop up will be of fundamental interest to up-to-date theories. It would be highly desirable if some scholars took the initiative of drawing up a well thought-out list ("une liste raisonnée") of all the points under discussion, and propose them to the community of intonologists for taking the right initiatives.

Acknowledgements

I would like to thank Antonis Botinis, and his colleagues Georgios Kouroupetroglou and George Carayannis, who invited me at the Athens symposium on Intonation to present a keynote paper whose development his my participation to this book. Their encouraging invitation was very helpful. I would like also to thank all the reviewers for their useful and constructive comments on the first version of my contribution.

References

Abe, I. 1955. Intonational patterns in English and Japanese. *Word* 11, 386-398.

Altmann, H. (ed.). 1988. *Intonationsforschungen*. Tübingen: Niemeyer.

Arvaniti, A. 1994. Acoustic features of Greek rhythmic structures. *Journal of Phonetics* 22, 239-268.

Aubergé, V. 1992. Developing a structured lexicon for synthesis of prosody. In Bailly *et al.* (eds.), 307-321.

Avesani, C. 1990. A Contribution to the Synthesis of Italian Intonation. *Proc. ICSLP* '90 (Kobe, Japan), vol. 2, 833-836.

Avesani, C., J. Hirschberg and P. Prieto. 1995. The intonation disambiguation of potentially ambiguous utterances in English, Italian, and Spanish. *Proc. 13th ICPhS* (Stockholm, Sweden), vol. 1, 144-177.

Bailly, G., C. Benoît and T. Sawallis (eds.). 1992. *Talking Machines: Theories, Models and Designs*. Amsterdam: Elsevier Science.

Bakenecker, G., U. Block, A. Batliner, R. Kompe, E. Nöth and P. Regel-Brietzmann. 1994. Improving parsing by incorporating 'prosodic clause boundaries' into grammar. *Proc. ICSLP* '94 (Yokohama, Japan), vol. 3, 1115-1118.

Bannert, R. 1983. Modellskizze für die Deutsche Intonation. *Zeitschrift für Literaturwissenschaft und Linguistik* 49, 9-34.

Bannert, R. 1995. Variations in the perceptual modelling of macro-prosodic organisation of spoken Swedish: prominence and chunking. (Reports from the Department of Phonetics, Umeå University) *PHONUM* 3, 31-53.

Bartkova, K. and C. Sorin. 1987. A model of segmental duration for speech synthesis in French. *Speech Communication* 6, 245-260.

Beckman, M.E. 1986. *Stress and Non-Stress Accent.* Dordrecht: Foris.

Beckman, M.E. 1993. Modelling the production of prosody. *Working Papers* 41, 258-263. Dept of Linguistics and Phonetics, Lund University.

Beckman, M.E. and J. Edwards. 1990. Lenghtenings and shortenings and the nature of prosodic constituency. In Kingston and Beckman (eds.), 152-178.

Beckman, M.E. and J.P. Pierrehumbert. 1986. Intonational structure in Japanese and English. *Phonology Yearbook* 3, 255-310.

Berkovits, R. 1993. Utterance-final lengthening and the duration of final stop closures. *Journal of Phonetics* 21, 479-489.

Bertinetto, P. M. 1979. *Aspetti Prosodici della Lingua Italiana.* Padova: Clesp.

Bierwish, M. 1966. Regeln für die Intonation deutscher Sätze. *Studia Grammatica* 7, 92-102.

Bloomfield, L. 1935. *Language.* London: George Allen & Unwin.

Bolinger, D.L. 1951. Intonation: levels versus configurations. *Word* 7, 199-210.

Bolinger, D.L. 1958. A theory of pitch accent in English. *Word* 14, 109-149.

Bolinger, D.L. 1965. *Forms of English: Accent, Morpheme, Order.* Harvard University Press.

Bolinger, D.L. 1970. Relative Height. Prosodic feature analysis. *Studia Phonetica* 3, 109-125.

Bolinger, D.L. (ed.). 1972. *Intonation.* Harmondsworth: Penguin Books.

Bolinger, D.L. 1985. Two views of accent. *Journal of Linguistics* 21, 79-123.

Booj G. and J. van Marle (eds.). 1993. *Yearbook of Morphology.* Dordrecht: Kluwer Academic Publishers.

Botinis, A. 1989. *Stress and Prosodic Structure in Greek.* Lund University Press.

Botinis, A. 1992. Accentual distribution in Greek discourse. *Travaux de l'Institut de Phonétique d'Aix* 14, 13-52.

Bouquet, S. 1997. *Introduction à la Lecture de Saussure.* Paris: Payot.

Brazil, D., M. Coulthard and C. Johns. 1980. *Discourse Intonation and Language Teaching.* London: Longman.

Bresnan, J. 1971. Sentence stress and syntactic transformations. *Language* 47, 357-281.

Bresnan, J. 1972. Stress and syntax: a reply. *Language* 48, 326-342.

Brown, G. 1983. Prosodic structure and the given/new distinction. In Cutler and Ladd (eds.), 66-77.

Bruce, G. 1977. *Swedish Word Accents in Sentence Perspective.* Lund: Gleerup.

Bruce, G. and E. Gåding. 1978. A prosodic typology for Swedish dialects. *Nordic Prosody*, 219-228. Dept of Linguistics and phonetics, Lund University.

Bruce, G., B. Granström, K. Gustafson and D. House. 1993. Phrasing strategies in prosodic parsing and speech. *Proc. EUROSPEECH* '93 (Berlin, Germany), 1205-1208.

Bruce, G., U. Willstedt and P. Touati. 1990. On Swedish interactive prosody: Analysis and synthesis. *Nordic Prosody* V, 36-48. Turku University.

Butterworth, B. (ed). 1980. *Language production 1, Speech and talk.* London: Academic Press.

Caelen, G. Forthcoming. *Prosodie et Sens*. Paris: Editions du CNRS.

Campbell, N. 1993. Automatic detection of prosodic boundaries in speech. *Speech Communication* 13, 343-355.

Carbonnel, N., J.P. Haton, F. Lonchamp and J.M. Pierrel. 1982. Elaboration expérimentale d'indices prosodiques pour la reconnaissance: application à l'analyse syntaxico-sémantique dans le système Myrtille II. *Actes du Séminaire Prosodie et Reconnaissance Automatique de la Parole* (Aix-en-provence), 59-91.

Chang, N.-C. 1958. Tones and intonation in the Chengdu dialect, Szechuan, China. *Phonetica* 2, 59-84.

Chao, Y.R. 1933. Tone and intonation in Chinese. *Bulletin of the National Research Institute of History and Philology of the Academia Sinica* 4, 124-134.

Chomsky, N. 1962. The logical basis of linguistic theory. *Proc. 9th ICL* (Cambridge, MA., USA), 914-1008.

Cohen P.R., J. Morgan and M.E. Pollack (eds.). 1990. *Intentions in Communication.* Cambridge, Mass.: MIT Press.

Cohen, A. and J. 't Hart. 1965. Perceptual analysis of intonation patterns. *Proc. 5th ICA* (Liège, Belgium) 1A, A16.

Cohen, A., and S.G. Nooteboom (eds.). 1975. *Structure and Process in Speech Perception.* New York: Springer-Verlag.

Collier, R. 1992. A comment on the prediction of prosody. In Bailly *et al.* (eds.), 205-208.

Collier, R. and J. 't Hart. 1975. The role of intonation in speech perception. In Cohen and Nooteboom (eds.), 107-121.

Cooper-Kuhlen, E. 1986. *An introduction to English Prosody.* London: Edward Arnold.

Cruttenden, A. 1986. *Intonation.* Cambridge University Press.

Cooper, W. and J. Paccia-Cooper. 1980. *Syntax and Speech.* Harvard University Press.

Cruttenden, A. 1992. The origins of nucleus. *Journal of the International Phonetics Association* 20, 1-9.

Crystal, D. 1969. *Prosodic Systems and Intonation in English.* Cambridge University Press.

Crystal, D. and R. Quirk. 1964. *Systems of Prosodic and Paralinguistic Features in English.* The Hague: Mouton.

Cutler, A., D. Dahan and W. van Danselaar. 1997. Prosody in the comprehension of spoken language: a literature review. *Language and Speech* 40, 141-201.

Cutler, A. and S.D. Isard. 1980. The production of prosody. In Butterworth (ed), 245-269.

Cutler, A. and D.R. Ladd (eds.). 1983. *Prosody: Models and Measurements.* Berlin: Springer-Verlag.

Daneš, F. 1960. Sentence intonation from a functional point of view. *Word* 16, 34-54.

Daneš, F. 1967. Order of elements and sentence intonation. In *To Honour of Roman Jakobson*, 499-512. The Hague: Mouton.

De Groot, A.W. 1945. L'intonation de la phrase néerlandaise et allemande considérée du point de vue de la linguistique structurale. *Cahiers Ferdinand de Saussure* 5, 17-31.

Delattre, P. 1962. A comparative study of declarative intonation in American English and Spanish. *Hispania* XLV, vol. 2, 233-241.

Denes, P. 1959. A preliminary investigation of certain aspects of intonation. *Language and Speech* 2, 106-122.

Di Cristo, A. 1975. *Soixante et Dix Ans de Recherche en Prosodie*. Editions de l'Université de Provence.

Di Cristo, A. 1981. L'intonation est congruente à la syntaxe: une confirmation. In Rossi *et al.* (eds.). 272-289.

Di Cristo, A. 1998. Intonation in French. In Hirst and Di Cristo (eds.), 195-218.

Di Cristo, A. and D.J. Hirst. 1993. Rythme syllabique. rythme mélodique et représentation hiérarchique de la prosodie du français. *Travaux de l'Institut de Phonétique d'Aix* 15, 9-24.

Di Cristo, A. and D.J. Hirst. Forthcoming. L'accentuation non emphatique en français: stratégies et paramètres. *Hommages I. Fónagy*. Paris: L'Harmattan.

Di Cristo, A., J.P Haton, M. Rossi and J. Vaissière (eds.). 1982. *Prosodie et Reconnaissance Automatique de la Parole* (Actes du Colloque Prosodie et reconnaissance, Aix-en Provence, France).

Di Cristo. A. and D.J. Hirst. 1996. Vers une typologie des unités intonatives du français. *XXIièmes Journées d'Études sur la Parole* (Avignon, France), 10-14.

Eco, U. 1988. *Trattato di Semiotica Generale*. Milano: Bompiani.

Fant, G. and A. Kruckenberg. 1994. Notes on stress and word accent in Swedish. *STL-QPSR* 2-3, 125-144.

Fant, G., A. Kruckenberg and L. Nord. 1991. Durational correlates of stress in Swedish, French and English. *Journal of Phonetics* 19, 351-365.

Farnetani, E. and S. Kori. 1983. Interaction of syntactic structure and rhythmical constraints on the realisation of word prosody. *Quaderni del Centro di Ricerca di Fonetica di Padova* 2, 287-318.

Faure, G. 1962. *Recherches sur les Caractères et le Rôle des Éléments Musicaux dans la Prononciation Anglaise*. Paris: Didier.

Féry, C. 1993. *German Intonational Patterns*. Tübingen: Niemeyer.

Fónagy, I. 1952. Accent et intonation du français moderne. *Grammaire Descriptive du Français Moderne*. Budapest, 62-82.

Fónagy, I. 1967. *L'intonation du Hongrois*. Budapest.

Fretheim, T. and R.A. Nilsen. 1993. How to tell H% from L% in right-detached expressions in Norwegian. *Working Papers* 41, 58-65. Dept of Linguistics and Phonetics, Lund University.

Fry, D.B. 1960. Linguistic theory and experimental research. *Transactions of the Philological Society*, 13-39. (Reprinted in Jones and Laver (eds.), 66-87).

Fry, D.B. 1967. The present-day tasks of the phonetic sciences. *Proc. 6th ICPhS* (Prague, Czechoslovakia), 87-89.

Fujisaki, H. 1981. Dynamic characteristics of voice fundamental frequency in speech and singing. *Proc. 4th FASE Symposium* (Venice, Italy), 57-70.

Fujisaki, H. 1997. Prosody model and spontaneous speech. In Sagisaka *et al.* (eds.), 27-42.

Fujisaki, H., K. Hirose and K. Ohta. 1979. Acoustic features of the fundamental frequency contours of declarative sentences in Japanese. *Ann. Bull. Res. Inst. Logopedics and Phoniatrics* 13, 163-173.

Gårding, E. 1983. A generative model of intonation. In Cutler and Ladd (eds.), 11-25.

Gårding, E. 1990. West Swedish and East Norwegian intonation. *Nordic Prosody* V, 111-129. Turku University.

Gårding, E. 1992. Focal domains and their tonal manifestations in some Swedish dialects. *Nordic Prosody* VI, 65-76.

Gårding, E. 1993. On parameters and principles in intonation analysis. *Working Papers* 40, 25-47. Dept of Linguistics and Phonetics, Lund University.

Gårding, E. 1994. Prosody in Lund. *Speech Communication* 15, 59-67.

Gårding, E. A. Botinis, and P. Touati. 1982. A comparative study of Swedish, Greek and French intonation. *Working Papers* 22, 137-152. Dept of Linguistics and Phonetics, Lund University.

Garrido, J.M. 1993. Analysis of global pitch contour domains at paragraph level in Spanish reading text. *Working Papers* 41, 104-107. Dept of Linguistics and Phonetics, Lund University.

Geoffrois, E. 1995. *Extraction Robuste de Paramètres Prosodiques pour la Reconnaissance de la Parole*. PhD dissertation, Université de ParisXI-Orsay.

Gibbon, D. 1976. *Perspectives of Intonation Analysis*. Bern: Lang.

Goldsmith, J.A. 1976. *Autosegmental Phonology*. PhD dissertation, MIT (published 1979, New York: Garland).

Goldsmith, J.A. (ed.), 1995. *The Handbook of Phonological Theory*. Oxford: Blackwell.

Grabe, E. 1998. *Comparative Intonational Phonology, English and German*. PhD dissertation, Katholieke Universiteit Nijmegen.

Granström, B. 1997. Applications of intonation – an overview. *Proc. ESCA Workshop on Intonation* (Athens, Greece), 21-24.

Grice, M. 1995. *The Intonation of Interrogation in Palermo Italian: Implications for Intonation Theory*. Tübingen: Niemeyer.

Grønnum (Thorsen), N. 1992. *The Groundworks of Danish Intonation*. Copenhagen: Museum Tusculanum Press.

Grønnum (Thorsen), N. 1995. Superposition and subordination in intonation. A non-linear approach. *Proc. 13th ICPhS* (Stockholm, Sweden), vol. 2, 124-131.

Grønnum (Thorsen), N. 1998. A critical remark on D.R. Ladd's *Intonational Phonology*. *Journal of Phonetics* 26, 109-112.

Grosjean, F. 1983. How long is the sentence? Prediction and prosody in the on-line processing of language. *Linguistics* 21, 501-529.

Grosz, B.J. and J. Hirschberg. 1992. Some intonational characteristics of discourse structure. *Proc. ICSLP* '92 (Banff, Canada), vol. 1, 429-432.

Grosz, B.J. and C.L. Sidner. 1986. Attention, intentions, and the structure of discourse. *Computational Linguistics* 12, 175-204.

Gussenhoven, C. 1983. Focus, mode and the nucleus. *Journal of Linguistics* 19, 377-417.

Gussenhoven, C. 1984. *On the Grammar and Semantics of Sentence Accents*. Dordrecht: Foris.

Gussenhoven, C. 1985. Two views of accent: a reply. *Journal of Linguistics* 21, 125-138.

Gussenhoven, C. 1990. Tonal association domains and the prosodic hierarchy in English. In Ramsaran (ed.), 27-37.

Gussenhoven, C., D.L. Bolinger and C. Kejsper. 1987. *On accent*. Bloomington: IULC.

Hadding-Koch, K. 1961. *Acoustico-phonetic Studies in the Intonation of Southern Swedish*. Lund: Gleerup.

Hadding-Koch, K. and M. Studdert-Kennedy. 1964. An experimental study of some intonation contours. *Phonetica* 11, 175-185.

Halliday, M.A.K. 1967. *Intonation and Grammar in British English*. The Hague: Mouton.

Harris, Z.S. 1951. *Structural Linguistics*. The University of Chicago Press.

Hart 't, J. Collier, R. and A. Cohen. 1990. *A Perceptual Study of Intonation*. Cambridge University Press.

Haugen, E. and M. Joos. 1953. Tone and intonation in East Norwegian. *Acta Philologica Scandinavica* 22, 41-64.

Hayes, B. and A. Lahiri 1991. Bengali intonational phonology. *Natural Language and Linguistic theory* 9, 47-96.

Hess, W. 1983. *Pitch Determination of Speech Signals*. Berlin: Springer-Verlag.

Hess, W., A. Batliner, A. Kiessling, R. Kompe, E. Nöth, A. Petzold, M. Reyelt and V. Strom. 1997. Prosodic modules for speech recognition and understanding in Verbomobil. In Sagisaka *et al.* (ed.), 361-382.

Heuft, B., T. Portele, J. Krämer, H. Meyer, M. Rauth and G. Sonntag. 1995. Parametric description of F_0-contours in a prosodic database. *Proc. 13th ICPhS* (Stockholm, Sweden), vol. 2, 378-381.

Hirschberg, J. and C. Nakatani. 1996. A prosodic analysis of discourse segments in direction-giving monologues. *Proc. 34th Annual Meeting of the Association for Computational Linguistics* (Santa Cruz, CA., USA), 286-293.

Hirschberg, J. and J.P. Pierrehumbert. 1986. The intonational structuring of discourse. *Proc. 24th Association for Computational Linguistics* (Santa Cruz, CA., USA), vol. 24, 136-144.

Hirst, D.J. 1991. Intonation models: towards a third generation. *Proc. 12th ICPhS* (Aix-en-Provence, France), vol. 1, 305-310.

Hirst, D.J. 1992. Prediction of prosody: an overview. In Bailly *et al.* (eds.), 199-204.

Hirst, D.J. 1993a. Peak, boundary and cohesion characteristics of prosodic grouping. *Working Papers* 41, 32-37. Dept of Linguistics and Phonetics, Lund University.

Hirst, D.J. 1993b. Detaching intonational phrases from syntactic structure. *Linguistic Inquiry* 24, 781-788.

Hirst, D.J. and A. Di Cristo. 1984. French intonation: a parametric approach. *Di Neueren Sprache* Bd 83, Heft 5, 554-569.

Hirst, D.J. and A. Di Cristo. 1998. A survey of intonation systems. In Hirst and Di Cristo (eds.), 1-44.

Hirst, D.J. and A. Di Cristo, (eds.). 1998. *Intonation Systems*. Cambridge University Press.

Hirst, D.J., A. Di Cristo and R. Espesser. Forthcoming. Levels of representation and levels of analysis for the description of linguistic systems. In Horne (ed.).

Hjelmslev, L. 1943. *Prolegomena to a Theory of Language*. Reproduced by University of Wisconsin Press, Madison. (Translated by F.J. Whitfield, 1963).

Horne, M.A. 1987. Towards a discourse-based model of English sentence intonation. *Working Papers* 32, 1-36. Dept of Linguistics and Phonetics, Lund University.

Horne, M. (ed). Forthcoming. Prosody: *Theory and Experiment*. Dordrecht: Kluwer Academic Publishers.

Horne, M., M. Filipsson, C. Johansson, M. Ljungqvist and A. Lindström. 1993. Improving the prosody in TTS systems. *Working Papers* 41, 208-211. Dept of Linguistics and Phonetics, Lund University.

House, D. 1990. *Tonal Perception in Speech*. Lund University Press.

Hunt, A. 1997. Training prosody-syntax recognition models without prosodic labels. *Proc. ICSLP* '94 (Yokohama, Japan), vol. 3, 1119-1122 (also in Sagisaka *et al.* (eds.), 309-326)

Inkelas, S. and D. Zec. 1990. *The Phonology-syntax Connection*. The University of Chicago Press.

Inkelas, S. and D. Zec. 1995. Syntax-phonology interface. In Goldsmith (ed.) 1995, 535-549.

Isacenko, A.V. and H.J. Shädlich. 1963. Erzeugung kûnstlicher deutscher Satzintonationen mit zwei kontrastierenden Tonstufen. *Monatsberichte der Deutschen Akademie der Wissenschaften su Berlin* 5-6, 365-372.

Jassem, W. 1952. *The Intonation of Conversational English*. Wroclaw: Wroclawskie Towarzystwo Naukow.

Jones W.E. and J. Laver (eds.). 1973. *Phonetics in Linguistics*. London: Longman.

Jun, S.-A. 1993. *The Phonetics and Phonology of Korean Prosody*. PhD dissertation. The Ohio State University (published 1996, New York: Garland).

Kaisse, E.M. 1985. *Connected Speech: The Interaction of Syntax and Phonology*. New York: Academic Press.

Kaisse, E.M. and P.A. Shaw. 1985. On the theory of lexical phonology. *Phonology Yearbook* 2, 2-30.

Karcevskij, S. 1931. Sur la phonologie de la phrase. *Travaux du Cercle Linguistique de Prague* 4, 188-227.

Keijsper, C.E. 1987. Two views of accent: a third opinion. In Gussenhoven *et al.* (eds.), 162-201.

Kingdon, R. 1958. *The Groundwork of English Intonation*. London: Longman.

Kingston, J. and M.E. Beckman (eds.). 1990. *Papers in Laboratory Phonology I*. Cambridge University Press.

Klatt, D.H. 1975. Vowel lengthening is syntactically determined in a connected discourse. *Journal of Phonetics* 3, 129-140.

Kohler, K. 1995. A language science in its own right? *Proc. 13th ICPhS* (Stockholm, Sweden), vol. 1, 10-17.

Kohler, K. 1997. Modelling prosody in spontaneous speech. In Sagisaka *et al.* (eds.), 187-210.

Lacheret, A. and Beaugendre, F. 1999. *La Prosodie du Français*. Paris: Editions du CNRS.

Ladd, D.R. 1996. *Intonational Phonology*. Cambridge University Press.

Ladd, D.R. 1998. A critical remark on Intonational Phonology: response to Nina Grønnum. *Journal of Phonetics* 26, 113-114.

Ladd, D.R. and N. Campbell. 1991. Theories of prosodic structure: evidence from syllable duration. *Proc. 12th ICPhS* (Aix en Provence, France), 290-293.

Laniran, Y.O. 1992. *Intonation in Tone Languages: The Phonetic Implementation of Tones in Yoruba*. PhD dissertation, University of Cornell.

Lea, W.A. 1972. *Intonational Cues to the Constituent Structure and Phonemics of Spoken English*. PhD dissertation, University of Purdue.

Lea, W.A. 1980. Prosodic aids to speech recognition. In Lea (ed.), 166-205.

Lea, W.A. (ed.). 1980. *Trends in Speech Recognition*. New Jersey: Prentice Hall.

Leben, W.R. 1973. *Suprasegmental Phonology*. PhD dissertation, MIT (published 1989, New York: Garland).

Lehiste, I. 1970. *Suprasegmentals*. Cambridge, Mass.: MIT Press.

Lehiste, I. 1973a. Phonetic disambiguation of syntactic ambiguity. *Glossa* 7, 107-121.

Lehiste, I. 1973b. Rhythmic units and syntactic units in production and perception. *J. Acoust. Soc. Am.* 54, 1228-1234.

Léon, P. 1971. Où en sont les études sur l'intonation?. *Proc. 7th ICPhS* (Montreal, Canada), 113-156.

Léon, P. and Ph. Martin. 1970. Prolégomènes à l'étude des structures intonatives. *Studia Phonetica* 2.

Liberman, M.Y. 1975. *The Intonational System of English*. PhD dissertation, MIT (published 1978 by IULC).

Liberman, M.Y. and A. Prince. 1977. On stress and linguistic rhythm. *Linguistic Inquiry* 8, 249-336.

Lieberman, Ph. and S.B. Michaels. 1962. Some aspects of fundamental frequency and envelope amplitude as related to the emotional content of speech. *J. Acoust. Soc. Am.* 34, 922-927.

Magno Caldognetto, E., F. Ferrero, C. Lavagnoli and K. Vagges. 1978. F_0 contours of statements, yes-no questions and WH-questions of two regional varieties of Italian. *Journal of Italian Linguistics* 3, 57-67.

Martin, Ph. 1981. Pour une théorie de l'intonation. In Rossi *et al.* (eds.), 234-271.

Martinet, A. 1961. *A Functional View of Language*. Oxford: Clarendon Press.

Mathesius, V. 1937. K teorii vetné intonace. *Slovo a Slovesnost* 3, 248-249.

McCarthy, J.J and A. Prince. 1993. Generalised alignment. In Booj and van Marle (eds.), 79-153.

Mennen, I. and E. den Os. 1993. Intonation of Modern Greek sentences. *Proc. Institute of Phonetic sciences* 17, 111-128. University of Amsterdam.

Mertens, P. 1993. Intonational grouping, boundaries, and syntactic structure in French. *Working Papers* 41, 156-159. Dept of Linguistics and Phonetics, Lund University.

Mettas, O. 1963. Etudes sur les facteurs ectosémantiques de l'intonation en français. *Travaux de Linguistique et de Littérature de Strasbourg* 1, 143-154.

Mixdorff, H. 1997. *Intonation Patterns of German-model-based Quantitative Analysis and Synthesis of F_0 Contours*. PhD dissertation. Technical University Dresden.

Mixdorff, H. and H. Fujisaki. 1995. Production and perception of statement, question, non-terminal intonation in German. *Proc. 13th ICPhS* (Stockholm, Sweden), vol. 2, 410-413.

Möbius, B. 1993. *Ein Quantitatives Modell der Deutschen Intonation. Analyse und Synthese von Grundfrequenzverläufen*. Tübingen: Niemeyer.

Möbius, B. 1995. Components of a quantitative model of German intonation. *Proc. 13th ICPhS* (Stockholm, Sweden), vol. 2, 108-115.

Nagano-Madsen, Y. 1992. *Mora and Prosodic Coordination*. Lund University Press.

Nespor, M. and I. Vogel. 1986. *Prosodic Phonology*. Dordrecht: Foris.

O'Connor, J.D. and J.F. Arnold. 1961. *Intonation of Colloquial English*. London: Longman.

O'Shaughnessy, D. and J. Allen. 1983. Linguistic modality effects on fundamental frequency in speech. . *J. Acoust. Soc. Am.*74, 1155-1171.

Ohala, J. 1991. The integration of phonetics and phonology. *Proc. 12th ICPhS* (Aix-en-Provence, France), 2-17.

Öhman, S.E.G. 1967. Word and sentence intonation: a quantitative model. *STL-QPSR* 2-3, 20-54.

Ostendorf, M.F. 1998. Linking speech recognition and language processing through prosody. *CCAI* 15 (3), 255-279.

Ostendorf, M.F., P.J. Price and S. Shattuck-Hufnagel. 1993. Combining statistical and linguistic methods for modelling prosody. *Working Papers* 41, 272-275. Dept of Linguistics and Phonetics, Lund University.

Palmer, H.E. 1922. *English Intonation with Systematic Exercises*. Cambridge: Heffer.

Pierrehumbert, J.B. 1980. *The Phonology and Phonetics of English Intonation*. PhD dissertation, MIT (published 1988 by IULC).

Pierrehumbert, J.B. and M.E. Beckman. 1988. *Japanese Tone Structure*. Cambridge, Mass.: MIT Press.

Pierrehumbert, J.B. and J. Hirschberg. 1990. The meaning of intonation contours in the interpretation of discourse. In Cohen, Morgan and Pollack (eds.), 271-311.

Pijper de, J.R. and A.A. Sanderman. 1994. On the perceptual strength of prosodic boundaries and its relation to suprasegmental cues. *J. Acoust. Soc. Am.* 96, 2037-2047.

Pike, K. 1945. *The Intonation of American English*. Ann Arbor: University of Michigan Press.

Pike, K. 1964. On the grammar of intonation. *Proc. 5th ICPhS* (Münster, Germany), 105-119.

Poser, W. 1984. *The phonetics and Phonology of Tone and Intonation in Japanese*. PhD dissertation, MIT.

Potapova, R.K. 1997. *Connotative Paralinguistics*. Moscow: Triada.

Price, P.J., M.F. Ostendorf, S. Shattuck-Hufnagel and C. Fong. 1991. The use of prosody in syntactic disambiguation. *J. Acoust. Soc. Am.* 90, 2956-2970.

Ramsaran, S. (ed.). 1990. *Studies in the Pronunciation in English*. London: Routledge.

Rossi, M. 1965. Contribution à l'étude des faits prosodiques dans un parler d'Italie du Nord. *Revue Langage et Comportement* 1, 5-29.

Rossi, M. 1985. L'intonation et l'organisation de l'énoncé. *Phonetica* 42, 135-153.

Rossi, M. 1993. A model for predicting the prosody of spontaneous speech (PPSS model). *Speech Communication* 13, 87-107.

Rossi, M. 1995. A Principle-based model for predicting the prosody of speech. In Sorin *et al.* (eds.), 159-170.

Rossi, M. 1995. The evolution of phonetics: a fundamental and applied science. *Speech Communication* 18, 96-102.

Rossi, M. 1997a. Intonation: past, present, future. *Proc. ESCA Workshop on Intonation* (Athens, Greece), 1-10.

Rossi, M. 1997b. Is syntactic structure prosodically recoverable? *Proc. EUROSPEECH* '97 (Rhodes, Greece), 1- 8.

Rossi, M. 1999. *L'Intonation, le Système du Français: Description et Modelisation.* Paris: Ophrys.

Rossi, M., A. Di Cristo, D. Hirst, Ph. Martin and Y. Nishinuma. 1981. *L'Intonation, de l'Acoustique à la Sémantique*. Paris: Klincksieck.

Rump, H.H. and R. Collier. 1996. Focus conditions and the Prominence of Pitch-Accented Syllables. *Language and Speech* 39, 1-17.

Safir, K. 1985. Binding in relatives and LF. *Glow Newsletter* 14, 77-79.

Sagisaka, Y., N. Campbell and N. Higuchi (eds.). 1997. *Computing Prosody*. New York: 9-Verlag.

Sanderman, A.A. 1996. *Prosodic Phrasing*. PhD dissertation, Eindhoven University of Technology.

Sanderman, A.A. and R. Collier. 1996. Prosodic rules for the implementation of phrase boundaries in synthetic speech. *J. Acous. Soc. Am.* 100, 3390-3397.

Schubiger, M. 1958. *English Intonation, its Form and Function.* Tübingen: Max Niemeyer.

Scripture, E.W. 1902. *The Elements of Experimental Phonetics*. London: Edward Arnold. (Reprinted by J.W. Black. Columbus: The Ohio state University, 1973).

Selkirk, E.O. 1984. *Phonology and Syntax: The Relation Between Sound and Structure.* Cambridge, Mass.: MIT Press.

Selkirk, E.O. 1990. On the nature of prosodic constituency: comments on Beckman and Edwards paper. In Kingston and Beckman (eds.), 179-200.

Selkirk, E.O. 1995. Sentence prosody: intonation, stress and phrasing. In Goldsmith (ed.), 550-569.

Shaumjan, S.K. 1968. *Problems of Theoretical Phonology*. The Hague: Mouton.

Shaumjan, S.K. 1971. *Principles of Structural Linguistics*. The Hague: Mouton.

Sluijter, A.M.C. and V. van Heuven. 1993. Perceptual cues of linguistic stress: intensity revisited. *Working Papers* 41, 246-249. Dept of Linguistics and Phonetics, Lund University.

Sorin, C., J. Mariani, H. Meloni and J. Schoentgen (eds.) 1995. *Levels in Speech Communication: Relations and Interactions*, A Tribute to Max Wajskop. Amsterdam: Elsevier Science.

Sosa, J.M. 1991. *Fonetica y Fonologia de la Entonacion del Espanol Hispanoamericano*. PhD dissertation, University of Massachusetts.

Stockwell, R.P. 1972. The role of intonation: reconsiderations and other considerations. In Bolinger (ed.), 87-109.

Strangert, E. 1990. Pauses, syntax and prosody. *Nordic Prosody* V, 294-305. Turku University.

Streefkerk, B.M. and L.C.W. Pols. 1996. Prominent accents and pitch movements. *Proc. Institute of Phonetic Sciences*, 111-120. University of Amsterdam.

Streeter, L. 1978. Acoustic determinants of phrase boundary perception. *J. Acoust. Soc. Am.* 71, 1582-1592.

Svetozarova, N.D. 1998. Intonation in Russian. In Hirst and Di Cristo (eds.), 261-274.

Swerts, M. 1994. *Prosodic Features of Discourse Units*. PhD dissertation, Eindhoven University of Technology.

Swerts, M. 1997. Prosodic features at discourse boundaries of different strength. *J. Acoust. Soc. Am.* 101, 514-521.

Terken, J.M.B. 1993. Issues in the perception of prosody. *Working Papers* 41, 228-233. Dept of Linguistics and Phonetics, Lund University.

Touati, P. 1987. *Structures Prosodiques du Suédois et du Français*. Lund University Press.

Trager, G.L. and B. Bloch. 1941. The syllable phonemes of English. *Language* 17, 223-246.

Trager, G.L. and H.L. Smith. 1951. *An Outline of English Structure*. Norman: Battenburg Press.

Truckenbrodt, H. 1995. *Phonological Phrases: Their Relation to Syntax, Prominence and Focus*. PhD dissertation, MIT.

Uhmann, S. 1988. Akzentöne, Grenztöne und Fokussilben: zum Aufbau eines phonologischen Intonationssystems für das Deutsche. In Altmann (ed.), 65-88.

Uldall, U. 1960. Attitudinal meanings conveyed by intonation contours. *Language and Speech* 3, 223-234.

Vaissière, J. 1983a. Language-independent prosodic features. In Cutler and Ladd (eds.), 53-66.

Vaissière, J. 1983b. Une composante suprasegmentale dans un système de reconnaissance: réduction du nombre d'hypothèses lexicales et détection des frontières majeures. *Recherches Acoustiques* (Lannion: CNET), vol. 7, 109-124.

van Donzel, M. and F.J. Koopmans-van-Beinum. 1997. Pitch accents, boundary tones, and information structure in spontaneous discourse in speech. 1997. *Proc. ESCA Workshop on Intonation* (Athens, Greece), 313-316.

Veilleux, N.M. and M.F. Ostendorf. 1993. Probabilistic parse scoring with prosodic information. *Proc. ICASSP* '93 (Minneapolis, MN), vol. 2, 51-54.

Verluyten, P. 1982. *Recherches sur la Prosodie et la Métrique du Français*. PhD dissertation, University of Antwerpen.

Wang, M.Q. and J. Hirschberg. 1992. Automatic classification of intonational phrase boundaries. *Computer Speech and Language* 6,175-196.

Ward, G. and J. Hirschberg. 1985. Implicating uncertainty: the pragmatics of fall-rise. *Language* 61, 747-776.

Wells, R.S. 1945. The pitch phonemes of English. *Language* 21, 27-39.

Whitney, W.D. 1867. *Language and the Study of Language*. London: N. Trübner. (Reprinted by Georg Olms, Hildesheim, 1973).

Wightman, C.W., S. Shattuck-Hufnagel, M.F. Ostendorf and P.J. Price. 1992. Segmental durations in the vicinity of prosodic phrase boundaries. *J. Acoust. Soc. Am.* 91, 1707-1717.

Wodarz, H.W. 1960. Uber vergleichende satzmelodische Untersuchungen. *Phonetica* 5, 75-98.

Wode, H. 1966. Englische Satzintonation. *Phonetica* 15, 3-4, 129-218.

Wunderlich, D. 1988. Der Ton macht die Melodie: zur Phonologie der Intonation der Deutschen. In Altmann (ed.), 1-40.

Section II

Prominence and Focus

3

Acoustic-phonetic Analysis of Prominence in Swedish

GUNNAR FANT, ANITA KRUCKENBERG AND JOHAN LILJENCRANTS

3.1 Introduction

The purpose of our contribution is to introduce subjectively scaled prominence as a reference for studies of the realisation of prosodic events, applicable to the relative weight of syllables and words as well as to the perceived distinctiveness of junctures and other grouping phenomena. The prominence scaling fulfils the need of a descriptive level interfacing prosodic phonological categories and acoustic correlates which allows the establishment of quantified relations between continuously scaled acoustic attributes and continuously scaled response categories. An example is to study how the prominence of a syllable increases with increments of each of a number of relevant acoustic parameters. This tool accordingly adds to the evaluation of prosodic features and has obvious applications in speech synthesis. Portele and Heuft (1997) have adopted the scaling method of Fant and Kruckenberg (1989) in the development of a prominence-based synthesis system. Some suggestions for prominence based synthesis-by-rule of Swedish appear in a recent report by Fant and Kruckenberg (1998).

A novel strategy in prosodic analysis and synthesis is accordingly to adopt a continuously graded prediction of word, syllable and boundary prominence from lexical, syntactic and pragmatic rules, which are to be transformed into corresponding acoustic parameter values. Our ambition is to study not only the traditional prosodic parameters of duration, F_0 and intensity but also parameters related to the respiratory system, the voice source and to the segmental composition, i.e. the entire production process.

The continuous prominence scaling is a prerequisite for quantified studies of parameter co-variations and their joint contribution, e.g. by

A. Botinis (ed.), Intonation, 55-86.

means of regression analysis of acoustic parameters and subjective response. An advantage of prominence encoding compared to traditional dichotomies such as stressed/unstressed, accented/unaccented, focal/non-focal is that it allows a finer grading. However, the prominence scaling does not replace a phonological categorisation but acts as an intermediate supporting level of analysis.

Our ambition has accordingly been to contribute to a maximally broad analysis of prosodic parameters, their co-variation and functional interrelation and of their perceptual impact and phonological significance. We may thus inquire about the role of voice source parameters as intensity determinants and their relation to the transglottal pressure drop. To what extent are stress and focal accentuation mirrored in the sub-glottal pressure profile? What is the relative salience of F_0, duration, intensity and source spectrum tilt as stress correlates? What other parameters should be considered? How is continuously scaled perceptual prominence related to phonological categories of stressed/unstressed and focal/non-focal accentuation? These are pertinent problems that deserve a more detailed penetration than what can be covered in the present report.

Swedish is a stress-timed language with temporal contrast between stressed and unstressed syllables which accounts for a rhythmical impression. In connected speech, function words usually reduce or even loose their stress patterns. Accentuation implies a typical F_0 modulation contour, a choice between accent 1 (acute) and accent 2 (grave). Except for the lowest degree of stress, all stressed words are accentuated (Gårding, 1989; Bruce & Granström, 1993).

We have not yet developed a complete system. Most of our work has been directed towards analysis of various components. Fant, Nord and Kruckenberg (1986) and Fant and Kruckenberg (1989) reported the basic methodology for prominence scaling. The latter is a major reference for our earlier work, mainly on duration. Fant and Kruckenberg (1994) reported preliminary data on the relation between incremental steps of parameter variations and incremental steps of perceived prominence. The emphasis was on duration and F_0 and realisations of the Swedish word accents 1 and 2. Intensity and voice source parameters were also treated, but in less detail. It was found that the semitone scale adopted for F_0-scaling tended to equalise female and male data.

In Fant and Kruckenberg (1994, 1998) we had the occasion to discuss the canonical notations and description of Swedish word accents offered by Bruce (1977). The basic distinction between the two word accents is that the primary accented vowel starts with an L* for accent 1 and with an H* for accent 2. We find that the transition HL* from the preceding syllable towards L* in accent 1, is highly variable and varies inversely with prominence. The main F_0 correlate of prominence in accent 1 is the relative increase following L*. In accent 2 the prominence is marked by the F_0 fall H*L and in addition by the height of an F_0 peak in a following, usually the next syllable. In focal accent 2 the rise towards the secondary peak has the same role as the rise after L* in accent 1. We find evidence for an identity of underlying F_0 control gestures even at lower prominence levels. Our observations support the suggestion of Elert (1995) and Engstrand (1995) that accent 1 should be regarded as unmarked.

The interaction of relative prominence and grouping is an important issue. To what extent can a high degree of prominence introduce a phrase boundary? The return of F_0 to a low level after the focal peak is also found inside a sentence indicating a potential phrase boundary (Bruce, Granström & House, 1992; Bruce & Granström, 1993).

We have devoted much work to duration. A major part has been concerned with the stressed/unstressed dichotomy. Consistent differences are found between the duration of stressed and unstressed syllables of the same number of phonemes per syllable, (Fant, Kruckenberg & Nord, 1991b; Fant & Kruckenberg, 1994; Kruckenberg & Fant, 1995). The contrast is speaker and speaking style dependent and also varies with language. More detailed systems for predicting syllable and phoneme duration are being developed. One of the objects of our present study will be to relate duration to the prominence parameter. We have also been engaged in studies of quasi-rhythmical properties of interstress intervals and tendencies of a quantal nature of the timing of speech events such as the duration of pauses and associated pre-pause lengthening (Fant & Kruckenberg, 1989; Fant & Kruckenberg, 1996a).

Another prominent area of investigations has been the nature of the voice source and how to parameterise source data (Fant, Liljencrants & Lin 1985; Fant, 1993, 1995, 1997; Fant & Kruckenberg, 1995). Earlier work within our department was reviewed in Fant (1993). Of special interest is the voice source spectral tilt as a prosodic parameter. Is it of major importance as suggested by Sluijter and van Heufen (1996) see also Strik and Boves (1992). In recent years we have initiated work directed to the role of subglottal pressure and other aspects of the respiratory system (Fant, Hertegård & Kruckenberg, 1996; Fant, Kruckenberg, Hertegård & Liljencrants, 1997). The ultimate goal of phonetics, to integrate all relevant knowledge concerning production, acoustic pattern, and perception within a linguistic-phonetic frame also applies to the domain of prosodic analysis.

3.2 Syllable and Word Prominence

3.2.1 Experimental Technique. Word Class Dependency

The technique for prominence rating was described in Fant and Kruckenberg (1989). A listener crew was engaged in the assessment of each syllable or word in a read text presented over a loudspeaker in repeated chunks of the size of a sentence. The direct estimate technique involved the setting of a pencil mark on a vertical line scaled from 0 to 30 for each syllable or word. This is a continuous interval scale, which we label Rs. As the only guideline, subjects are told that typical values for unstressed syllables would be Rs=10 and for stressed syllables Rs=20.

The speech material originated from a session (Fant *et al.*, 1996) in which simultaneous measures of sub- and supra glottal pressure had been recorded. The speaker, SH, was a medical doctor specialising in voice research with standard Swedish pronunciation. Fifteen staff members and students graded the entire corpus of 213 syllables within the nine-sentence paragraph of our standard text. Each sentence was played twenty times in succession over a loudspeaker system. The whole test took about an hour to complete. It incorporated a larger corpus of sentences than that of Fant and Kruckenberg (1989) and should be more representative. The standard deviation among the 15 subjects in our listening crew was of the order of 3 Rs-units only, which implies an uncertainty of the means, $0.7\sigma/(N^{0.5})$, of the order of 0.4 units. The greatest spread was found in the secondarily stressed syllables of accent 2 words and compounds.

Our earlier tests from the 1989 study showed that word prominence assessments closely follow those of the dominating syllable in the word, i.e. the syllable carrying maximum stress in isolated lexical pronunciation. Word prominence has accordingly been quantified indirectly as that of the dominant syllable of the word. These data are shown in Table 3.1.

Table 3.1. Syllable prominence Rs versus word class. N = number of words.

Content words	Rs	N	*Function words*	Rs	N
Numerals	22.8	1	Pronouns	12.5	22
Nouns	19.8	31	Prepositions	11.1	18
Adjectives	18.2	5	Auxiliary verbs	10.7	8
Verbs	17.1	22	Others	9.4	18
Adverbs	17.0	6			
Weighted mean	18.6	65	*Weighted mean*	11.0	64

Content words are usually stressed and function words unstressed. However, content words may be de-emphasised and function words may be raised in prominence, but we have also to consider extra high emphasis on some content words and a far going reduction of function words. Averages of Rs=18.6 for content words and Rs=11.0 for function words depart from the expected Rs=20 respectively 10. The reduced contrast may to some extent reflect an uncertainty in stress judgements. In view of the small standard deviation in individual ratings this should not be a major problem.

3.3 Prominence and Duration

3.3.1 Syllable Duration

It has been shown (Fant & Kruckenberg, 1994; Fant, Kruckenberg & Nord, 1992) that average values of syllable duration are highly predictable from a binary stressed/unstressed classification and the number of phonemes in a syllable. Stressed and unstressed syllable duration increases at about the same rate with an increase of the number of phonemes in the syllable. The difference in duration is close to 100 ms. It varies with individual and occasional speaking style, increasing with the distinctiveness of speech. It also serves as a language-contrasting feature (Fant, Kruckenberg & Nord, 1991a; Kruckenberg & Fant, 1995). We find that the duration of unstressed syllables show less variability with speaking style than stressed syllables.

Figure 3.1 presents the data for two subjects, our speaker SH of the present study and speaker ÅJ whose speech is well documented through a larger database of prose reading which has been used for predicting duration from lexical and syntactic structure (Fant *et al.*, 1992). The close similarity of the two data sets is apparent.

In the present study with SH as a speaker, we have set the boundary between the two prominence categories at Rs=15. The duration of syllables in the low prominence, "unstressed", group averaging Rs=11 increased linearly with the number of phonemes N as D = 1.5+ 57N. For syllables in the high prominence, "stressed", group averaging Rs = 18.6, in the range of N equal to or greater than 2 we derived a regression D=104+57N. Phrase final and phrase initial syllables have been excluded from these data.

A first order prediction of syllable duration from Rs-values or the inverse, predicting Rs from duration, may be obtained by interpolation or extrapolation within Figure 3.1 where the two lines for subject SH pertain to Rs=11 and Rs=18,6. These are 100 ms apart which implies an Rs increase of the order of 0.08 units per ms. A comparison of experimentally determined Rs values with those predicted from syllable duration is shown in Figure 3.2.

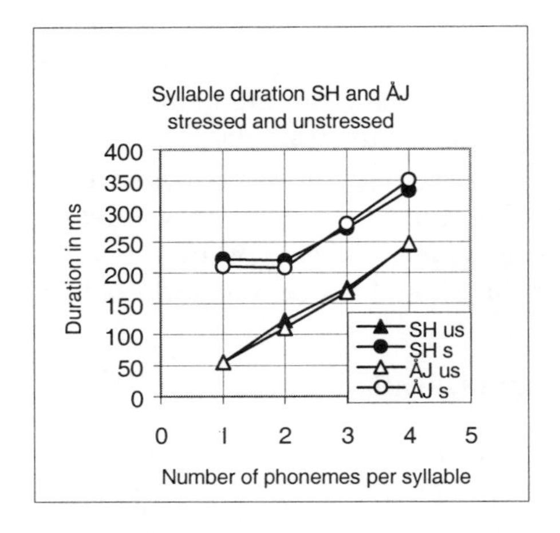

Figure 3.1. Stressed and unstressed syllable duration as a function of phonemes per syllable. Subjects SH & ÅJ.

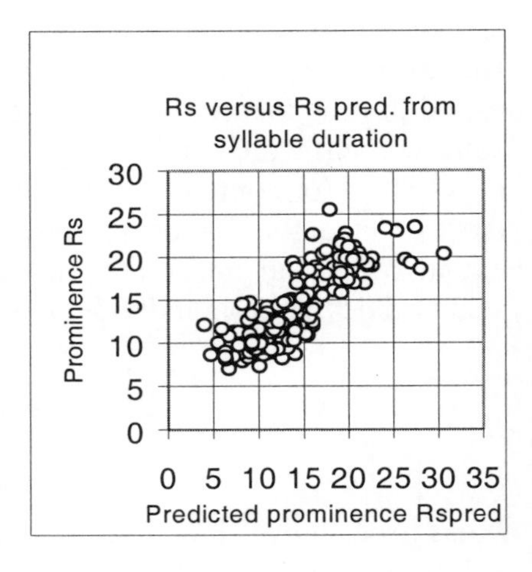

Figure 3.2. Measured versus first order predicted prominence determined from syllable duration.

3.3.2 Vowel Duration

Vowel duration also showed a substantially linear increase with Rs. Correlation coefficients were of the order of r = 0.8-0.9. Data for the short vowel [a] is shown in Figure 3.3.

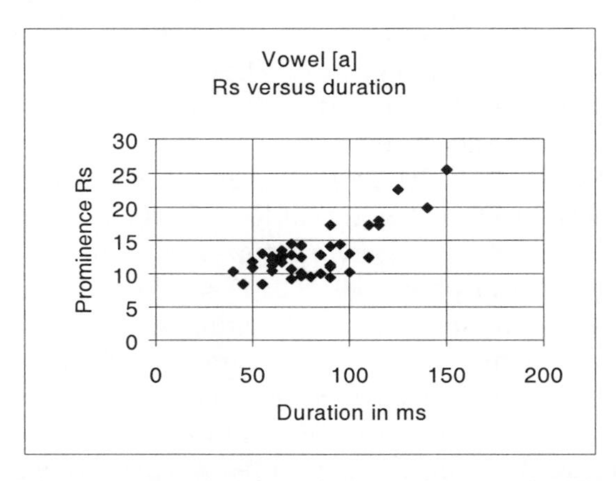

Figure 3.3. Prominence versus duration of all vowels [a] in the SH corpus.

Regression graphs relating prominence to duration were produced for all vowels. Durations corresponding to Rs=10 and Rs=20 were determined and are tabulated below.

Table 3.2. Vowel duration from linear regression.

	Short vowels		*Long vowels*	
	Unstressed	*Stressed*	*Stressed*	
Vowel	Rs=10	Rs=20	Vowel	Rs=20
[ɪ]	30	116	[iː]	140
[ɔ]	51	128	[oː]	220
[ɛ]	43	127	[eː]	155
[a]	53	144	[ɑː]	175
[ɑ]	70		[uː]	200

The long/short ratio of stressed vowels is of the order of 0.7 as expected. (Elert, 1964). Note the relatively small duration of unstressed short vowels which conforms with our earlier studies (Fant & Kruckenberg, 1989).

3.4 Subglottal Pressure

Our subject SH, a medical doctor, served as a subject for measurement of sub- and supraglottal pressures during the recording of a 1 minute long paragraph of a novel, the corpus referred to in the previous sections. A number of contrasting sentences were also included. Subglottal pressure was measured through a tracheal puncturing probe and supraglottal pressure through a probe inserted through the nasal pathways (Fant *et al.*, 1996; Fant *et al.*, 1997).

In order to attain a complete view of prosodic parameters we have combined our standard graphics of oscillogram, spectrogram, F_0, and intensity displays with continuous records of subglottal pressure, Psub and supraglottal pressure, Psup. Examples are shown in Figures 3.4 and 3.5. The assessed prominence Rs of successive syllables is displayed at the top. The bottom graph provides two intensity curves, the sound pressure level in dB with flat frequency weighting, SPL, and above it the sound pressure level with high frequency preemphasis, SPLH, see section 3.5.

The subglottal pressure in connected speech is usually confined to a range below 10 cm H_2O. Higher values are encountered in singing (Sundberg, Andersson & Hultqvist, 1999). The onset of voicing after a pause starts at a threshold Psub around 3-4 cm H_2O, but once voicing is established it can continue down to about 1 cm H_2O. The time constant of the initial rise is of the order of 150 ms. The termination is more variable. Voicing usually starts with a gesture of adduction and ends with an abduction gesture. The declination of Psub within a breath group, excluding a final rapid decay, is of the order of 1.5-2 cm H_2O and is approximately independent of the duration. It reflects the decreasing elastic forces of the respiratory system during the exhalation process.

The supraglottal pressure is close to zero at open articulations and approaches Psub at closure. The net driving pressure for phonation at narrowed articulations is the transglottal pressure, Psub-Psup. A prominent factor determining local minima of Psub within a phrase is glottal abduction as in a voiced /h/ or in a transitional interval towards an unvoiced consonant or an abducted termination of a phrase. During a following articulatory constriction Psub is restored.

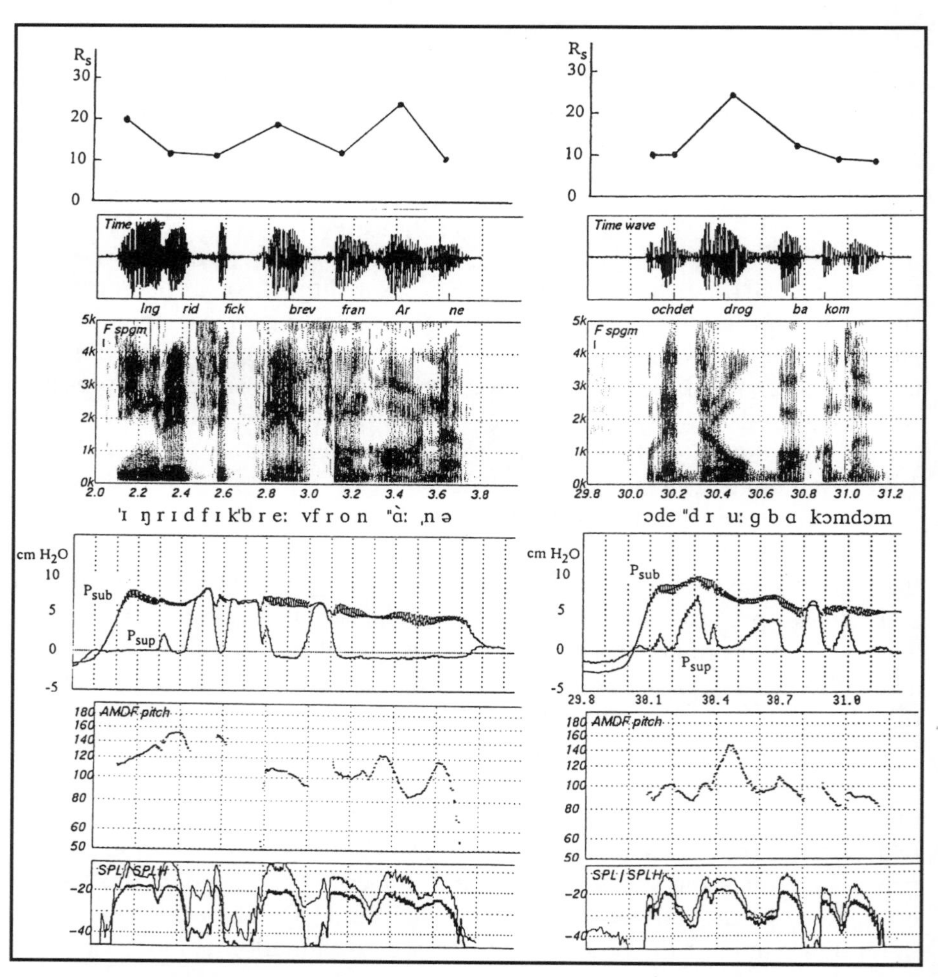

Figure 3.4. "Ingrid fick brev från Arne" Figure 3.5. "Å de drog bakom dom#"

These passive articulatorily induced variations should be kept in mind when sorting out the active respiratory gestures associated with stress, accentuation and emphasis. Our main conclusion is that a gesture of active pulmonary force is usually apparent at higher levels of prominence only and need not occur at moderate stress levels.

An interesting observation is that of a subglottal pressure build up ahead of a sufficiently accented syllable or word towards a maximum at the onset of the primary syllable, i.e. earlier than the F_0 and intensity peaks. In this respect it marks the P-center (Rapp, 1971; Marcus, 1981) of the syllable boundary. Figure 3.4 pertains to a neutral declarative sentence initiating a paragraph: "Ingrid fick brev från Arne" (Ingrid received a letter from Arne). The Psub contour is on the whole evenly declining, but drops somewhat faster in the first stressed syllable, "Ing". Figure 3.5 shows a typical example of focal accentuation with a high degree of prominence. It is a truncated sentence: "Å de drog bakom dom # ..(And there was a draught behind the ..). The Psub contour shows a peak at the onset of "drog" followed by a marked decay within the word. The typical rise-fall F_0 peak centred in the middle of the vowel is thus produced within a falling Psub. This pattern also contributes to marking a boundary between the verb "drog" and the following prepositional phrase.

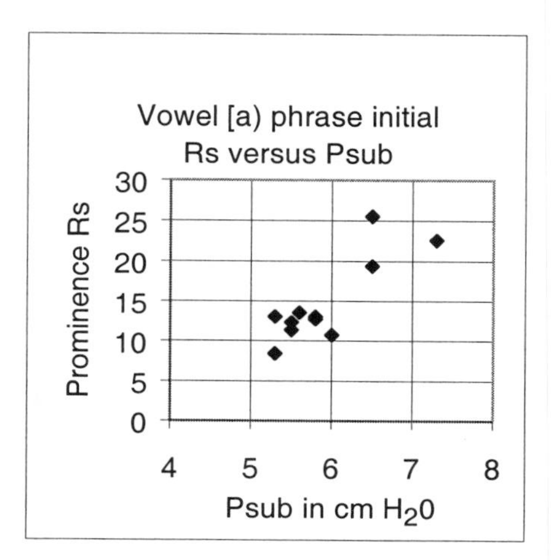

Figure 3.6. Prominence Rs versus Psub. 28.1 + 7.25Psub, (r=0.84).

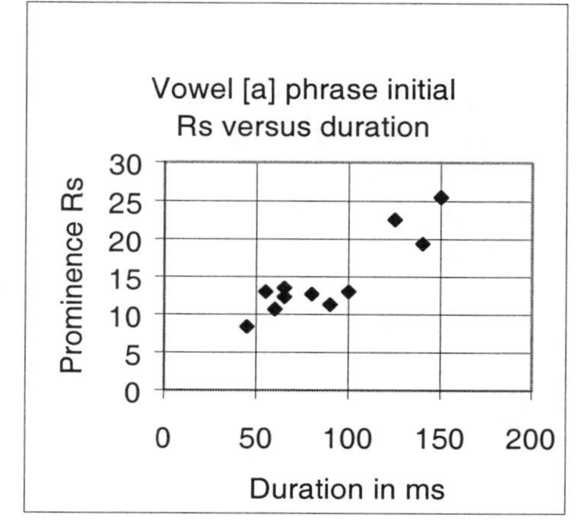

Figure 3.7. Prominence Rs versus duration Rs = Rs = 3.0 + 0.13D, (r=0.89).

The correlation between Psub and Rs based on the entire corpus of the 213 syllables of the prose reading was very weak. However, within a predefined context the situation improves. Figure 3.6 shows prominence Rs versus Psub for vowel [a] in phrase initial position. Here we find Rs=10 at Psub=5.2 and Rs=20 at Psub=6.6. The stressed/unstressed range in Psub is apparently rather small. Other co-varying parameters have to be considered. An example is the graph of Rs versus duration in Figure 3.7 of the same set of [a] vowels as in Figure 3.6. The similarity is apparent.

3.5 Intensity and Source Parameters

The intensity of a speech segment is a function of both a source and a filter, the latter also referred to as the vocal tract transfer function. In a frequency domain view, we can consider the sound spectrum to be a sum of the source spectrum and the vocal tract transfer function. The source supplies the raw material, which is shaped by the transfer function (Fant, 1960, 1973).

Open vowels possess an inherently greater intensity than close vowels, which is related to the transfer function and basically due to the higher frequency location of the first formant of the transfer function which raises the level of the entire spectrum envelope (Fant, 1960, 1973). In addition to the particular pattern of formant frequencies we also have to consider formant bandwidths. In a breathy voice the first formant bandwidth is increased which lowers the amplitude of the first formant peak (Fant, 1993, 1995, 1997; Fant & Kruckenberg, 1996b). The underlying mechanism is a glottal abduction, which affects both the source properties and the transfer function. A detailed mathematical treatment of source-filter decomposition may be found in Fant and Lin (1988).

Much work has been devoted to the analysis of the voice source (Fant, 1993, 1997; Fant *et al.*, 1985; Fant & Kruckenberg, 1996b). The overall intensity level is primarily determined by the slope of the glottal volume velocity air flow pulse at glottal closure, more specifically at the time of maximum rate of change of the slope in the transitional interval towards closure. In the LF model (Fant *et al.*, 1985) this maximum flow derivative is labelled Ee. It serves as a scale factor for formant amplitudes. In a breathy voice the point in time of maximum discontinuity, where Ee is determined, is displaced to a position in the falling branch of the glottal flow earlier than that of apparent closure.

The LF-model has additional parameters for the overall shape of the glottal pulse. One of these is the effective duration of the closing corner of the pulse. The shorter this transitional interval, the greater is the high frequency gain. We quantify this aspect of the source spectrum by the parameter Fa, which is inversely related to the duration of the transitional interval. At frequencies above Fa the source spectrum gradually attains a 6dB/octave steeper slope than at lower frequencies. Fa varies between 100 and 4000 Hz and averages 500-1000 Hz in normal phonation.

An additional source parameter, essentially determined by the overall pulse shape, is the difference in amplitude between the fundamental and the second harmonic, H1-H2, (Stevens, 1994; Stevens & Hanson, 1994). A true measurement of H1-H2 requires an inverse filtering to remove the influence of the transfer function. H1-H2 has now become a popular object for source studies (see Swerts & Veldhuis, 1997). However, the concept of spectral tilt as a candidate of prominence can not be attached to H1-H2 alone. The H1-H2 parameter is indicative of vocal cord abduction and thus of breathiness. It increases with the glottal open quotient. However, as a prominence cue, the Fa parameter or equivalent measures of the source slope, i.e. the "spectral tilt", are probably of greater perceptual importance. In practice, direct time-domain estimates of Fa are time consuming and subject to considerable uncertainty (Fant, 1995; Fant & Kruckenberg, 1996b). We therefore attempt to quantify Fa after having performed a frequency domain analysis by synthesis of all LF parameters. However, because of spectrum irregularities it is often difficult to find a unique match (see also Fant & Lin, 1988). Analytical derivations of H1-H2 and Fa within the constraints of the LF model appear in Fant (1995, 1997). A measure of spectral tilt related to the Fa parameter is the relation of second or third formant amplitude to the amplitude of the voice fundamental. (Stevens & Hanson, 1994; Hanson, 1997a, 1997b). A prerequisite is inverse filtering to remove the formant structure or to restrict the use of the measure to one and the same vowel in contrastive analysis.

As intensity correlates to prominence we have now established two measures. One is the sound pressure level in dB with flat frequency weighting, SPL. The other measure, which we label SPLH, differs from SPL by a high frequency pre-emphasis which introduces a gain G(f) which gradually increases with frequency:

$$G(f) = 10\log10\{(1+f^2/200^2)/(1+f^2/5000^2)\} \quad \text{dB} \quad (1)$$

This is our standard pre-emphasis function. At 200 Hz the gain is 3 dB, at 1000 Hz 14 dB, at 5000 Hz 25 dB.

SPLH is more sensitive to variations in the region of the second and the third formant, F2 and F3, than is SPL and should accordingly provide a better match to the concept of sonority. Moreover, the difference SPLH-SPL is related to the contribution of formants above F1. We will adopt SPLH-SPL as a measure of spectral tilt, but with an understanding that it is influenced not only by the source but also by the filter function, i.e. by Fa as well as by the particular formant pattern.

Graphs of SPL, SPLH and SPLH-SPL of the same set of [a] vowels as in Figures 3.6 and 3.7, are shown in Figures 3.8-3.10.

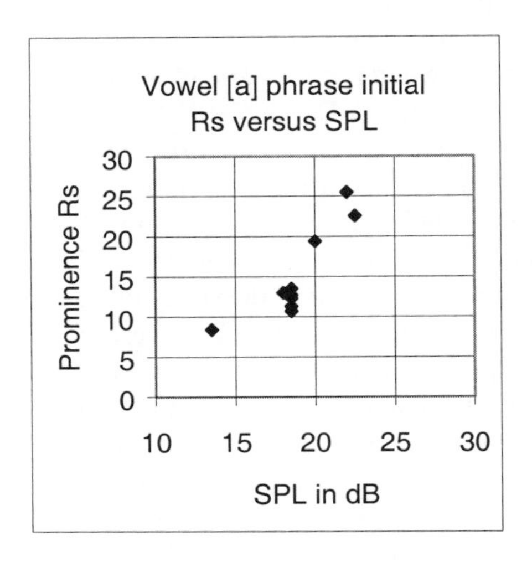

Figure 3.8. Prominence Rs versus SPL. Rs = Cspl + 1.8SPL (r=0.79)

Figure 3.9. Prominence Rs versus SPLH. Rs = Csplh + 1.3SPLH (r=0.91)

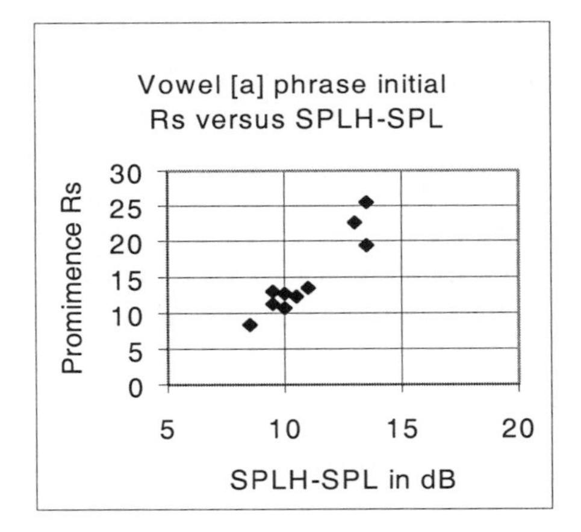

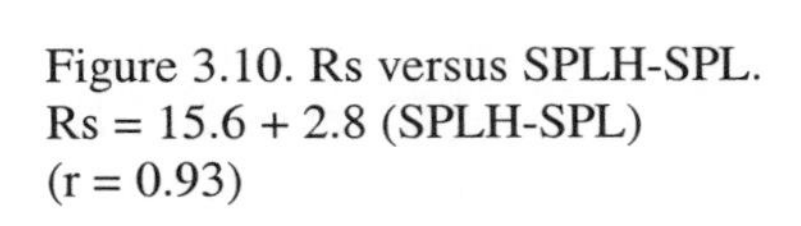

Figure 3.10. Rs versus SPLH-SPL.
Rs = 15.6 + 2.8 (SPLH-SPL)
(r = 0.93)

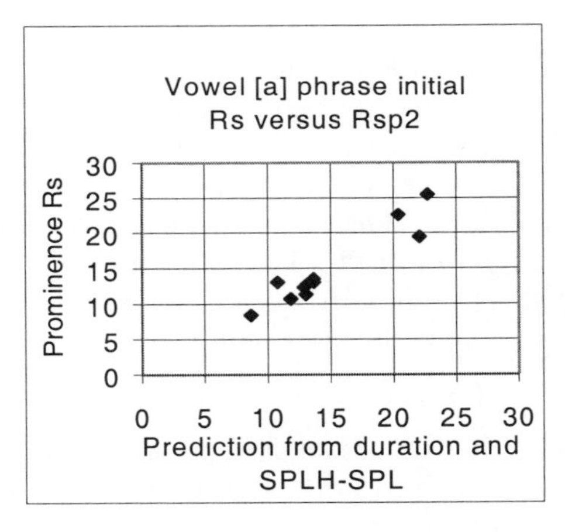

Figure 3.11. Rs versus Rs predicted from duration and SPLH-SPL
Rs = -1.1 + 1.07Rsp2 (r =0.95)

Rs may be predicted from the combined data on duration and SPLH-SPL. As noted in Figure 3.11, a correlation coefficient of r = 0.95 was achieved. The gain is rather small as could be expected in view of the already high correlation, r=0.89 for duration and r=0.93 for SPLH-SPL.

3.6 Fundamental Frequency

In dealing with the F_0 correlates of prominence we adopt the canonical description and notations of Bruce (1977) of the Swedish word accents 1 and 2. A few minor additions and a more specific interpretation of the L* of accent 1 is introduced. According to Bruce the basic accent 1 constituent is a HL* fall from the preceding syllable to the accented syllable, whereas accent 2 has an H*L fall within the primary stressed syllable. H* is located close to the left boundary of the stressed vowel. In the early synthesis rules of Carlson and Granström (1973) the rise up to H* starts from an F_0 minimum in the preceding accentuation. We have adopted a more detailed view by incorporating a starting point labelled Lo situated about 50 ms ahead of H* with an F_0 level about half way between H* and L, see Figures 3.4 and 3.18.

The addition of a sentence (focal) accent adds an F_0 rise to a value which we have denoted as Ha for accent 1 (acute) and Hg for accent 2 (grave). The Ha is usually located at the end of the accent 1 primary syllable, but the high F_0 may carry over to the next unstressed syllable indicating coherence (Bruce & Granström, 1993). In accent 2 the Hg is located in the following syllable or, in case of a compound accent 2, in a later syllable carrying secondary stress. The Hg, when prominent, produces a \secondary peak of accent 2. Our more specific notations Hfg and Hfa imply focal prominence, see Figure 3.18. An accent 1 or accent 2 word of sufficiently high prominence is usually terminated by an F_0 fall towards a low boundary location which we label Lt. The presence of the Lt introduces a grouping which may add to the prominence. In summary we will use the following notations.

Accent 1: (HL*) (Ha) (Lt) Accent 2: (LoH*L) (Hg) (Lt)

The three components we have put into parentheses above, the basic accent contour, the sentence accent (focal peak) and a terminating low tone reflect more or less independent production components, which are superimposed. The resulting blending has been pointed out by Bruce (1977). A quantitative decomposition will not be attempted here. We accordingly apply the symbols above to actually observed F_0 values in speech. However, a support for a local superposition view (Fant &

Kruckenberg, 1998) to be elaborated here, is that Ha and Hg exhibit quantitative equivalence in a range of prominence levels, not only focal, reflecting identical production gestures.

Increasing prominence of an accent 1 word introduces an increase of the sentence accent component, here represented by Ha. The rise of F_0 up to Ha has an early start and may elevate the observable L* to an F_0 level equal to or even higher than that of the preceding H (Fant & Kruckenberg, 1994). At low and moderate prominence levels the F_0 contour is usually falling within the accent 1 stressed syllable. All the same, in order to secure a consistent data analysis, our convention has been to locate L* at the left boundary of the stressed vowel irrespective of the following F_0-contour being rising, flat or falling.

Within a breath group, usually coinciding with a sentence, a clause or a major phrase, the declination in F_0 was of the order of 4 semitones in F_0 and tended to be rather independent of the duration. The corresponding declination in Psub was of the order of 2 cm H_2O. F_0 accentuation data were sampled at five positions within a clearly bounded group, usually a breath group. Position 1 pertains to a stressed syllable initiating a group and position 5 implies a group final syllable carrying main or secondary stress. Positions 2 and 4 were next to terminal locations. Position 3 included all intermediate positions.

In order to exclude terminal specific effects and to maintain a sufficient number of observations, accent 1 words were pooled from the locations, 2,3,4. Figure 3.12 shows prominence Rs as a function of H-L*.

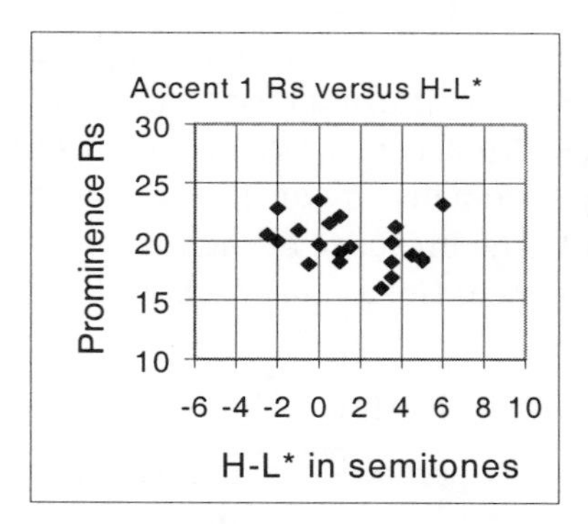

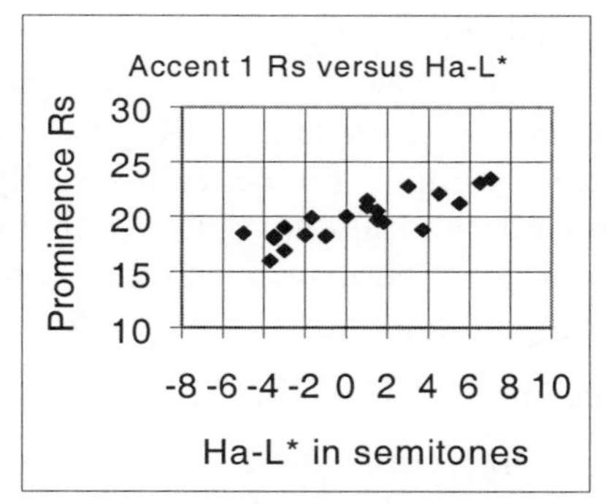

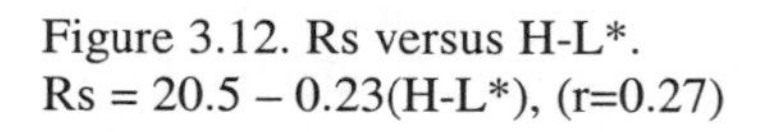
Figure 3.12. Rs versus H-L*.
Rs = 20.5 – 0.23(H-L*), (r=0.27)

Figure 3.13. Rs versus Ha-L*.
Rs = 19.6 + 0.48 (Ha-L*), (r=0.86)

At high prominence levels H-L* is predominantly negative, L* exceeds H as discussed above. The spread in H is considerable. An H-L* fall, when present, signals accent 1 but is not useful as a prominence parameter. A better correlate, see Figure 3.13, is the step from L* to the following Ha. which implies that the transition from L* to Ha tends to be flat in the region of Rs=20. At higher Rs values the sentence accent sets in by a positive slope, Ha>L*. The slope tends to be negative at low stress levels. In the initial position, 1, the accent 1 rise from L* to Ha of subject SH achieved a boost of 2-4 semitones which derives from the phrase contour.

The critical Rs at which the sentence accent Ha approaches L* appears to be speaker and dialect dependent. In the study of Fant and Kruckenberg (1994) the critical Rs level observed (subject ÅJ) was lower than in the present study (subject SH), i.e. the L* to Ha transition more often showed a positive slope.

The presence of accent 2 is signalled by the typical (H*-L) fall. In connected speech it also serves as an Rs correlate supported by the following rise from L to Hg, see Figures 3.14 and 3.15 which pertain to final positions. However, as will be discussed in connection with Figure 3.18, at higher prominence levels above Rs=22 typical of focal accentuation, the (H*-L) measure saturates and the Hg peak takes over the role of major Rs correlate.

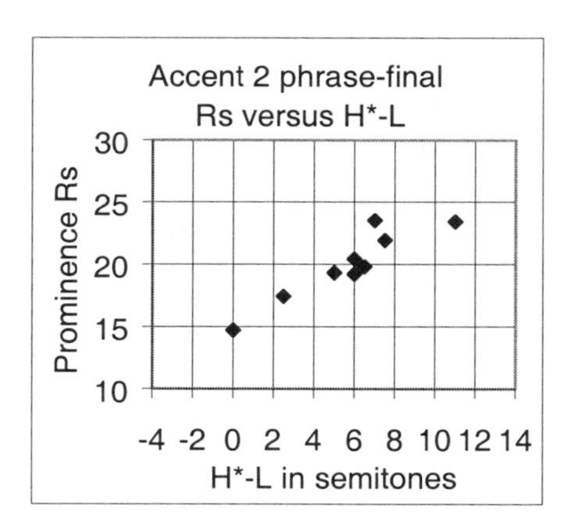

Figure 3.14. Rs versus H*-L.
Rs = 15,1 + 0.84(H*-L), (r=0,92)

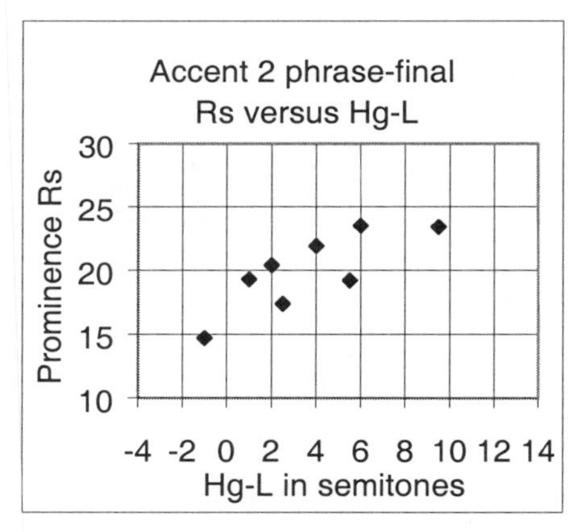

Figure 3.15. Rs versus Hg-L
Rs = 17.2 + 0.74(Hg-L), (r=0.84)

As seen in Figure 3.15 the relative level of the secondary peak Hg also provided a good correlation to Rs. In non-final positions the accent 2 data showed greater spread and a weaker correlation of Rs with (H*-L) and

(Hg-L). This is mainly due to infrequent occurrence of sentence accent. We had no examples of accent 2 in the first word of an utterance.

The degree of correlation between F_0-parameters and prominence observed here is greater than what was found by Strangert and Heldner (1995) who adopted a simplified measure of prominence based on average listener response within a four-level scale (0,1,2,3).

The variation of the accent 1 parameters H and L* and of the accent 2 H* and L as a function of location within an utterance is shown in Figure 3.16. The data points pertain to average values at Rs=20. The accent 2 data dominate the F_0 span. The upper boundary is set by H* and the lower by L. The size of the H*-L drop is close to 6 semitones independent of the position. The accent 1 parameters H and L* are situated in the middle of the graph. The distance between H and L* is rather small and of the order of 0-2 semitones. However, these average values obscure the large spread with incidental occurrences of (H-L*) drops of a greater magnitude as well as incidents of negative values. A large variability, specially at low Rs values, originates from the superimposed phrase or sentence intonation.

Unaccented syllables attain F_0 values close to L and were on the average 0.5 semitones above L. The graph also shows the overall declination, which is of the order of 0.75 semitones per position step.

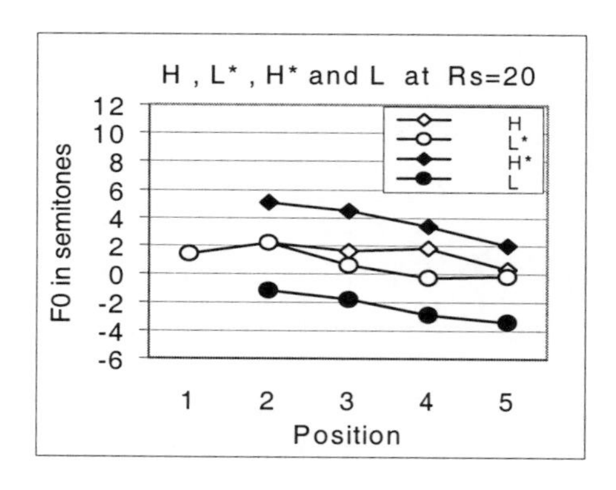

Figure 3.16. Positional variation of accent 1 and 2 parameters.

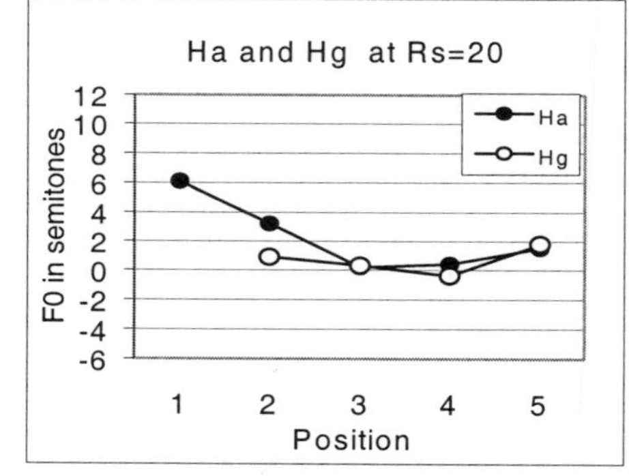

Figure 3.17. Positional variation of Ha and Hg

Figure 3.17 illustrates the quantitative equality of Ha and Hg in locations 3,4 and 5 at Rs=20. In positions 1 and 2 the phrase initial rising F_0 contour affects the observed Ha level. The equivalence was also found to hold for lower and higher values of Rs, which supports the identity in contexts other than focal accentuation.

3.7 Focal Accentuation

The concept of focal accentuation is not entirely a matter of physical prominence. Semantic factors, the relative novelty of a word and a potential presence of multiple focal accentuation may obscure a straightforward definition of focal accentuation. Still, we may expect that a prediction of relative prominence will preserve essential requirements for synthesis.

In addition to our present corpus of prose reading, which is now being extended to other speakers than SH, we have data from specially constructed lab sentences. Our collected experience supports the validity of the semitone scale which preserves relational measures and provides a reliable normalisation of female and male data.

3.7.1 F0 Correlates

Figure 3.18 illustrates a lab sentence, which is varied with respect to focal versus non-focal accent 1 and accent 2. It was spoken by a male subject SH and a female subject AK.

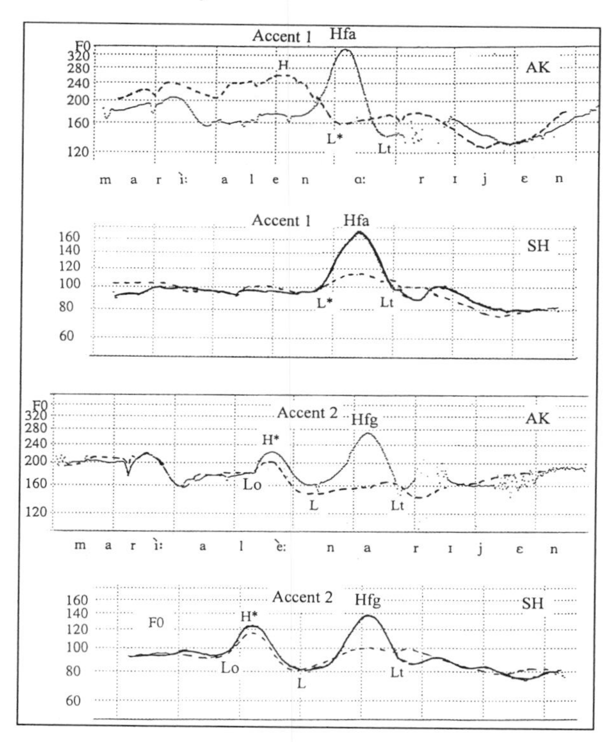

Figure 3.18. Accent 1 and accent 2. Male subject SH, female subject AK. Solid lines focal accentuation, broken lines neutral version. F_0 in log scale.

The accent 1 version was “Å Maria Lenar igen”, [ɔmaɹ̆iːalɛ"nɑːɹɪ'jɛn], “And Maria Lenar again”. The sentence stress, here denoted Hfa, is located on the second syllable of “Lenar”. The accent 2 version was “Å Maria lenar igen”, [ɔmaɹ̆iːa"lɛ̀ːˌnaɹɪ'jɛn], “And Maria is soothing again”. Here, the sentence stress Hfg falls on the second syllable of the word “lenar” and the primary stress, H*, on the first syllable. Observe the rather large height of the Hfa peak, 12 semitones for AK and 10 semitones for SH. The Hfg peak was 9 semitones for both AK and SH. These focal accentuations are realised as single peaks, pointed hat-patterns (Collier, 1991), of an Hf rise and a Lt fall with a base of 200 ms. They are absent or very weak in the non-focal versions. For both subjects the H*L fall in accent 2, of the order of 5 semitones, is only marginally greater in focal than in non-focal accentuation. Observe how subject SH produces an almost complete reduction of F_0 modulations before and after the focally stressed syllable.

We shall now return to the prose text and show how five subjects, 2 females and 3 males, executed focal accentuation of varying positions within the initial sentence of a paragraph: “Ingrid fick brev från Arne” ['ɪŋɹɪdfɪk'bɹeːvfɹon'ɑːɳɛ] “Ingrid received a letter from Arne”. This is the sentence already shown in Figure 3.4. All words are lexically accent 1 except the last word, “Arne” which carries accent 2. We do not yet have Rs data for this corpus. Our object will be to demonstrate F_0, duration, intensity and spectral slope as acoustical correlates of the simple dichotomy, neutral (default) compared to focal contrastive accentuation.

In order to simplify the analysis we have represented F_0 by maximum values in each syllable. This will in effect bring out the Ha of accent 1 and the H* and the Hg of accent 2. In order to concentrate on relational properties we shall mainly study changes in parameter values. This can be achieved in several ways. One is to demonstrate how a specific focal position affects the parameter values of each syllable in the sentence compared to the values observed in a neutral reading without intended focus. This will bring out focal emphasis as well as de-emphasis of preceding and following parts of the utterance.

However, the intended neutral version is not free from focus. There is default prominence in the first word related to its novelty and there is a sentence final default prominence of the word “Arne”. One way to single out this initial bias is to compute a neutral non-focal reference. For each syllable we have accordingly calculated the mean of a group of five conditions, that of the neutral version and the values the syllable obtains when focus is placed on the other four words of the utterance. We shall refer to this as the weighted neutral reference as an alternative to the neutral reading alone.

In the following graphs we have denoted the seven syllables of the sentence as 1=Ing, 2=rid, 3=fick, 4=brev, 5=från, 6=A, 7=rne. The six focal conditions are denoted as 1=neutral, 2=Ingrid, 3=fick, 4=brev, 5=från, 6=Arne.

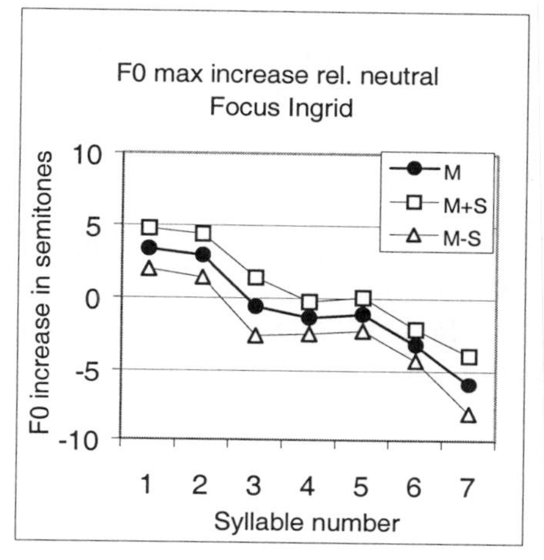

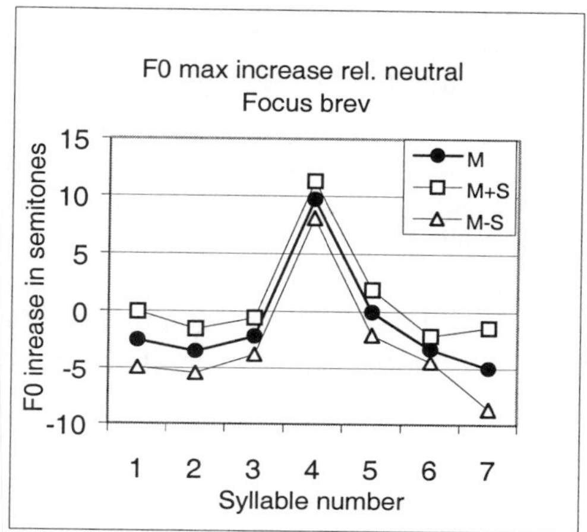

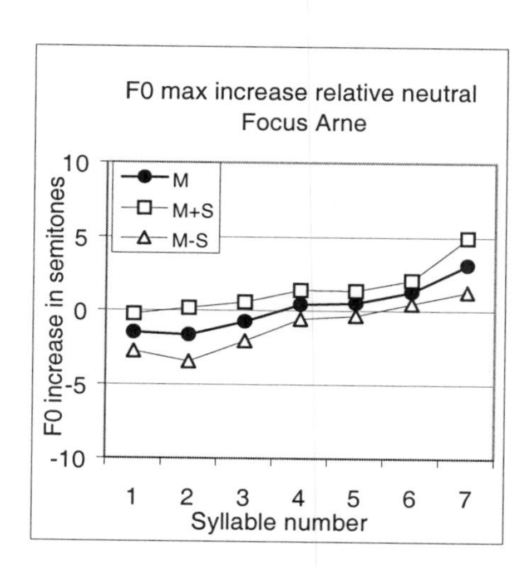

Figure 3.19. Focally induced F_0 change at successive syllables in the sentence "Ingrid fick brev från Arne" for three locations of the focus, *Ingrid, brev and Arne.* Means plus and minus one standard deviation. Five subjects.

Figure 3.19 shows the contours of F_0 shift of each syllable of the sentence when focus is laid on “Ingrid”, “brev“ and “Arne” respectively. The ordinate is the difference between F_0 max in the particular utterance and in the neutral utterance. The three curves represent the mean of the five speakers and the mean plus and minus a standard deviation. The individual spread is moderate. The focal regions, intonation pivots in the terminology of Gårding (1981) are apparent. There is a clear de-emphasis of F_0, not only after but also before the focal maxima, which increases with the distance from the focal word. The post-focal reduction is somewhat more apparent than the pre-focal reduction. This is a well known effect (Bruce, 1982). De-emphasis is also associated with a reduced local modulation depth, a compressed grid, Gårding (1981). It can be quantified by our measures Ha-L* and H*-L versus prominence Rs, see Figures 3.13 and 3.14. With decreasing Rs the lower bounds L* and L vary less than Ha and H*.

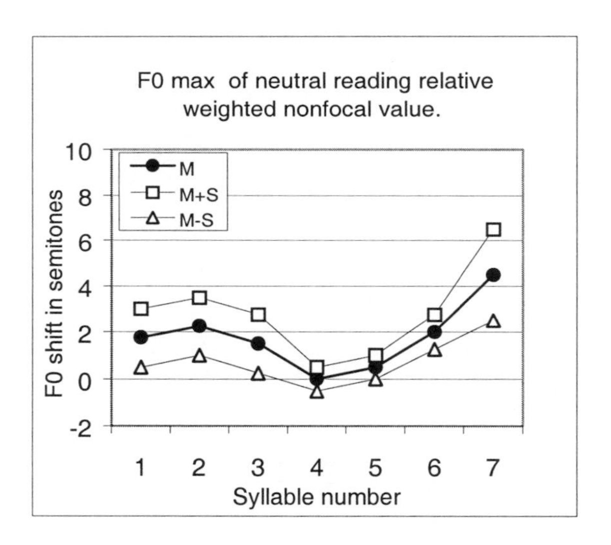

Figure 3.20. The difference between the neutral and the weighted neutral F_0 of each successive syllable. Mean of five subjects plus and minus a standard deviation.

Figure 3.20 shows the difference between the neutral and the weighted neutral mean of each successive syllable in the sentence. The relative prominence of the first and the final word of the neutral sentence indicating a default focus is apparent.

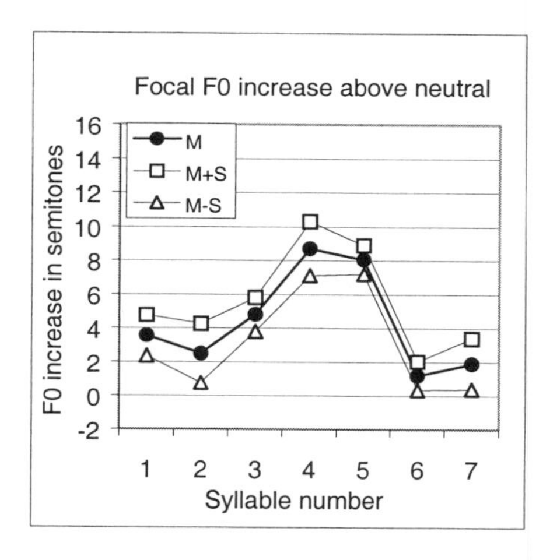

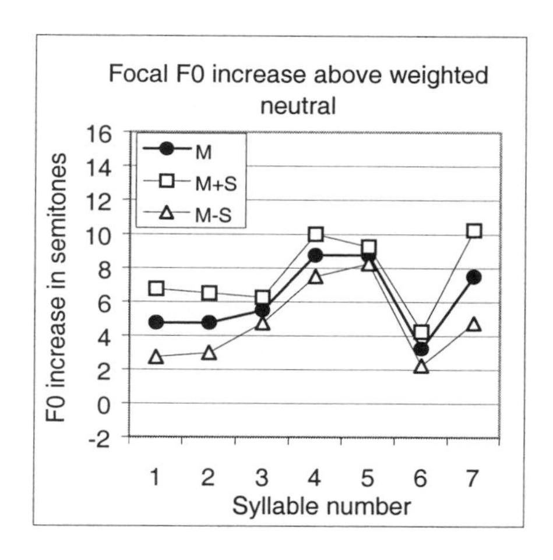

Figure 3.21. Focal F_0 increase of each syllable above its neutral reference, left, and above its weighted reference, right. Means plus and minus one standard deviation, five subjects.

Figure 3.21 shows the focal gain (F_0focus – F_0neutral) of each successive syllable within the sentence, and to the right the gain with the weighted neutral value as reference. The latter provides a more direct measure of focal increase above an essentially unaccented level.

3.7.2 Duration Correlates

The five speakers showed a high degree of uniformity with respect to syllable duration. The inter-subject standard deviation averaged 20 ms for the neutral readings. Average duration also showed a substantial agreement with lexically and contextually predicted values. As with the F_0 data above, the analysis will be directed to variations rather than to absolute values.

Figure 3.22 shows how much the duration of each successive syllable of the sentence is increased relative to the neutral reading when in focus. Mean values, M, and means plus and minus one standard deviation, M+S and M-S, are included. Syllable 5, the preposition "från", attains the highest focal increase. In view of its low prominence in neutral context this is to be expected. We noted that the focal version sounded somewhat exaggerated.

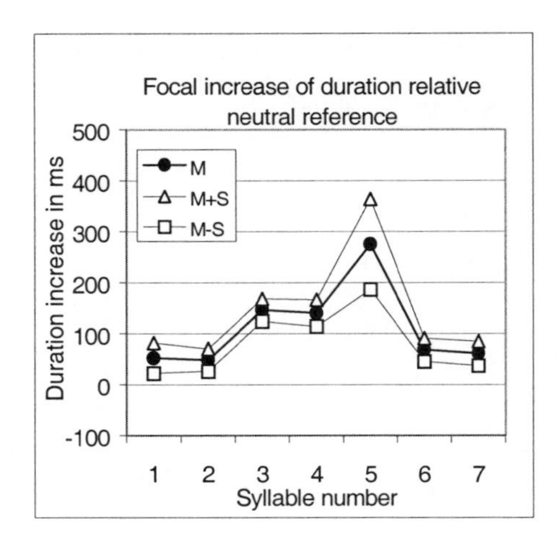

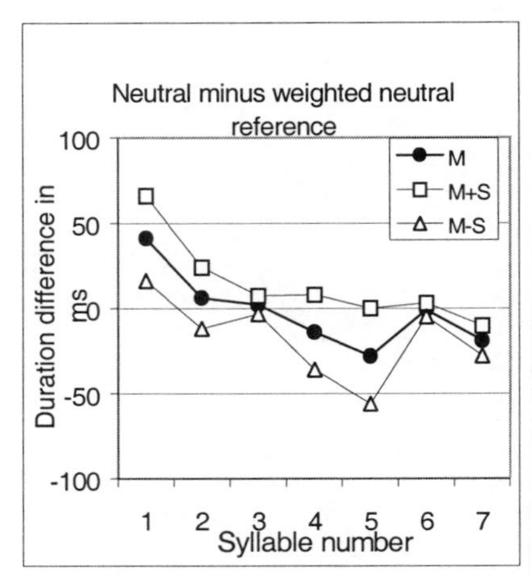

Figure 3.22. Focal increase of duration with the neutral reading as reference, left. Neutral minus weighted neutral reference, right. Mean values plus and minus a standard deviation, five subjects.

The initial noun "Ingrid" and the final noun "Arne" both have a fairly high prominence in the "neutral" reading. The duration increase with added prominence is here more modest and of the order of 50-70 ms in the primary as well as in the secondary syllables. We lack Rs data for our five subjects' readings in the variable focus test. However, the Rs values obtained from an earlier recording of the male subject SH, Figure 3.4, may serve as a provisional reference for the "neutral" versions of Rs=20 for "Ing" and Rs=24 for "A". An approximate prediction of the increase in Rs with focus may be obtained from the general relation between syllable prominence and duration, see the comments to Figure 3.1. With 0.08 Rs units per ms syllable increase we may thus predict an Rs increase of 60x0.8 = 5 for syllable nr 1, "Ing" of Ingrid and 70x0.08 = 5.5 Rs units for syllable nr 6, "A", of Arne. The resulting "focal" Rs-values should be of the order of 20+5=25 for "Ing" and 24+5.5=29.5 for "A". These rather approximate predictions will be followed up by perceptual assessments.

The difference between the neutral and the weighted, "non-focal" duration references in the right hand part of Figure 3.22 supports the view that the "neutral" version of "Ingrid" possesses an inherent default prominence, backed up by a 40 ms difference in the first syllable of Ingrid. No duration difference was found in the "A" of the final word "Arne".

3.7.3 Intensity Correlates

The gain in SPL of each syllable when in focus compared to the SPL in neutral reading is shown in Figure 3.23. Averaged over the five subjects it is of the order of 4-6 dB and does not show any large positional variations.

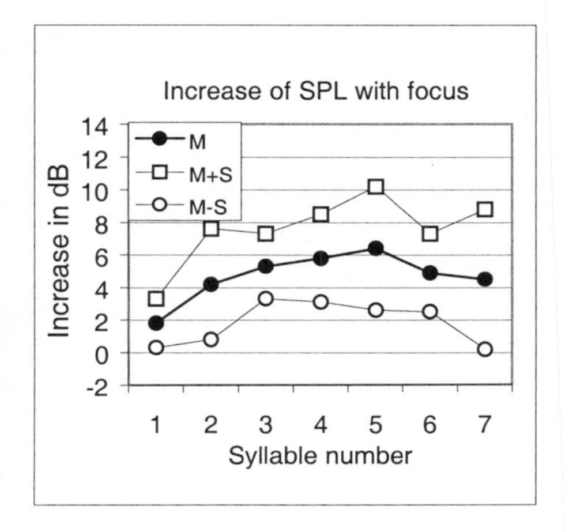

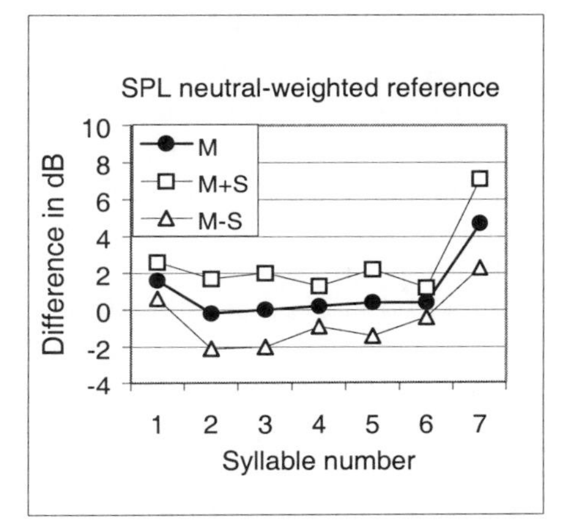

Figure 3.23. Increase of sound pressure level, SPL, with focus, left, and the difference between the neutral and the weighted reference of each successive syllable, right. Means plus and minus a standard deviation of five subjects.

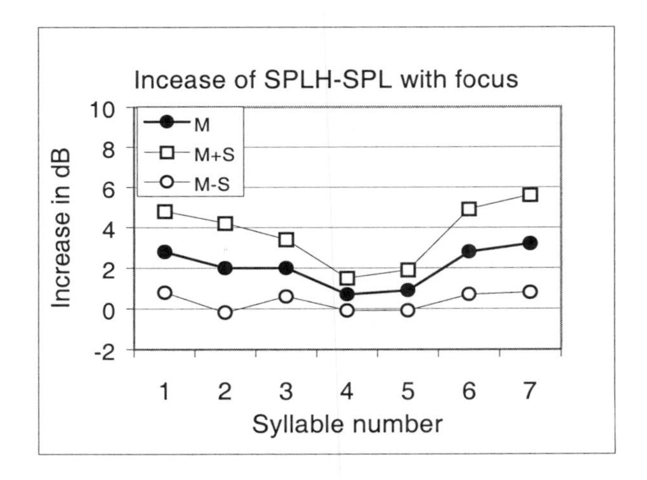

Figure 3.24. Increase of SPLH-SPL with focus in each successive syllable.

As can be seen in Figure 3.23 the SPL data from the neutral reading do not depart much from the non-focal weighted reference. Exceptions are the first and last syllables of the sentence where the neutral reading exceeds the weighted reference by 2 and 4 dB respectively.

The spectral tilt parameter, SPLH-SPL, see Figure 3.24, is a fairly consistent focal stress correlate. It is of the order of 2-3 dB, except for syllables 4 and 5, i.e. “brev” and “från” where it is of the order of 0-1 dB only. As earlier discussed the spectral tilt depends on both the source spectrum and possible changes in the formant pattern. We do not yet have sufficient insight into the relative role of the two components in this example, but it seems probable that the low sensitivity to SPLH-SPL of syllables 4 and 5 , “brev” and “från” is due to a constancy of their formant patterns with increased articulatory effort. Both vowels, [e:] and [o:], are long and non-open and thus in the boundary region towards vowels which attain a more closed articulation with increasing effort.

A conclusion from our collected data is that the SPLH-SPL parameter usually increases by about 0.5 dB per 1 dB increase of SPL. This has been verified by a number of regression studies for all of our subjects. Individual variations exist but the examples shown in Figure 3.25 pertaining to our reference female subject AÖ and the male subject GF are representative.

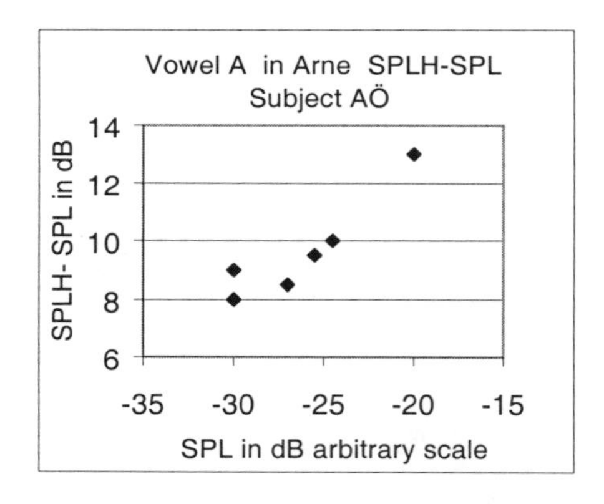

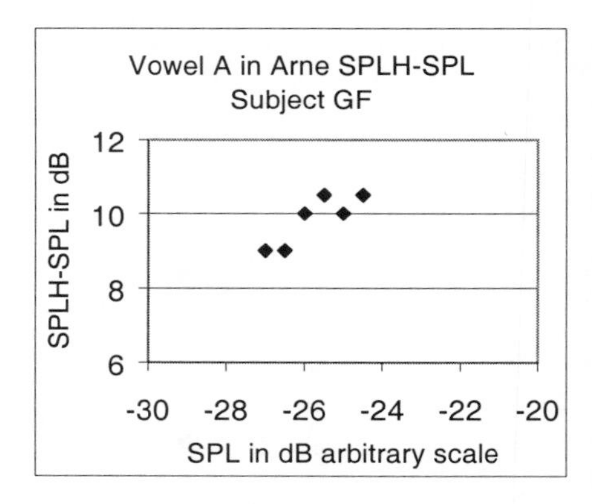

Figure 3.25. The growth of SPLH-SPL with increasing SPL. Female subject AÖ, left, male subject GF, right.

In the AÖ data the slope is 0.43 dB (SPLH-SPL) per dB increase of SPL and the correlation coefficient is r=0.92. Corresponding values for GF are: slope 0.63 and r=0.86. Subject AÖ apparently covered a greater intensity range than GF.

3.7.4 Additional Prominence Correlates

A single word or a prosodic word of high focal prominence is often prompted by a short pause and terminated by an F_0 fall which has the combined function of marking a juncture and adding prominence. As already discussed, the resulting F_0 contour is a single LHL peak where the H is the Hfa of accent 1 or the Hfg of accent 2. In accent 1, at a moderate stress level, F_0 tends to stay high in an unstressed syllable following Ha, which signals coherence, but when the prominence is raised above a certain level F_0 shows a fall which generally starts already in the stressed syllable. This was observed in the syllable "Ing" in focus compared to the neutral reading. Another example was presented in Figure 3.5 where the verb "drog" (*There was a draught*) is produced with a high prominence. The (L* Ha Lt) peak in the vowel [u:] is combined with a decaying subglottal pressure contour which adds to marking the juncture towards the following prepositional phrase.

The intensity contour associated with the high prominence F_0 peak has interesting properties described in a few earlier publications (Fant & Kruckenberg, 1996b; Fant *et al.*, 1996; Fant, 1997). At a critical high F_0 the normal rise of intensity with F_0 was found to level off in the peak region. If combined with falling subglottal pressure there may also appear a minimum in the top of the peak area. Another instance of a juncture adding to prominence is the marking of a voiced word boundary by a glottal stop. This can be seen in the spectrogram of Figure 3.4 as a narrow intensity dip in the boundary between the words "från" and "Arne". The intensity and spectral contrast between a consonant and a following vowel within a stressed syllable increases with prominence. A typical example is that the local intensity minimum associated with an [ɹ] dip increases in depth. Similarly, the contrasts in SPLH and SPLH-SPL between a nasal or liquid and a following vowel are observed to increase with Rs.

These prominence sensitive contrast effects may be related to changes in formant patterns towards those of more extreme articulations. But the relations are fairly complex. Open and short vowels tend to be articulated more open with emphasis, while long close vowels and consonants tend to approach more close targets. A typical example is the Swedish long vowel [i:] which when highly stressed may approach a target of complete closure. The prominence is then conveyed by the apparent mode of production rather than by the intensity. Another example is that increased emphasis may change an [h] segment from voiced to unvoiced by increased vocal cord abduction. These phenomena are all part of the hypo/hyper dimension in articulatory control (Lindblom, 1990).

3.8 Summary of Acoustical Correlates of Stress and Focus

At various stages of our studies we have dealt with two different systems for classifying prominence. We have used a simple auditory stressed/unstressed labelling as the base for studies of syllable duration as exemplified in Figure 3.1 and in studies of vowel and consonant duration (see also Fant & Kruckenberg, 1989). This binary categorisation has influenced our frame for the subjective prominence scale by our choice of Rs=10 for unstressed and Rs=20 for stressed syllables as initial guidelines for the scaling in perceptual assessments. In a more differentiated system, see Table 3.3, five levels of accented syllables are specified.

In retrospect we now are in a position to project the Rs scale onto established prosodic categories as suggested in table 3.3. There are no sharp acoustic boundaries.

Table 3.3. The prominence parameter Rs and prosodic categories.

Category	Rs
A. Binary system	
Lower stress ("unstressed")	Rs<15, average Rs=11
Higher stress ("stressed")	Rs>15, average Rs=19
B. Differentiated system	
Unstressed	Rs = 5-12
Stressed unaccented	Rs =12-15
Stressed accented	
non-focal accentuation	Rs =15-20
focal accentuation	Rs =20-25
higher degrees of prominence	Rs =25-35

As a rule of thumb the step from Rs=11 to 19 of the binary system above is associated with an increase of syllable duration of 100 ms irrespective of the number of phonemes in the syllable, see Figure 3.1. By extrapolation we may expect a step of 10 Rs units, e.g. from Rs=10 to 20 or from Rs=15 to 25, to provide a difference in syllable duration of 125 ms. The vowel part of a syllable accounts for about 80 ms.

The focal lengthening of the secondary syllables of the disyllabic first and last words of the test sentence, see Figure 3.2.2, was of the same order of magnitude as that of the primary syllables. The domain of duration increase in focal accentuation is accordingly the whole word (Strangert & Heldner, 1998).

The step from the lower bound of accentuation, Rs=15, to the upper bound of focal not extreme prominence, Rs=25, is associated with about 8 semitones rise in the "sentence accent" parameters Ha and Hg. The H*-L accent 2 fall shows a smaller range of variation in this Rs interval, an increase of the order of 4 semitones from 4 to 8 semitones. At higher prominence levels, Rs=25-35, the H*-L fall tends to be constant and increasing prominence is carried by Hg alone.

Intensity, in terms of SPL, shows about 6 dB increase from Rs=15 to Rs=25. At the same time the high frequency weighted SPLH increases with about 9 dB which implies an increase of the spectral tilt parameter (SPLH-SPL) of the order of 3 dB. It is speaker and context dependent. An observed increase of (SPLH-SPL) is in part a matter of a more efficient voice source with a more abrupt vocal fold closure but, as we have already noted, it is also influenced by a shift in the formant pattern towards that of an inherent higher overall intensity (Sluijter & van Heufen, 1996; Hanson, 1997b). We have found the influence of the formant pattern to be especially apparent comparing accented and unstressed versions of the Swedish short vowel [a], the latter approaching a neutral schwa when de-emphasised.

There is a need for detailed inverse filtering studies to sort out the source and filter contributions and the particular influence of the F_0 level. Anyhow, irrespective of underlying mechanisms, SPLH is more closely related to sonority than is SPL, and the (SPLH-SPL) serves the function of an easily accessible measure of spectral tilt. The intensity contrast between voiced consonants and vowels is much more apparent in the SPLH than in the SPL measure.

In summary, an increase of Rs of an accented syllable by 10 units from 15 to 25 is associated with 125 ms increase of duration, about 4-8 semitones in F_0, 6 dB in SPL and 9 dB in SPLH, thus 3 dB in the increase of SPLH-SPL. Our studies of prose reading as well as the study of focus variation support these data. It should be kept in mind that the vowel specific inherent values of SPLH-SPL vary between 4-15 dB, the lowest for [u:], the highest for none-close front vowels.

There is evidence that the assumed stressed unaccented category, see Table 3.3 has an unclear identity. The regression graphs of prominence versus acoustic parameters all show some tendency towards a clustering of the Rs<15 data in one group clearly separated from the high prominence data, which indicates a tendency of categorisation. However, accent 2 modulation is already well developed at Rs=15 and extends to a lower region of Rs values.

Our general impression from the varying focus study is that of a fairly consistent co-variation of F_0, duration, subglottal pressure, SPL and SPLH-

SPL within the 5 talkers of which 2 were females. Our data reveal the existence of a default prominence of the first and the last words within the intended neutral utterance, an instance of "sentence accent", which could be expected (Bruce, 1982; Horne & Filipsson, 1996). Some of our graphs of focally induced parameter increase are accordingly based on the weighted neutral reference. The graphs of neutral minus weighted neutral reference illustrate clearly the extent of this default bias.

We lack data as to what extent a low value of one parameter, e.g. F_0, can be compensated for by other parameters, e.g. duration and intensity. The conclusion of Heldner and Strangert (1997), that an F_0-rise is neither necessary nor sufficient to perceive a word as focused, needs to be verified. Although F_0 appears to be the major focal accent correlate, the normal covariation of F_0 with all other parameters should be studied in greater detail. There is evidence that the durational component at times can be rather small while the F_0 component is perceptually dominant.

Synthesis experiments are needed in order to study the relative salience of prominence parameters, but in order to avoid unnatural combinations they should be carried out with due regard to inherent constraints in the production mechanism. An important issue is to gain experience concerning individual variations and speaking style.

There are physical limits to the degree that intensity can be varied independently of F_0. Increasing the subglottal pressure by one cm H_2O causes a passive increase of F_0 of the order of 4 Hz only, but this is considerably lower than the normal covariation of Psub and F_0 in speech which is determined by physiological ease and amounts to F_0 increasing in proportion to $Psub^{1.4}$, i.e. Psub is proportional to $F_0^{0.7}$. (Fant, 1997). Furthermore, the voice source amplitude Ee was found to be proportional to $(Psub^{1.1})(F_0^{1.35})$ or to approximately F_0^{2}.

This means that half an octave, i.e. 6 semitones increase in F_0, is associated with a doubling of Ee and thus a 6 dB increase in SPL of which 4 dB is the contribution from F_0 at constant Psub and 2 dB is the contribution from Psub at constant F_0. In addition, since a larger number of excitation per second at a higher F_0 implies an increase of speech power, there is an extra 1.5 dB gain in SPL to account for in the half octave rise in F_0 and thus in all a 7.5 dB gain. However, at a high F_0 above the middle of a speaker's available F_0-range, there is a saturation effect limiting the growth of SPL with increasing F_0 which should be taken into account (Fant, 1997; Fant *et al.*, 1996, 1997). Here, in the upper F_0-range, a more conservative prediction of the order of 6 dB increase in SPL with a 6 semitones rise in F_0 seems reasonable and is close to what is found above.

In conclusion, Psub and F_0 have a primary role in determining SPL. A glottal adduction contributes to raising the efficiency of the voice source and thus an increase of both SPL and the spectral tilt (SPLH-SPL) at constant Psub. In focal accentuation a change in formant pattern towards a more sonorous sound also contributes to SPL and to (SPLH-SPL). As pointed out by Hanson (1997b), glottal adduction is potentially a means of executing stress without an increase of subglottal pressure whilst focal accentuation usually requires an active P-sub increase. In singing the subglottal pressure attains a greater importance for control of loudness and varies over a much larger range than in speech (Sundberg *et al.*, 1999).

The role of spectral tilt as a parameter contributing to signal emphasis in focal accent is by now well established but opinions differ as to whether it also contributes to the stressed/unstressed contrast at lower prominence levels. Sluijter and van Heuven (1997) report positive observations while Campbell and Beckman (1977) state that a change in spectral tilt requires an accent contrast. In our data we have frequent examples of stressed/unstressed contrasts at lower prominence levels without a spectral tilt component. On the other hand our data in Figures 3.10 and 3.25 indicate that the (SPLH-SPL) parameter is effective at moderate Rs and SPL levels.

Acknowledgements

This work has been supported by grants from the Bank of Sweden Tercentenary foundation and the Swedish Research Council for Engineering Sciences, TFR. The authors are indebted to Kjell Gustafson and Eva Gårding for rewarding discussions.

References

Bailly, G., C. Benoît and T. Sawallis (eds.). 1992. *Talking Machines: Theories, Models and Designs*. Amsterdam: Elsevier Science.

Bruce, G. 1977. *Swedish Word Accents in Sentence Perspective*. Lund: Gleerup.

Bruce, G. 1982. Developing the Swedish intonation model. *Working Papers 22*. 51-116. Dept of Phonetics and Linguistics, Lund University.

Bruce, G. and B. Granström. 1993. Prosodic modelling in Swedish speech synthesis. *Speech Communication* 13, 63-74.

Bruce, G., B. Granström and D. House. 1992. Prosodic phrasing in Swedish speech synthesis. In Bally *et al.* (eds.), 307-321.

Campbell, N. and M.E. Beckman. 1997. Stress, prominence and spectral tilt. *Proc. ESCA Workshop on Intonation* (Athens, Greece), 18-21.

Carlson, R. and B. Granström. 1973. Word accent, emphatic stress, and syntax in a synthesis-by-rule scheme for Swedish. *STL-QPSR* 2-3, 31-35.

Collier, R. 1991. Multi-language intonation synthesis. *Journal of Phonetics* 19, 61-73.

Elert, C.-C. 1964. *Phonological Studies of Quantity in Swedish.* Stockholm: Almqvist & Wiksell.

Elert, C.-C. 1995. *Allmän och Svensk Fonetik* (7th edition). Stockholm: *Almqvist & Wiksell.*

Engstrand, O. 1995. Phonetic interpretation of the word accent contrast in Swedish. *Phonetica* 52, 171-179.

Fant, G. 1960 (2nd edition, 1970). *Acoustic Theory of Speech Production.* The Hague: Mouton.

Fant, G. 1973. *Speech Sounds and Features.* Cambridge, Mass.: MIT Press.

Fant, G. 1993. Some problems in voice source analysis. *Speech Communication* 13, 7-22.

Fant, G. 1995. The LF-model revisited. Transformations and frequency domain analysis. *STL-QPSR* 2-3, 119-155.

Fant, G. 1997. The voice source in connected speech. *Speech Communication* 22, 125-139.

Fant, G. and A. Kruckenberg. 1989. Preliminaries to the study of Swedish prose reading and reading style. *STL-QPSR* 2, 1-83.

Fant, G. and A. Kruckenberg. 1994. Notes on stress and word accent in Swedish. *Proc. International Symposium on Prosody* (Yokohama, Japan). Also published in *STL-QPSR* 2-3, 125-144.

Fant, G. and A. Kruckenberg. 1995. The voice source in prosody. *Proc. 13th ICPhS* (Stockholm, Sweden), vol. 2, 622-625.

Fant, G. and A. Kruckenberg. 1996a. On the quantal nature of speech timing. *Proc. ICSLP* '96 (Philadelphia, USA), 2044-2047.

Fant, G. and A. Kruckenberg. 1996b. Voice source properties of the speech code. *TMH-QPSR* 4, 45-46.

Fant, G. and A. Kruckenberg. 1998. Prominence and accentuation. Acoustical correlates. *Proc. Swedish Phonetics Conference Fonetik* '98 (Stockholm, Sweden), 142-145.

Fant, G., A. Kruckenberg and S. Hertegård. 1996. Focal accent and subglottal pressure. *Proc. Swedish Phonetics Conference Fonetik* '96 (Stockholm, Sweden), *TMH-QPSR* 2, 29-32.

Fant, G., A. Kruckenberg, S. Hertegård and J. Liljencrants. 1997. Accentuation and subglottal pressure in Swedish. *Proc. ESCA Workshop on Intonation* (Athens, Greece), 111-114.

Fant, G., A. Kruckenberg and L. Nord 1991a. Prosodic and segmental speaker variations. *Speech Communication* 10, 521-531.

Fant, G., A. Kruckenberg and L. Nord. 1991b. Durational correlates of stress in Swedish, French and English. *Journal of Phonetics* 19, 351-365.

Fant, G., A. Kruckenberg and L. Nord. 1992. Prediction of syllable duration, speech rate and tempo. *Proc. ICSLP* '92 (Banff, Canada), vol. 1, 667-670.

Fant, G., A. Kruckenberg and L. Nord. 1997. Some studies of accentuation and juncture in Swedish. *Proc. Swedish Phonetics Conference Fonetik* '97 (Umeå, Sweden), *PHONUM* 4, 157-160.

Fant, G., J. Liljencrants and Q. Lin. 1985. A four-parameter model of glottal flow, *STL-QPSR* 4, 1-13.

Fant, G. and Q. Lin 1988. Frequency domain interpretation and derivation of glottal flow parameters, *STL-QPSR* 2-3, 1-21.

Fant, G., L. Nord and A. Kruckenberg. 1986. Individual variations in text reading. A data-bank pilot study, *STL-QPSR* 4, 1-17 (also in *RUUL* 17, 1987, 104-114).

Fujimura O. and M. Hirano (eds.). 1994. *Vocal Fold Physiology*. Singular Publishing Group.

Gårding, E. 1981. Contrastive prosody: a model and its application. *Studia Lingvistica* 35, 146-166.

Gårding, E. 1989. Intonation in Swedish. *Working papers* 35, 63-88. Dept of Linguistics and phonetics, Lund University.

Hanson, H. 1997a. Glottal characteristics of female speakers. Acoustic correlates. *J. Acoust. Soc. Am.* 101, 466-481.

Hanson, H. 1997b. Vowel amplitude variation during sentence production. *Proc. ICASSP* '97 (München, Germany), vol. 3, 1627-1630.

Hardcastle W.J. and A. Marchal (eds.), *Speech Production and Modelling*. Dordrecht: Kluwer Academic Publishers.

Heldner, M. 1996. Phonetic correlates of focus accents in Swedish. *TMH-QPSR* 2, 1-4.

Heldner, M. and E. Strangert. 1997. To what extent is perceived focus determined by F_0 cues? *Proc. EUROSPEECH* '97 (Rhodes, Greece), 875-877.

Horne, M. and M. Filipsson. 1996. Implementation and evaluation of a model for synthesis of Swedish intonation. *Proc. ICSLP* '96 (Philadelphia, USA), vol. 3, 1848-1851.

Horne, M., E. Strangert and M. Heldner. 1995. Prosodic boundary strength in Swedish: final lengthening and silent interval duration. *Proc. 13th ICPhS* (Stockholm, Sweden), vol. 1, 170-173.

Kruckenberg, A. and G. Fant. 1995. Notes on syllable duration in French and Swedish. *Proc. 13th ICPhS* (Stockholm, Sweden), vol. 1, 58-161.

Lindblom, B. 1990. Explaining phonetic variation: a sketch of the H&H theory. In Hardcastle and Marchal (eds.), 403-439.

Marcus, S.M. 1981. Acoustic determinants of perceptual center (P-center) location. *Perception and Psychophysics* 30, 247-256.

Portele, T. and B. Heuft. 1997. Towards a prominence-based synthesis system. *Speech Communication* 21, 1-72.

Rapp, K. 1971. A study of syllable timing. *STL-QPSR* 1, 14-19.

Sluijter, A.M.C. and V.J. van Heuven. 1996. Spectral balance as an acoustic correlate of linguistic stress. *J. Acoust. Soc. Am.* 100, 2471-2484.

Stevens, K.N. 1994. Prosodic influences on glottal waveform: preliminary data. *International Symposium on* Prosody (Yokohama, Japan), 53-63.

Stevens, K.N. and M. Hanson 1994. Classification of glottal vibrations from acoustic measurements. In Fujimura and Hirano (eds.), 147-170.

Strangert, E. and M. Heldner, 1995. The labelling of prominence in Swedish by phonetically trained transcribers. *Proc. 13th ICPhS* (Stockholm, Sweden), vol 4, 204-207.

Strangert, E. and M. Heldner. 1998. On the amount and domain of focal lengthening in two-syllable and longer Swedish words. *Proc. Swedish Phonetics Conference Fonetik* '98 (Stockholm, Sweden), 154-157.

Strik, H. and L. Boves. 1992. On the relation between voice source parameters and prosodic features in connected speech. *Speech Communication* 1, 167-174.

Sundberg, J., M. Andersson, and C. Hultqvist. 1999. Effects of subglottal pressure variations on professional baritone singers' voice sources. *J. Acoust. Soc. Am.* 105, 1965-1971.

Swerts, M. and R. Veldhuis. 1977. Interactions between intonation and glottal-pulse characteristics. *Proc. ESCA Workshop on Intonation* (Athens, Greece), 297-300.

4

Prosodic Disambiguation in English and Italian

JULIA HIRSCHBERG AND CINZIA AVESANI

4.1 Introduction

It is widely believed that syntactic and semantic ambiguities, such as the scope of negation and quantifiers, the association of focus sensitive operators, and the attachment of prepositional phrases, adverbials, and relative clauses can be disambiguated intonationally (Jackendoff, 1972; Bolinger, 1989; Renzi, 1988). While there has been some experimental investigation of the role prosody plays in influencing hearers' interpretation of certain syntactic ambiguities in English (Altenberg, 1987; Beach, 1991; Price, Shattuck-Hufnagel & Fong, 1991), there has been little empirical study of these phenomena in other languages, and even less cross linguistic research. In this paper we present results of a study designed to compare the mechanisms speakers employ to disambiguate syntactically and semantically ambiguous utterances in English and Italian. We wanted to discover, first, whether phenomena believed to be intonationally disambiguable would be so disambiguated by naive subjects. Second, we wanted to see whether speakers of two languages in which prosodic features such as phrasing and pitch accent can be freely varied to convey differences in meaning would use those features similarly or not.

In an earlier production study comparing English, Italian, and Spanish (Avesani, Hirschberg & Prieto, 1995) we found that native speakers of these languages did differentiate among some types of syntactic and scopal ambiguity intonationally, employing differences in intonational phrasing and prominence. Their strategies differed among languages, with Spanish and Italian patterning together more often than either patterned with English. English, Spanish, and Italian speakers were most similar in their

A. Botinis (ed.), Intonation, 87-95.

disambiguation of the scope of negation employing variation in prosodic phrasing to distinguish wide from narrow scope productions, with wide scope utterances produced as a single phrase and narrow produced as two phrases. Italian and Spanish speakers also differentiated wide from narrow scope by similar variation in phrasing; however, they also placed nuclear stress on the verb to indicate wide scope negation, while English speakers located nuclear stress later in the utterance. Also, English speakers further distinguished wide from narrow scope by utterance final tonal variation, with continuation rise employed for wide scope readings and falling intonation for narrow. While our Italian speakers consistently used phrasing variation to indicate differences in PP attachment (between NP and VP attachment), English and Spanish subjects were inconsistent in this regard. For quantifier disambiguation, the picture was more complex: For Italian and Spanish speakers, renditions of sentences containing scope ambiguous negative quantifiers were disambiguated by variation in placement and in prosodic phrasing; for two English speakers, accent placement served to disambiguate these utterances. In cases of association with focus, *only/solo/sólo* was treated inconsistently by speakers of all three languages.

4.2 Current Study

In order to test the validity of our previous results, we conducted a larger study of English and Italian. We constructed materials for a production study to test intonational variation in sentences that contained the following potential ambiguities: *pp* and adverbial attachment, relative clauses focus sensitive operators (*only* and *even*), the scope of negation and quantifier scope (*none*). We embedded these sentences in paragraphs designed to disambiguate the potential ambiguity. For example, "*William isn't drinking because he's unhappy*" was embedded in paragraphs in which context favoured (1a) the reading in which William drinks, but not because he is unhappy and (1b) the reading in which William does not drink, because of his unhappiness.

1a. William is a hopeless case. There's nothing anyone can do to make him stop drinking. For a while, his friends thought he only drank to forget his troubles. But that doesn't seem to be true. William isn't drinking because he's unhappy. He drinks because he's an alcoholic.

1b. When my friend William is happy, he loves to go to parties, and he can drink more than anyone I know. But lately he never wants to come to parties. When he does, he only drinks water. I think I understand. William isn't drinking because he's unhappy. And I don't know how long this will last.

There were a total of 21 pairs of paragraphs, three for each of the phenomena under examination.

Twelve subjects (six native speakers of standard American English and six of Tuscan Italian) were given each pair of paragraphs, asked to read them over for understanding and then were recorded reading them aloud. After reading each pair, speakers were then asked to answer series of questions designed to elicit their interpretation of the target sentences in the differing contexts. We wanted to make sure that the interpretations intended for each sentence in each context were in fact the interpretations the subjects understood and had been trying to convey in reading the paragraphs aloud. First, subjects were asked to explain in their own words the difference in meaning of the target sentence in each context. A typical answer from one subject for paragraph (1a) was "It is not the case that William's unhappiness is causing him to drink"; for (1b) the same subject wrote "William is unhappy so he isn't drinking." After completing this task, subjects were then asked to answer a forced choice question for each paragraph; the following was asked for paragraphs (1a) and (1b): "Does William drink?". Both the free and forced choice tasks were scored by one of the authors to determine whether the subject had interpreted the task as intended for the purpose of the experiment or not.

Subjects' speech productions were excised from the disambiguating contexts, pitch tracked using Entropic WAVES+ software, and labelled by someone who had not heard the utterance in context, using the ToBI labelling scheme. Pitch accents, phrase accents, boundary tones, and relative prominence of accents were marked.

4.3 Results

Preliminary analysis of our data indicate, first, that both English and Italian subjects had no difficulty in understanding the differences in interpretation that we hoped to convey with our disambiguating contexts. Nearly all explained the differences as we had intended to convey them in the free form condition, and similarly answered the questions posed in ways that confirmed their interpretation of the paragraphs was the same as the intended interpretation. Of the six English subjects' written responses to the two conditions for each of the 42 paragraphs (N=252), only six responses indicated that the subject had understood a different interpretation of the target utterance from the intended one; four of these were from one of our speakers. For the six Italian speakers, only two responses showed a different interpretation from the intended one.

The speech productions from the English speakers show clear trends for all speakers in their production of sentences ambiguous with respect to scope of negation and the focus of the focus sensitive operators, *only* and *even*. Results are less clear for sentences where the ambiguity derived from attachment distinctions and there are no discernible patterns in the quantifier scope productions.

The English speakers disambiguate scope of negation, in paragraphs like (a) and (b) above, primarily by varying the phrasing of the ambiguous sentence. Target utterances produced in the wide scope condition (1a) rarely (2/18) contain internal phrase boundaries, while utterances produced in the narrow scope context (1b), usually (12/18) exhibit major or minor prosodic phrase boundaries before the subordinate conjunction ('because' in all cases in our study). Wide scope productions often end in 'continuation rise' (LH%) (10/18), while narrow scope productions usually were falling contours (LL%) (15/18). Examples of typical productions for (1a) and (1b) are (2a) and (2b), respectively.

2a. H* William isn't !H* drinking because he's H*unhappy LH%

2b. H* William isn't !H* drinking LH% because he's H* unhappy LL%

These findings are similar to those of our earlier study (Avesani *et al.*, 1995).

These speakers also display consistency in their production of variation in the focus of *even* and *only*. Examples of such sentences include "*Harold even telegraphed the paper*", where contexts vary the focus of *even* from the verb to the direct object, and "*He only wounded Anne*", where *wounded* and *Anne* alternate as foci. In 64/72 target utterances in this category, the focus of the operator represents the nuclear stress of the utterance; 30/36 utterances with *even* exhibit this pattern and 34/36 with *only*, as illustrated in (3):

3a. He H* only H* wounded Anne LL%

3b. He H* only !H* wounded !H* Anne LL%

Attachment ambiguities produced different results, depending upon syntactic category: English speakers make no clear prosodic distinctions in their production of sentences ambiguous with respect to PP attachment. So, sentences such as "*He managed to find the woman with the binoculars*" are produced similarly, whether the VP or the NP attachment of the prepositional phrase is favoured by context. However, speakers do exhibit some regularities in their disambiguation of adverbial and relative clause attachment, in producing sentences such as "*He had spoken to her quite clearly*" and "*The professor who loves jelly beans died in terrible pain*". In contexts favouring the S attachment of the adverbial expression (e.g. "It was quite clear that he had spoken to her"), speakers produce the target sentence with an internal prosodic boundary separating the adverb from the remainder of the sentence (11/18) (e.g. '*He had L+H* spoken to her L-H*quite !H* clearly LL%*) while in contexts favouring a VP attachment (e.g. "He had spoken to her in a clear manner.") only 1/18 productions exhibit such an internal boundary. For contexts favouring a non restrictive reading of target sentences containing a relative clause, speakers produce utterances in which the clause represented a separate prosodic phrase in 9/18 cases (e.g. '*The H* professor LH% who loves H* jelly beans LH% H* died in !H* terrible H* pain LL%*'; only four productions of the restrictive reading exhibit such phrasing. So, there is weak evidence that speakers employ phrasing to signal attachment differences for some phenomena. Speakers' productions of sentences containing the quantifier *none* do not show any consistent distinctions for the wide *vs.* narrow scope contexts, despite the fact that speakers **do** distinguish between the two readings. In most cases, speakers vary the accenting of the quantifier itself; 13/18 production pairs in this category showed such differences, which ranged from deaccenting the quantifier in the narrow scope condition and accenting in the wide in a few (3) cases, to varying the type of pitch accent between the two conditions. However, there are no clear differences in choice of accent for either of the conditions (e.g. 7/18 of the narrow scope cases and 11/18 of the wide bore **L+H*** accents. Nor is there a distinction in relative prominence between the two; in 10/18 cases the quantifier receives nuclear stress in the narrow condition, and similarly in 12/18 of the wide. It might be argued that the distinction between readings might have been more difficult for speakers to comprehend than for some of the other phenomena, thus leading to less consistency in productions. Paragraphs such as (4) certainly seem more artificial than (1), for example.

4a. It is really hard to say who you should vote for the town council. All of the Democratic candidates are pretty mediocre, but none of them is actually corrupt. *The election of none of these candidates would be a disaster.* Any of them would probably do a decent enough job.

4b. It's absolutely essential that the Democratic party win some seats in the coming election, or nothing is ever going to get better in our state. They are the only party who can save us from bankruptcy. *The election of none of these candidates would be a disaster.* Unless at least some of them are elected we will really be in trouble.

However, only one of the speakers exhibited any interpretation problems with these sentences in the written condition. And the lack of consistency we find here is consistent with our earlier results for English speakers, reported in (Avesani *et al.*, 1995).

A systematic use of intonational phrasing appears to be the main strategy adopted by our Italian speakers in disambiguating the scope of negation. All but one instance of the narrow scope utterances (16/17)[1] is uttered in two intermediate phrases, with a high (10/16) or low (6/16) phrase accent marking the syntactic boundary between the two clauses. All instances of the wide scope reading are uttered as a single intonational phrase, with a nuclear pitch accent on the VP of the main clause and deaccenting the remainder of the utterance. The intonational contours of both the wide (5a) and the narrow scope (5b) readings end in a fall, as it is illustrated in the following example:

5a. H* Guglielmo non H+L* beve perché è infelice LL%

5b. H* Guglielmo non H* beve H- perché è H+L*infelice LL%

The difference in the scope of focus sensitive operators such as *solo* (*only*) and *anche* (*even*) is generally conveyed by accent placement. The alternative readings of 7 out of 36 ambiguous sentences in this category did not show any prosodic evidence of disambiguation by the speakers. Of the 58 target utterances (of 29 ambiguous pairs) that did show evidence of disambiguation, 28 display a pitch accent associated both with the focus sensitive operator and with the focused item. The item not in focus is deaccented. So, in (6a) only Anna was wounded and no one else, and in (6b), Anna was only wounded, but not, say, killed.

6a. Ha H* solo ferito H+L* Anna LL%

6b. Ha H* solo H+L* ferito Anna LL%

The focus of the operator in these cases represents the nuclear stress of the utterance. In 15/30 remaining utterances, the focused word occurs in sentence final position and is preceded by a prenuclear accent, as in "*Ha H* anche H* abbracciato il H+L* poliziotto LL%*". The possible ambiguity caused by accenting both candidates for operator focus is solved by speakers attributing to the focused item a noticeably higher degree of prominence, attained through a higher F_0 and/or a longer vowel duration.

Ambiguous attachment of prepositional phrases, adverbials and relative clauses is resolved by Italian speakers with less consistency. Only 11/18 sentences with ambiguous attachment of the prepositional phrase, 13/18 sentences with ambiguous adverbial attachment and 8/18 sentences with embedded relative clauses are disambiguated prosodically. Globally, the preferred strategy used by our speakers in attachment disambiguation is to vary intonational phrasing. Among the two possible attachments a PP or ADVP and a relative clause can have in our sentences, if a phrase attaches to the syntactically higher node, it is prosodically separated from the remainder of the utterance via a H- or L- phrase boundary. (10/11 for PP attachment, 12/13 for adverbial attachment, 6/8 for relative clauses). For example, in a sentence such as *Ha provato a mettersi in contatto col suo oculista a Roma*, the PP is set off in an intermediate phrase if the intended meaning is that "the subject was in Rome when he tried to get in touch with his ophthalmologist" (VP attachment); and it is part of a larger intermediate phrase - together with the phrase it attaches to - in the case of NP (or PP) reading ("the ophthalmologist and not the subject was in Rome"). The prosodic boundary is usually marked by a constellation of different cues in addition to F_0 movement. When the phrase accent is not used, the intended attachment is conveyed by relative prominence of the pitch accents across the boundary, preboundary final lengthening and other sandhi phenomena (3/11). In "*Lui le aveva parlato chiaramente*", the adverbial phrase is set off as a separate intermediate phrase if the adverbial is intended to attach to S ("it was obvious that he spoke to her"), and is part of a larger phrase otherwise (VP attachment, i.e. "he spoke to her in a clear manner"). In "*La ragazza che mi ha fregato il taxi era una testimone dell' omicidio*", a L- or H- phrase boundary sets off the relative clause from the NP *the girl* when the clause has a non restrictive meaning. Despite the fact that the same prosodic phenomenon (phrasing) is used in almost all cases of attachment that have been disambiguated, only 60% of attachment ambiguities have shown evidence of prosodic disambiguation.

Ambiguity in the scope of the negative quantifier (e.g., for "*La presenza di nessuno dei professori la metterebbe in imbarazzo*" the readings "the absence of all the professors would embarrass her", and "there are no professors whose presence could embarrass her") is distinguished by Italian speakers in 16/18 of such sentences. Two different strategies are used. Both the quantifier and the NP bound by it (*nessuno dei professori*) are accented in one reading (4a) and either the quantifier is accented and the NP deaccented (4b) or the quantifier deaccented and the NP accented in the alternative reading.

4a. La H* presenza di H* nessuno dei H* professori la metterebbe in H+L* imbarazzo LL%

4b. La H* presenza di H* nessuno dei professori la metterebbe in imbarazzo LL%

Other speakers (11/18 productions) accent both the quantifier and the bound NP in both readings, but the pitch accents differ in prominence. The two readings are distinguished by: (a) assigning more prominence (higher F_0) to either the quantifier or to the noun phrase in one reading and reversing that order in the alternative one. (b) Changing the difference in prominence between the quantifier and the bound noun phrase. The difference is higher in one reading and lower in the alternative one. Only two of our speakers use one of the above strategies consistently, however.

4.4 Conclusion

Comparing the production of our English and Italian speakers, we see that, for both, semantic ambiguities were disambiguated more consistently than were syntactic ambiguities. So, the scope of negation and of focus sensitive operators were reliably distinguished by both groups, both using phrasing to disambiguate scope of negation and accent placement to distinguish focused items. However, another semantic ambiguity, quantifier scope, was **not** consistently disambiguated by speakers of either language. These results are consistent with our previous findings for scope of negation, for both English and Italian speakers. However, our earlier studies showed no clear patterns for the disambiguation of focus ambiguities but did find consistent behaviour in Italian speakers' disambiguation of quantifier scope.

Syntactic ambiguities were in general much less clearly disambiguated by speakers of either language. PP attachment, in fact, was disambiguated consistently by neither English nor Italian speakers, while only English speakers showed clear trends in the disambiguation of ambiguously attached adverbials and results for ambiguous relative clause attachment is quite mixed. However, in all cases, speakers who **did** consistently disambiguate syntactic ambiguities did so by varying intonational phrasing.

While much has been proposed for the role of prosodic phenomena as contributing additional information essential for the processing of syntactic and semantic constructions, our findings suggest that this role is not a simple one. Phenomena such as attachment decisions, in particular, which seem intuitively easy to disambiguate intonationally, rarely **were** so disambiguated, in context, even by subjects who understand their ambiguity, in contrast to phenomena such as focus location and scope of negation. One possible explanation for this is that our "disambiguating contexts" were, in some cases, too successful. Often phenomena disambiguable by prosodic variation have a "neutral" (or "unmarked") production, which is felicitous for either interpretation. For example, sentences in which attachment may be signalled by the variation of internal phrase boundaries may usually be produced without internal boundaries in either condition, if the ambiguity is resolvable by other means. Thus the disambiguating contexts in which target sentences were embedded may in fact have allowed subjects felicitously to produce intonationally ambiguous utterances. While this explanation might account for the lack of prosodic distinctions for some sentences, however, it would not explain those productions of sentences also in clearly disambiguating contexts which **did** exhibit consistent intonational variation, except by the hypothesis that, in such cases, a 'neutral' production was less available.

Notes

1. One production is missing.

References

Altenberg, B. 1987. *Prosodic Patterns in Spoken English.* Lund University Press.

Avesani, C., J. Hirschberg and P. Prieto. 1995 The intonational disambiguation of potentially ambiguous utterances in English, Italian, and Spanish. *Proc. 13th ICPhS* (Stockholm), vol. 1, 174–177.

Beach, C. 1991. The interpretation of prosodic patterns at points of syntactic structure ambiguity: Evidence for cue trading relations. *Journal of Memory and Language* 30, 644–663.

Bolinger, D.L. 1989. *Intonation and Its Uses: Melody in Grammar and Discourse.* London: Edward Arnold.

Jackendoff, R.S. 1972. *Semantic Interpretation in Generative Grammar.* Cambridge, Mass.: MIT Press.

Price, P.J., S. Shattuck-Hufnagel and C. Fong. 1991. The use of prosody in syntactic disambiguation. *J. Acoust. Soc. Am.* 90: 2956–2970.

Renzi, L. (ed.) 1988. *Grande grammatica di consultazione* (vol. 1). Bologna: Il Mulino.

5

Contrastive Tonal Analysis of Focus Perception in Greek and Swedish

ANTONIS BOTINIS, ROBERT BANNERT AND MARK TATHAM

5.1 Introduction

The present study is an experimental investigation of Greek and Swedish tonal perception of focus. Focus is a complex category involving a variety of linguistic correlates, among them, phonetic, syntactic, semantic and pragmatic ones (Halliday, 1967; Bresnan, 1971; Chomsky, 1971; Jackendoff, 1972; Rossi, Di Cristo, Hirst, Martin & Nishinuma, 1981; Gussenhoven, 1983; Ladd, 1996; Lambrecht, 1996; Cruttenden, 1997). The study is concentrated on the phonetics of focus and in particular on tonal perception of voice fundamental frequency (F_0), or pitch. Focus, and fairly synonymous terms such as nucleus and sentence stress, have been studied extensively and the results indicate both local and global tonal realisation (e.g. Bruce, 1977; Pierrehumbert, 1980; Gårding, Botinis & Touati, 1982; Botinis, 1989; Hirst & Di Cristo, 1998).

Five main questions in relation to focus perception are addressed: (1) what is the effect of local tonal range (2) what is the effect of poststressed tonal flattening, (3) what is the effect of tonal shift, (4) what is the effect of tonal neutralisation, and (5) what is the effect of tonal defocalisation? The identification of focus is also considered and the conflicting cues issue is discussed. The results of the present study are based on one experiment in each Greek and Swedish and the relevant discussion is primarily addressed to these two languages[1].

Phonetic studies report F_0, duration, intensity and vowel quality (or timbre) as acoustic correlates of prominence distinctions in different languages (Fry, 1955, 1958; Bolinger, 1958, 1972; Lieberman, 1960; Bruce, 1977; Cooper, Eady & Mueller, 1985; Beckman, 1986; Botinis,

A. Botinis (ed.), Intonation, 97-116.

1989; Fant, Kruckenberg & Liljencrants, this volume). Although the acoustics of focus has been investigated intensively for Greek and Swedish, there is very little published on perception. Thus, the present investigation aims to carry out a basic analysis of tonal perception and examine the effects of tonal manipulations on focus perception in Greek and Swedish.

In Greek, the acoustic correlates of focus include a local tonal range expansion, associated with the stressed syllable of the word in focus, in combination with a global tonal structure compression. This compression is mostly related to postfocus material, the tonal structure of which may be fairly flattened, but may be evident even across prefocus material (Botinis, 1982, 1989, 1998). Duration and intensity may also correlate with focus, although less invariably than F_0 does. Focus may also have an effect on vowel quality, modifying the formant structure and expanding the acoustic space of the vowel of the stressed syllable in focus (Botinis, Fourakis & Katsaiti, 1995; Fourakis, Botinis & Katsaiti, 1999).

In Swedish, the tonal correlates of focus include a local tonal gesture, associated with the poststressed (i.e. postaccented) syllable of the word in focus, in combination with a global tonal structure compression. This compression is mostly related to postfocus material, the tonal structure of which may be downstepping (or fairly flattened), but may be evident even across prefocus material (Bruce, 1977; Bruce, Filipsson, Frid, Granstöm, Gustafson, Horne & House, this volume; Fant *et al.*, this volume). Duration, unlike in Greek, is also a constant acoustic correlate of focus in Swedish (Bannert, 1979, 1986; Botinis, Erkenborn, Isacsson & Westin, 1999; Fant, Kruckenberg & Nord, 1991) whereas intensity may correlate with focus. Prominence, and thus focus, may also have an effect on vowel quality modifying the acoustic space and the formant structure of the prominent vowel (Engstrand, 1988).

Thus, focus, apart from other grammatical markers such as clefting and topicalisation, has fairly constant phonetic correlates in production and, presumably, in perception. The general consensus is, at least for the languages investigated in this study, that focus has both local and global tonal correlates. However, the relative perceptual salience of local *vs.* global tonal effects is hardly understood. Neither are the conflicting cues of F_0 *vs.* duration and intensity or the significance of F_0 well understood, especially from a contrastive perspective. Furthermore, it should be noted that perception studies of focus investigation in its right dimension are essentially missing but, instead, prominence distinctions, usually word stress, are investigated in *focus position*. This is a major scientific issue to be further considered in the discussion section (5.4).

5.2 Experimental Methodology

5.2.1 Speakers and Speech Material

The speech material was produced by two male professional phoneticians[2], who read the test sentence "Mona saw Molly in London", in standard orthography, in Greek and Swedish. These productions were either with no context, i.e. focus-neutral, or in a question-answer context with focus elicitation in different positions. The question was textually variable whereas the answer was textually constant and varied exclusively in prosodic structure in accordance with the corresponding question. Focus productions with focus-initial (i.e. Mona), focus-medial (i.e. Molly) and focus-final (i.e. London) were thus answers to the questions "Who saw Molly in London?", "Who did Mona see in London?" and "Where did Mona see Molly?" respectively (Table 5.1). Greek and Swedish refer to standard Athenian Greek and standard Stockholm Swedish respectively.

Table 5.1. Contextual frames and test sentence productions with focus-neutral as well as focus-initial, focus-medial and focus-final (underlining and capital letters) in Greek and Swedish.

	Contextual frames	Test sentence productions
	Greek	
0		[i ˈmona ˈiðe ti ˈmoli sto lonˈðino] 'Mona saw Molly in London.'
1	[pça ˈiðe ti ˈmoli sto lonˈðino] 'Who saw Molly in London?'	[i <u>ˈmona</u> ˈiðe ti ˈmoli sto lonˈðino] 'MONA saw Molly in London.'
2	[pça ˈiðe i ˈmona sto lonˈðino] 'Whom did Mona see in London?'	[i ˈmona ˈiðe ti <u>ˈmoli</u> sto lonˈðino] 'Mona saw MOLLY in London.'
3	[pu ˈiðe i ˈmona ti ˈmoli] 'Where did Mona see Molly?'	[i ˈmona ˈiðe ti ˈmoli sto <u>lonˈðino</u>] 'Mona saw Molly in LONDON.'
	Swedish	
0		[ˈmoːna sog ˈmɔlːy i ˈlɔndɔn] 'Mona saw Molly in London.'
1	[vem sog ˈmɔlːy i ˈlɔndɔn] 'Who saw Molly in London?'	[<u>ˈmoːna</u> sog ˈmɔlːy i ˈlɔndɔn] 'MONA saw Molly in London.'
2	[vem sog ˈmoːna i ˈlɔndɔn] 'Whom did Mona see in London?	[ˈmoːna sog <u>ˈmɔlːy</u> i ˈlɔndɔn] 'Mona saw MOLLY in London.'
3	[var sog ˈmoːna ˈmɔlːy] 'Where did Mona see Molly?'	[ˈmoːna sog ˈmɔlːy i <u>ˈlɔndɔn</u>] 'Mona saw Molly in LONDON.'

5.2.2 Acoustic Analysis of Original Test Sentences

The acoustic analysis was carried out using the Entropic ESPS/*waves+*™ signal processing package at Umeå University Phonetics Laboratory. Initial Tonal Boundary (ITB), Maximum Tonal Top (MTP), Final Tonal Boundary (FTB) as well as tonal range values (i.e. the difference between MTP and FTB) are basic reference units of the analysed speech material.

5.2.2.1 Greek

Figure 5.1 (left) shows the tonal structure of the Greek test sentence productions. The focus-neutral production (0) has a tonal gesture associated with the stressed syllable of lexical words; the tonal range is 110Hz (IB=100Hz, MTT=185Hz, FB=75Hz). The focus productions have a local tonal range expansion associated with the stressed syllable of the respective words in focus whereas non-focus material has a compressed tonal range, especially after focus. The focus-initial (1), focus-medial (2) and focus final (3) tonal range is 145Hz (IB=100Hz, MTT=215Hz, FB=70Hz), 155Hz (IB=110Hz, MTT 225Hz, FB=70Hz) and 140Hz (IB=100Hz, MTT 211Hz, FB=71Hz) respectively. The tonal structure of the present material is fairly regular, according to studies in Greek prosody (e.g. Botinis, 1982, 1989, 1998).

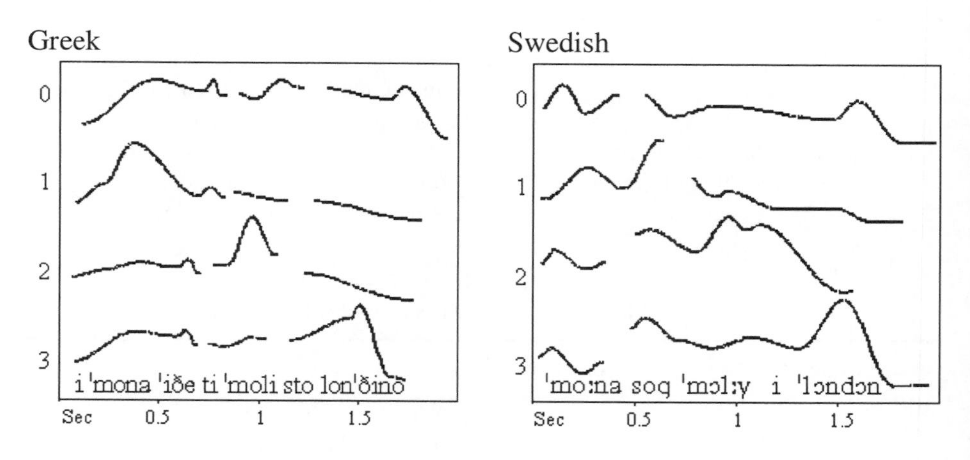

Figure 5.1. Greek (left) and Swedish (right) tonal contours of test sentence productions with focus-neutral (0) as well as focus-initial (1), focus-medial (2) and focus-final distribution (3).

5.2.2.2 Swedish

Figure 5.1 (right) shows the tonal structure of the Swedish test sentence productions. The focus-neutral production (0) has a tonal gesture associated with the accented syllable of lexical words; the tonal range is 110Hz (ITB=110Hz, MTT=170Hz, FTB=60Hz). The focus productions have a local tonal gesture associated with the poststressed (postaccented) syllable of the respective words in focus (except for focus-final) whereas the non-focus material is compressed, especially after focus with the downstepping tonal structure of standard Swedish. The focus-initial (1), focus-medial (2) and focus-final (3) tonal range is 155Hz (ITB=100Hz, MTT=215Hz, FTB=60Hz), 140Hz (ITB=110Hz, MTT=198Hz, FTB=58Hz) and 162Hz (ITB=109Hz, MTT=222Hz, FTB=60Hz) respectively. The focus-neutral and the focus-final productions have noticeable differences, especially with reference to tonal range, although, however, they are the two most similar of the four productions with respect to tonal structure. The present material is fairly regular and in general accordance with studies of intonation and prosodic structure in Swedish (e.g. Bruce, 1977; Bruce *et al.*, this volume; Fant *et al.*, this volume).

5.2.3 Tonal Manipulations and Test Stimuli

Analysed tonal contours of the four original test sentence productions were stylised on a frame-by-frame smooth F_0 interpolation and constitute the "reference" tonal contours (alias "reference stimuli/sentences", etc.). Tonal manipulations were carried out and 15 resynthesised stimuli in each language were constructed, including the reference sentence stimuli, in an LPC/PSOLA software environment.

The concept of the tonal manipulation methodology was from pattern A to pattern B, where A and B are different focus structures, in either local or global tonal realisations, in accordance with the experimental targets and the corresponding questions outlined at the Introduction section (5.1). The difference between the reference stimuli and the manipulated stimuli is exclusively based on tonal differences (shown in thick and thin lines at the figures respectively). No other prosodic or acoustic parameter of the reference sentence productions was manipulated.

The test stimuli in each language were set up in ten blocks, in random order, resulting in ten repetitions for each stimuli and thus in 150 stimuli. Ten university students participated in the experiments in each language and the results are thus based on 100 counts per stimulus (i.e. 10 stimulus repetitions x 10 listeners), which totals to 1.500 stimuli per language. There was a pause of 2 seconds between the stimuli and 5 seconds between the blocks and each experiment lasted approximately 10 minutes. The listeners had four answer options: mark one word in focus, i.e. "Mona", "Molly" or "London", or not mark any word, i.e. "neutral" production.

The perceptual experiments were conducted in Athens University, Greece, and Umeå University, Sweden, by the authors of the present study.

5.2.3.1 Greek

The tonal structure of the 15 Greek stimuli, 11 manipulated as well as 4 reference ones, is shown in Figure 5.2a-d (left). Figure 5.2a shows four manipulated stimuli (0.1-0.4), associated with focus-neutral production (0). In stimulus 0.1, the local tonal range associated with the stressed syllable of the word 'Mona' has been expanded by 30Hz, in accordance with the tonal contour of the focus-initial production with 'Mona' in focus. In stimulus 0.2, the local tonal range associated with the stressed syllable of the word 'Molly' has been expanded by 40Hz, in accordance with the tonal contour of the focus-medial production with 'Molly' in focus. In stimulus 0.3, the tonal structure of the poststressed material with reference to word 'Molly' has been manipulated in accordance with the flattened tonal contour of the focus-medial production with 'Molly' in focus. In stimulus 0.4, the tonal structure of the poststressed material with reference to the word 'Mona' has been manipulated in accordance with the flattened tonal contour of the focus-initial production with 'Mona' in focus.

Figure 5.2b shows three manipulated stimuli (1.1-1.3), associated with focus-initial production (1). In stimulus 1.1, the tonal structure of the reference focus-initial production has undergone a tonal shift manipulation, in accordance with the tonal structure of the focus-medial production. In stimulus 1.2, the focus tonal gesture associated with the stressed syllable of the word 'Mona' has been neutralised at a low tonal contour. In stimulus 1.3 the tonal structure of the focus-initial production has been manipulated in accordance with the focus-neutral production.

Figure 5.2c shows three manipulated stimuli (2.1-2.3), associated with focus-medial production (2). In stimulus 2.1, the tonal structure of the reference focus-medial production has undergone a tonal shift manipulation, in accordance with the tonal structure of the focus-initial production. In stimulus 2.2, the focus tonal gesture associated with the stressed syllable of the word 'Molly' has been neutralised at a low tonal contour. In stimulus 2.3 the tonal structure of the focus-medial production has been manipulated in accordance with the focus-neutral production. Figure 5.2d shows one manipulated stimulus (5.1), associated with focus-final production (3), the global tonal structure of which has been manipulated in accordance with the focus-neutral production.

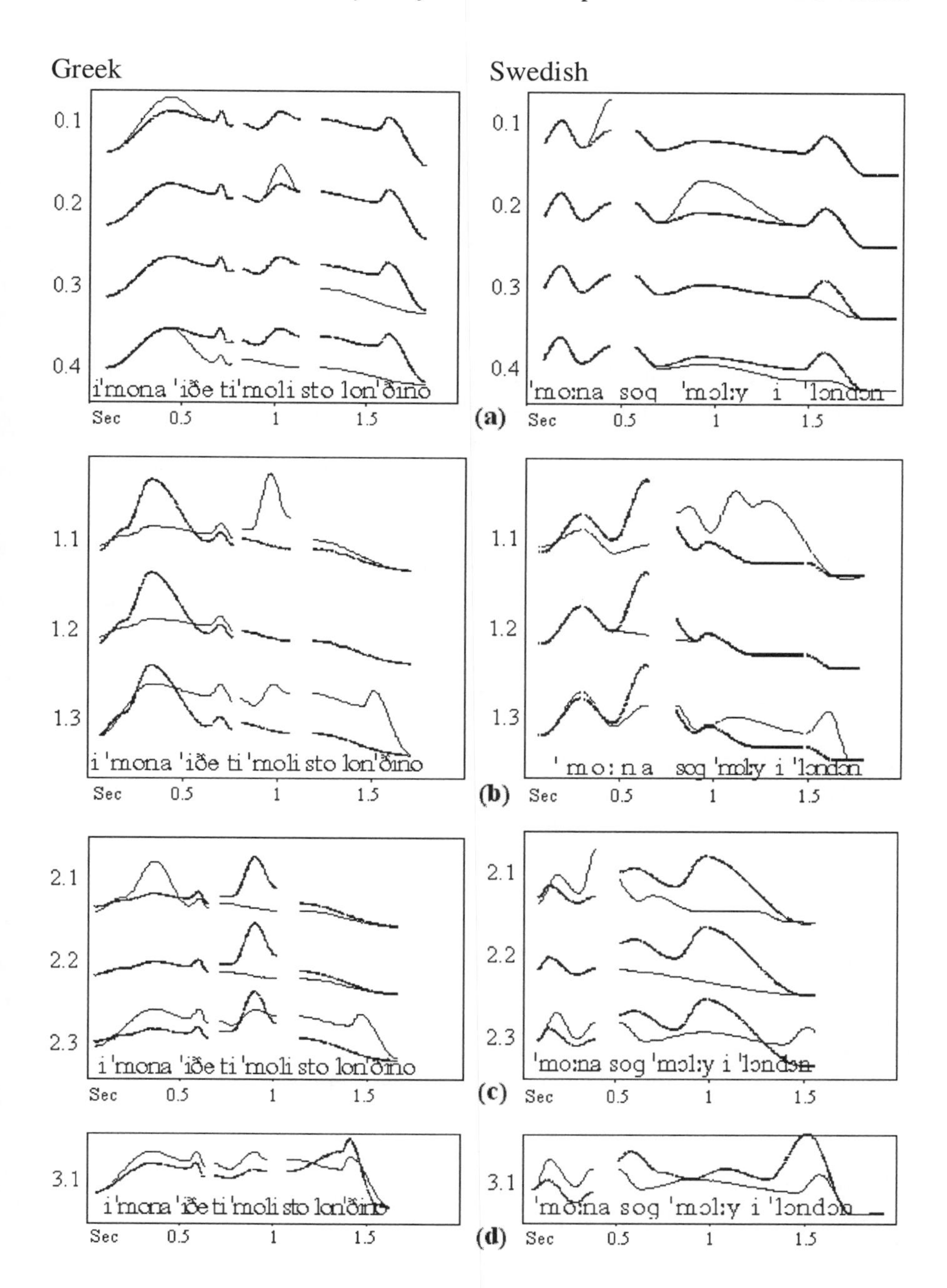

Figure 5.2. Greek stimuli on the left and Swedish stimuli on the right (see text).

5.2.3.2 Swedish

The tonal structure of the 15 Swedish stimuli, 11 manipulated as well as 4 reference ones, is shown in Figure 5.2a-d (right). Manipulations in Swedish are fairly analogous to manipulations in Greek. Figure 5.2a shows four manipulated stimuli (0.1-0.4), associated with focus-neutral production (0). In stimulus 0.1, the local tonal range associated with the stressed syllable of the word 'Mona' has been expanded by 65Hz, in accordance with the tonal contour of the focus-initial production with 'Mona' in focus. In stimulus 0.2, the local tonal range associated with the stressed syllable of the word 'Molly' has been expanded by 68Hz, in accordance with the tonal contour of the focus-medial production with 'Molly' in focus. In stimulus 0.3, the tonal structure of the poststressed material with reference to word 'Molly' has been manipulated in accordance with the compressed/downstepping tonal contour of the focus-medial production with 'Molly' in focus. In stimulus 0.4, the tonal structure of the poststressed material with reference to the word 'Mona' has been manipulated in accordance with the compressed/downstepping tonal contour of the focus-initial production with 'Mona' in focus.

Figure 5.2b shows three manipulated stimuli (1.1-1.3), associated with focus-initial production (1). In stimulus 1.1, the tonal structure of the reference focus-initial production has undergone a tonal shift manipulation, in accordance with the tonal structure of the focus-medial production. In stimulus 1.2, the focus tonal gesture associated with the stressed syllable of the word 'Mona' has been neutralised at a low tonal contour. In stimulus 1.3 the tonal structure of the focus-initial production has been manipulated in accordance with the focus-neutral production.

Figure 5.2c shows three manipulated stimuli (2.1-2.3), associated with focus-medial production (2). In stimulus 2.1, the tonal structure of the reference focus-medial production has undergone a tonal shift manipulation, in accordance with the tonal structure of the focus-initial production. In stimulus 2.2, the focus tonal gesture associated with the stressed syllable of the word 'Molly' has been neutralised at a low tonal contour. In stimulus 2.3 the tonal structure of the focus-medial production has been manipulated in accordance with the focus-neutral production.

Figure 5.2d shows one manipulated stimulus (3.1), associated with focus-final production (3), the global tonal structure of which has been manipulated in accordance with the focus-neutral production.

5.3 Results

The results are based on comparisons of perception response percentages for each focus production and its manipulated versions, using the chi-square statistic. Thus, the response percentages for S0 were compared to those for S0.1, S0.2, S0.3, and S0.4. Response percentages for S1 were compared to those for S1.1, S1.2, and S1.3, and so on. Thus, there were thirteen comparisons. The significance level was adjusted to compensate for the large number of comparisons and was set to .005.

5.3.1 Perception of Focus Productions

Figure 5.3 presents the percentage of focus perception responses given to each focus production for Greek and Swedish. The responses to the focus-neutral production (S0) for each language were distributed over the four response options (Fn = neutral, Fi = initial, Fm = medial, and Ff = final in this and all other figures). In contrast the focus-initial production (S1) shows mainly Fi perception responses, the focus-medial production (S2) mainly Fm responses, and the focus-final production (S3) mainly Ff responses for both languages. These results indicate that, although an utterance may be produced with no specific focus, focus-neutral prosodic structure does not constitute a perceptual invariance. No remarkable focus-final perception dominance is either observed, especially in Swedish.

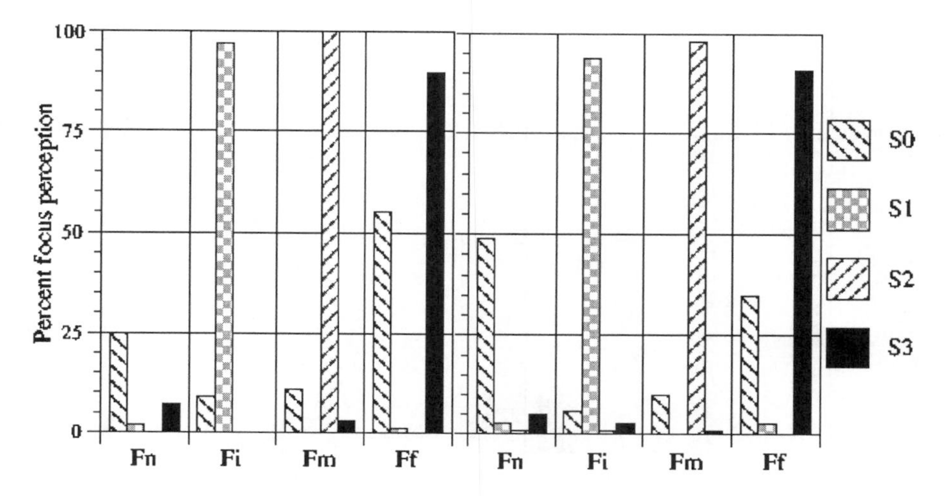

Figure 5.3. Greek (left) and Swedish (right) percentage of focus perception on the ordinate and focus response options on the abscissa (see text).

5.3.2 Perception of Focus-neutral Manipulations

Figure 5.4 shows the perception percentages for the focus-neutral production (S0) and its manipulated versions for Greek and Swedish. For Greek, there was no significant difference between the responses to S0 and S0.1 (chi-square = 5, df = 3, p>.1). All other comparisons were significant at the p<.0001 level. For Swedish all comparisons were significant at the p<.0001 level.

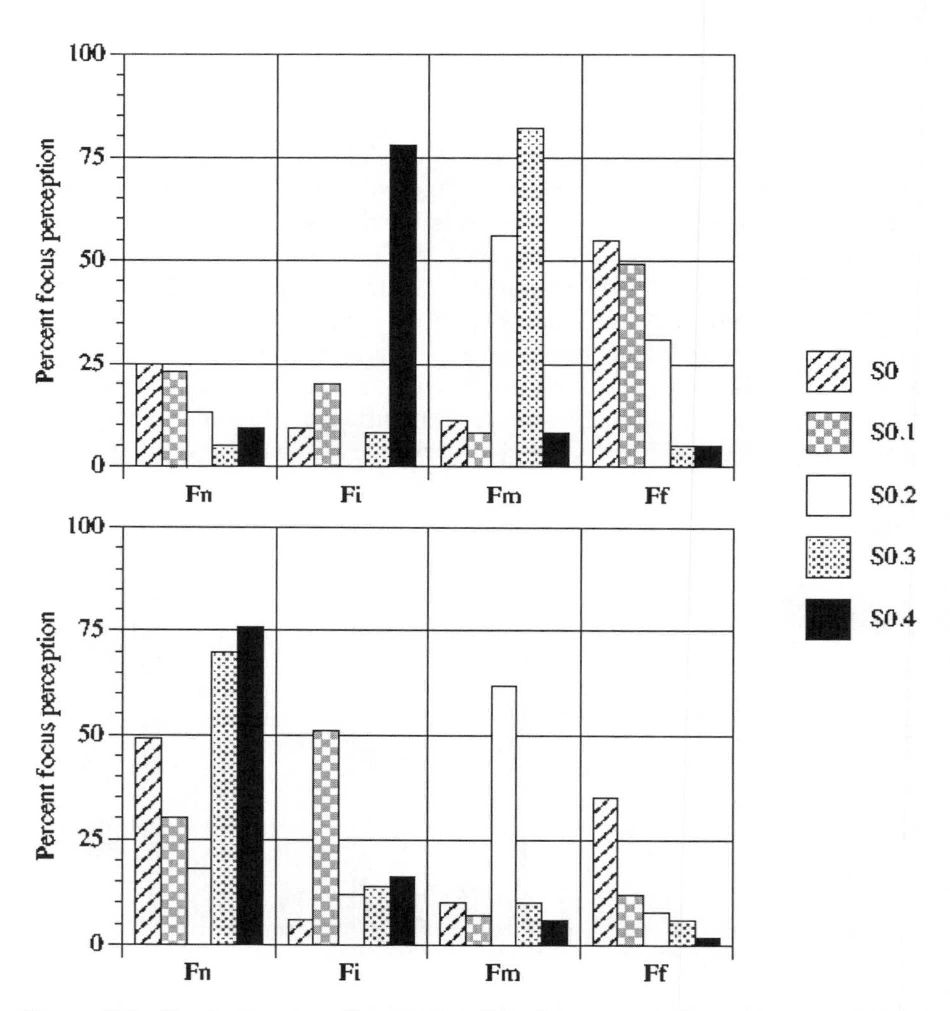

Figure 5.4. Greek (top panel) and Swedish (bottom panel) percentage of focus perception on the ordinate and focus response options on the abscissa (see text).

The results presented in Figure 5.4 indicate that poststressed tonal flattening (S0.3 & S0.4) may have a first-order perceptual effect for focus in Greek, assigning over 75% focus responses to this manipulation. In Swedish, however, poststressed tonal flattening favours the focus-neutral option.

Tonal range (S0.1 & S0.2) may have a second-order effect in both Greek and Swedish, assigning focus responses far below the 75% level in both languages.

5.3.3 Perception of Focus-initial Manipulations

Figure 5.5 shows the perception percentages for the focus-initial production (S1) and its manipulated versions for Greek and Swedish. All comparisons were significant at the p<.0001 level for both languages.

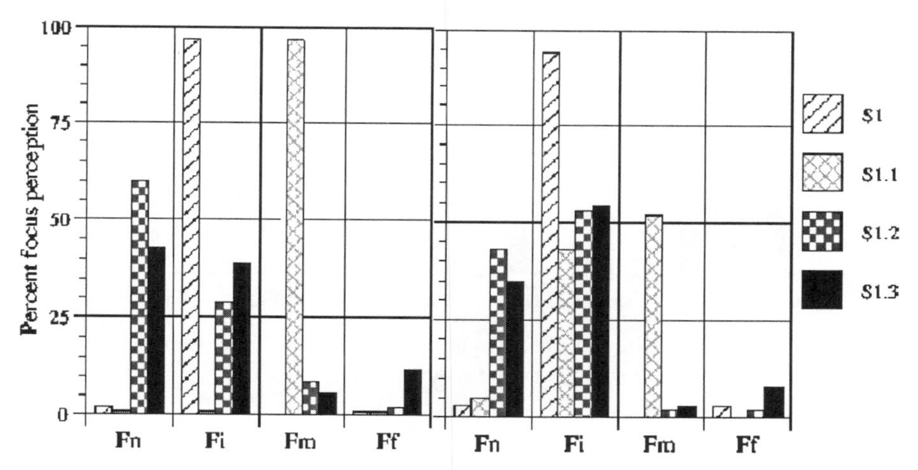

Figure 5.5. Greek (left) and Swedish (right) percentage of focus perception on the ordinate and focus response options on the abscissa (see text).

The results presented in Figure 5.5 indicate that tonal shift manipulation (S1.1) may cause a complete perception change in Greek, shifting focus from focus-initial to focus-medial, in accordance with the tonal shift manipulation. In Swedish, however, the corresponding tonal shift manipulation does not have similar effects but rather causes a fairly even distribution between the focus-initial and focus-medial options.

Tonal neutralisation manipulation (S1.2) may cause a perceptual defocalisation in both Greek and Swedish, although in a different pattern among the four alternative options in the two languages.

Tonal defocalisation manipulation (S1.3) may also cause a perceptual defocalisation in both Greek and Swedish, although in a different pattern among the four alternative options in the two languages.

5.3.4 Perception of Focus-medial Manipulations

Figure 5.6 shows the perception percentages for the focus-medial production (S2) and its manipulated versions for Greek and Swedish. All comparisons were significant at the p<.0001 level for both languages.

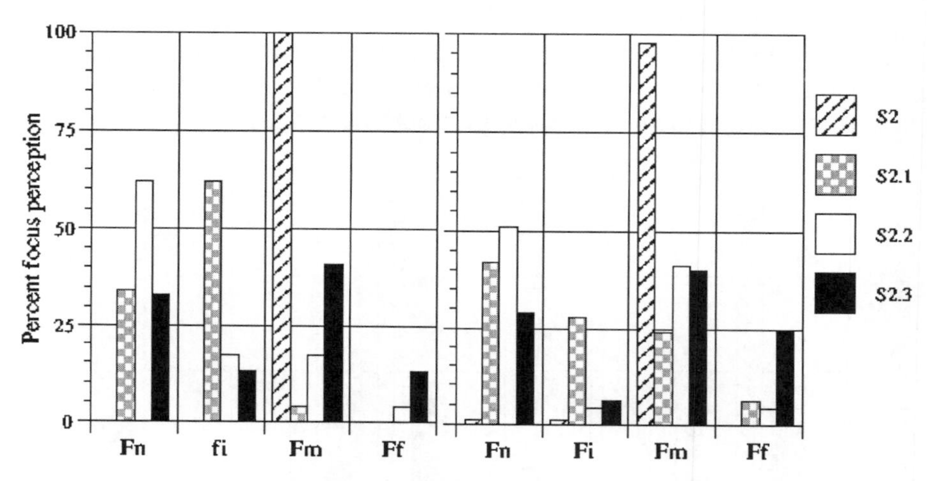

Figure 5.6. Greek (left) and Swedish (right) percentage of focus perception on the ordinate and focus response options on the abscissa (see text).

The results presented in Figure 5.6 are fairly similar in structure to the results presented in Figure 5.5. Thus, tonal shift manipulation (S2.1) may cause a complete perception change in Greek, although to a less degree than the one described in Figure 5.5, shifting focus from focus-medial to focus-initial, in accordance with the tonal shift manipulation. In Swedish, however, the corresponding tonal shift manipulation does not have similar effects but rather causes a distribution among the four options.

Tonal neutralisation manipulation (S2.2) may cause a perceptual defocalisation in both Greek and Swedish, although in a different pattern among the four alternative options in the two languages.

Tonal defocalisation manipulation (S1.3) may also cause a perceptual defocalisation in both Greek and Swedish, although in a different pattern among the four alternative options in the two languages.

5.5.5 Perception of Focus-final Manipulations

Figure 5.7 shows the perception percentages for the focus-final production (S3) and its manipulated version for Greek (left) and Swedish (right). For Greek, there was no significant difference between the responses to S3 and its manipulated version S3.1 (chi-square = .583, df = 3, p>.1). For Swedish, there was a marginally significant difference between the responses to S3 and S3.1 (chi-square = 12.8, df = 3, p=.005).

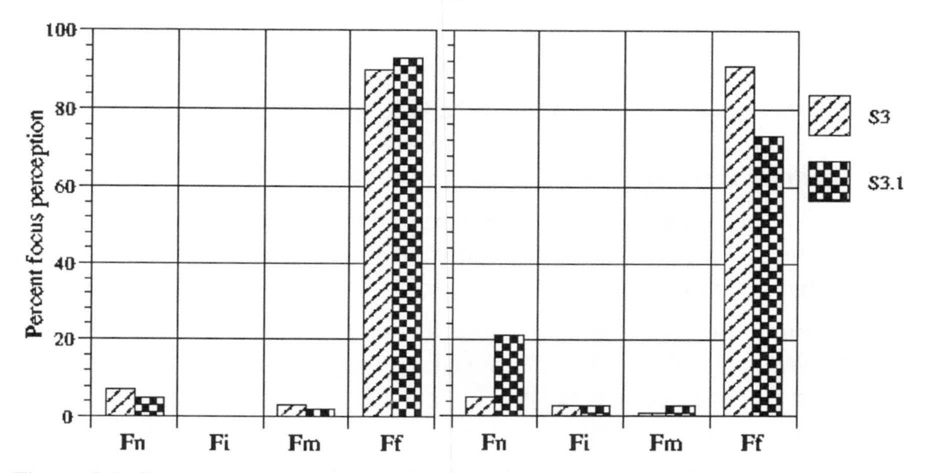

Figure 5.7. Greek (left) and Swedish (right) focus identification with identification rate on the ordinate and focus distribution responses on the abscissa (see text).

The results presented in Figure 5.7 indicate that tonal defocalisation of final-focus productions may not have a substantial perceptual effect in either Greek or Swedish. These results are not similar to the results of focus-initial defocalisation (see Figure 5.5, S1.3). It rather seems that postfocus tonal compression/flattening is a major perceptual correlate of focus and its tonal restructuring has bigger perceptual effects than analogous manipulations to other parts of the corresponding stimulus.

In summary, the present results indicate that focus perception is highly determined by tonal variations as evidenced by the significance of the overwhelming majority of the manipulated stimuli. Still, analogous tonal manipulations did not have similar effects on Greek and Swedish, an indication that the other prosodic correlates as well as the prosodic structure of each language have a substantial interference on tonal perception.

5.4 Discussion

Focus-neutral productions did not have the corresponding perception in either Greek or Swedish whereas focus-initial, focus-medial and focus-final productions were correspondingly perceived in both languages. With reference to tonal manipulations, Greek and Swedish have shown basic similarities but striking differences on tonal perception of focus as well, which may be language-specific characteristics of the two languages. The main points summarising the results, in accordance with the questions addressed at the Introduction, are the following:

1. Tonal range manipulations had a substantial effect but did not cause a complete perceptual change in accordance with the focus-target manipulation in either Greek or Swedish.
2. Poststressed tonal flattening manipulations had a major perceptual effect on both Greek and Swedish and caused a complete perceptual change in Greek but not in Swedish.
3. Tonal shift manipulations had a major perceptual effect on both Greek and Swedish and caused a complete perceptual change in Greek but not in Swedish.
4. Tonal neutralisation manipulations had a major perceptual effect and caused a defocalisation in both Greek and Swedish.
5. Tonal defocalisation manipulations had variable perceptual effects, associated with different focus productions in both Greek and Swedish.

5.4.1 Perception Effects of Tonal Range

Tonal range manipulations had a substantial effect in both Greek and Swedish. This is in accordance with our earlier assumptions for Greek (Botinis, 1989, 1998), according to which focus has a complex realisation within a global domain. Thus, in addition to the local contribution of tonal gestures and prosody in general both prefocal and especially postfocal tonal distribution is highly relevant for focus assignment. Although a great number of studies in different languages, including Swedish (see Bruce, 1977), describe the global effects of focus, no particular attention is paid with reference to these global realisations as tonal correlates of focus *per se*. They are rather assumed to be a by-product of focus, the main correlates of which are restricted to the local domain, i.e. syllable or stress group.

5.4.2 Perception Effects of Tonal Flattening

Tonal flattening manipulations caused a complete perceptual change in Greek but not in Swedish. In Greek, focus responses increased categorically in favour of the word just before the onset of the tonal flattening whereas, in Swedish, focus responses favoured the focus-neutral option. This is a striking difference of tonal perception of focus between Greek and Swedish. These results may be attributed to the assumption with reference to the global acoustic and perceptual correlates of focus in Greek as well as the contribution of a combination of prosodic parameters for focus perception in Swedish. In Greek, the poststressed tonal compression functions as a main perceptual correlate of focus; other combinations of prosodic parameters for focus perception are only marginally necessary and the global tonal structure rather than the local tonal range determines the focus distribution (see also Botinis, 1989, 1998). On the contrary, in Swedish, the poststressed tonal compression has not a decisive effect on the perception of focus as a combination of other prosodic parameters, probably duration in the first place, is a necessary condition for focus perception (see also Bruce *et al.*, this volume; Fant et *al.*, this volume). However, no grounded hypothesis may be postulated with reference to the substantial increase of focus-neutral responses in Swedish associated with tonal flattening.

5.4.3 Perception Effects of Tonal Shift

Tonal shift manipulations caused a complete perceptual change in Greek but not in Swedish. These results, in combination with the tonal neutralisation results (see below) support the assumption that F_0 is an all-or-none perceptual correlate of focus in Greek. A tonal shift to a focus target has a decisive effect as there are hardly any conflicting cues for focus perception left, since duration and intensity are primarily acoustic and perceptual correlates of the word stress of the reference material. However, the tonal shift from focus-initial to focus-medial had a considerably bigger effect than the reverse tonal shift, a difference for which only tentative speculations may be provided. Similar results, with reference to the effects of tonal shift manipulations, have been also reported for word stress perception in Greek but have been attributed to tonal structure interpretation on behalf of the listener and not to word stress perception *per se* (see Botinis, 1989, 1998). On the other hand, a tonal shift in Swedish has not a decisive effect as the other prosodic parameters of the reference focus may function as conflicting cues and thus block a complete identification change in favour of the focus target.

5.4.4 Perception Effects of Tonal Neutralisation

Tonal neutralisation manipulations had a major perceptual effect and caused a defocalisation in both Greek and Swedish. This is direct evidence that F_0 is a basic prosodic correlate for focus perception not only in Greek, the focus realisation of which has hardly any duration correlates, but also in Swedish, the focus realisation of which has constant duration correlates. Recently, it has been assumed that F_0 may not be an absolute necessity for focus perception in Swedish (Heldner, 1998), but still, in the present study, the perception of focus with a neutralised tonal contour is drastically reduced. It should be noted that Greek does not carry any quantity distinction whereas Swedish does have short *vs.* long vowels, in complementary distribution with post-vocalic consonants, and thus duration in Swedish seems more multi-functional than in Greek, at least as far as the present topic of discussion is concerned. In a perceptual study of word stress in Greek (Botinis, 1989, 1998), tonal neutralisations had no substantial effect and the manipulated stimuli were thus identified in accordance with the original word stress distribution (88% and 91% identification for the first and second element of the word stress minimal pair respectively). The conclusions reached in this study were that duration combined with intensity are the main acoustic and perceptual correlates of word stress in Greek whereas F_0 functions for focus distinctions and semantic weighting in general. These conclusions seem to be corroborated in general by the results of the present study.

5.4.5 Perception Effects of Tonal Defocalisation

Tonal defocalisation had fairly similar effects in both Greek and Swedish. There was, however, a basic difference in the perception pattern between the manipulated stimuli stemming from focus-initial and focus-final reference stimuli. Tonal defocalisation of focus-initial reference production (Figure 5.5, S1.3) had a defocalisation perception effect whereas tonal defocalisation of focus-final reference production (Figure 5.7, S3.1) had only a marginal effect. It might be argued that tonal structure is a basic perceptual correlate, in the absence of which no focus distinctions are perceived, with reference to the former case but this very argument is contradicted with reference to the latter case. On the other hand, no final-focus perception bias have been observed in other stimuli sets in the present study and thus no grounded hypothesis can be provided for the above perceptual asymmetry. Obviously, more research is required on this point as well as on focus perception in general.

5.4.6 Theoretical Considerations

In general, with reference to the present experiments, Greek seems more sensitive to tonal manipulations in comparison to Swedish. The present results are in accordance with and support previous results on the acoustics of focus, according to which F_0 has been found an invariant acoustic correlate of focus in Greek (e.g. Botinis, 1989; Botinis *et al.*, 1995; Fourakis *et al.*, 1999). In Swedish, however, a combination of the prosodic parameters seems to be required to a greater degree than in Greek, in both production and perception, although F_0 may be a primary phonetic correlate of focus, as referred to in the established Swedish literature (e.g. Bruce *et al.*, this volume).

To our knowledge, no perceptual study has explicitly compared focus-neutral to other focus productions. Even though these structures are encountered in the literature (e.g. Botinis, 1989) no comprehensive analysis has been carried out. Thus, in Greek, there is hardly any established consensus about focus-neutral *vs.* focus-final productions. On the other hand, there is an established consensus for Swedish (and even English and other European languages) that focus is by default assigned to the right end constituent, if not specifically assigned to an earlier constituent. Swedish is exemplary, where an extra tonal gesture may be assigned to the right of the word accent of the last constituent (Bruce, 1977). It has thus been inferred that focus-neutral and the focus-final productions should not be distinguished. Acoustic as well as perceptual evidence is however provided in the present experiment that these two prosodic structures may be distinct (see Figure 5.1). A right constituency dominance in perception, with reference to the focus-neutral productions, is barely noticeable in Greek whereas, with reference to the focus productions, no right constituency bias is observed. On the contrary, focus-final perception has the lowest scores in both Greek and Swedish whereas focus-medial perception has the highest scores in both languages.

It should be pointed out that tonal and prosodic manipulations with reference to focus perception are not straightforward. In the international literature, e.g. in English, numerous studies have concentrated on prosodic perception in a "nuclear stress" context (see Beckman, 1986 and references therein). However, the majority of these studies hardly address the appropriate question but, instead, the perception of *word stress* is investigated in *focus position* (i.e. nucleus) and the test sentences which the listeners are usually asked to identify consist of minimal word stress pairs such as OBject *vs.* obJECT. Thus, the results in the nucleus context reported often in the English literature refer in reality to word stress perception and not to focus perception *per se.*

Beyond the results of the present study, it should be emphasised that different languages do have different options of focus signalling, which may be carried out by a variety of linguistic means. Thus, focus by prosodic means may have a considerable variability of functional load among languages and, consequently, the functional specifications of focus should be reflected in prosodic perception as well. In conclusion, focus is a very complex concept and much research is required before its wider dimensions from phonetic, linguistic and communicative points of view are fairly understood. Focus applications are also among central areas in speech and language technology, for which the main aspects outlined in the present study, and even more, must be taken into consideration.

Notes

1. Work on focus perception is in progress by the present research group (see also Botinis, Fourakis and Gawronska, 1999).
2. The Swedish tonal structure presented here is slightly different from the one presented in Botinis and Bannert (1997). The main difference between the two is that the speaker of the former made hardly any distinction between the neutral and the final focus productions whereas the latter did.

Acknowledgements

We would like to thank Thierry Deschamps for his help with the stimuli generation at Umeå University Phonetics Laboratory as well as Marios Fourakis and Thanasis Protopapas for statistical evaluation. Thanks also to our reviewers, Gösta Bruce, Gunnar Fant and Mario Rossi, for insightful comments and much useful feedback.

References

Bannert, R. 1979. The effect of sentence accent on quantity. *Proc. 9th ICPhS* (Copenhagen, Denmark), vol. 2, 253-259.

Bannert, R. 1986. Independence and interdependence of prosodic features. *Working Papers* 29, 31-60. Dept of Linguistics and Phonetics, Lund University.

Bannert, R. 1995. Variations in the perceptual modelling of macro-prosodic organisation of spoken Swedish: prominence and chunking. (Reports from the Department of Phonetics, Umeå University), *PHONUM* 3, 31-53.

Beckman, M.E. 1986. *Stress and Non-Stress Accent*. Dordrecht: Foris

Bolinger, D.L. 1958. A theory of pitch accent in English. *Word* 14, 109-49.

Bolinger, D.L. 1972. Accent is predictable (if you're a mind-reader). *Language* 48, 633-44.

Botinis, A. 1982. Stress in Modern Greek: an acoustic study. *Working Papers* 22, 27-38. Dept of Linguistics and Phonetics, Lund University.

Botinis, A. 1989. *Stress and Prosodic Structure in Greek.* Lund University Press.

Botinis, A. 1998. Intonation in Greek. In Hirst and Cristo (eds.), 288-310.

Botinis, A. and R. Bannert. 1997. Tonal perception of focus in Greek and Swedish. *Proc. ESCA Workshop on Intonation* (Athens, Greece), 47-50.

Botinis, A., S. Erkenborn, C. Isacsson and P. Westin. 1999. Prosodic variability and segmental durations in Greek and Swedish. *Proc. Swedish Phonetics Conference Fonetik* '99 (Gothenburg, Sweden), 27-30.

Botinis, A., M. Fourakis and M. Katsaiti. 1995. Acoustic characteristics of Greek vowels under different prosodic conditions. *Proc. 13th ICPhS* (Stockholm, Sweden), vol. 4, 404-407.

Botinis, A., M., Fourakis and B. Gawronska. 1999. Focus identification in English, Greek and Swedish. *Proc. 14th ICPhS* (San Francisco, CA, USA), vol. 2, 1557-1560.

Bresnan, J. 1971. Sentence stress and syntactic transformations. *Language* 47, 257-280.

Bruce, G. 1977. *Swedish Word Accents in Sentence Perspective*. Lund: Gleerup.

Bruce, G., M. Filipsson, J. Frid, B. Granstöm, K. Gustafson, M. Horne and D. House. This volume. Modelling of Swedish discourse intonation in a speech synthesis framework.

Chomsky, N. 1971. Deep structure, surface structure and semantic interpretation. In Steinberg and Jakobovits (eds.), 183-216.

Cooper, W.E., S.J. Eady and P.R. Mueller. 1985. Acoustical aspects of contrastive stress in question-answer contexts. *J. Acoust. Soc. Am.* 77, 2142-55.

Cruttenden, A. 1997 (2nd edition). *Intonation.* Cambridge University Press.

Engstrand, O. 1988. Articulatory correlates of stress and speaking rate in Swedish VCV utterances. *J. Acoust. Soc. Am.* 83, 1863-1875.

Fant, G., A. Kruckenberg and J. Liljencrants. This volume. Acoustic-phonetic analysis of prominence in Swedish.

Fant, G., A. Kruckenberg and L. Nord. 1991. Durational correlates of stress in Swedish, French and English. *Journal of Phonetics* 19, 351-365.

Fourakis, A., A. Botinis and M. Katsaiti. 1999. Acoustic characteristics of Greek vowels. *Phonetica* 56, 28-43.

Fry, D.B. 1955. Duration and intensity as physical correlates of linguistic stress. *J. Acoust. Soc. Am.* 27, 765-8.

Fry, D.B. 1958. Experiments in the perception of stress. *Language and Speech* 1, 126-52.

Gårding, E., A. Botinis and P. Touati. 1982. A comparative study of Swedish, Greek and French intonation. *Working Papers* 22, 147-153. Dept of Linguistics and Phonetics, Lund University.

Gussenhoven, C. 1983. Focus, mode and the nucleus. *Journal of Linguistics* 19, 377-417.

Halliday, M.A.K. 1967. Notes on transitivity and theme in English. *Journal of Linguistics* 3, 199-244.

Heldner, M. 1998. Is an F_0 rise a necessary or a sufficient cue to perceived focus in Swedish. *Nordic Prosody* VII, 109-125.

Hirst, D.J. and A. Cristo (eds.). 1998. *Intonation Systems*. Cambridge University Press.

Jackendoff, R.S. 1972. *Semantic Interpretation in Generative Grammar*. Cambridge, Mass.: MIT Press.

Ladd, D.R. 1996. *Intonational Phonology*. Cambridge University Press.

Lambrecht, K. 1996. *Information Structure and Sentence Form*. Cambridge University Press.

Lieberman, Ph. 1960. Some acoustic correlates of word stress in American English. *J. Acoust. Soc. Am.* 32, 451-454.

Pierrehumbert, J.B. 1980. The *Phonology and Phonetics of English Intonation*. PhD dissertation, MIT (published 1988 by IULC).

Rossi, M. This volume. Intonation: past, present, future.

Rossi, M., A. Di Cristo, D.J. Hirst, Ph. Martin and Y. Nishinuma. 1981. *L' Intonation: De l'Acoustique à la Sémantique*. Paris: Klincksieck.

Steinberg D.D. and L.A. Jakobovits (eds.). 1971. *Semantics: An Interdisciplinary Reader in philosophy, Linguistics and Psychology*. Cambridge University Press.

Section III

Boundaries and Discourse

6

Phonetic Correlates of Statement versus Question Intonation in Dutch

VINCENT J. VAN HEUVEN AND JUDITH HAAN

6.1 Introduction

6.1.1 Basic Considerations

In recent years, the formal elements of Dutch intonation have been laid down in two comprehensive models ('t Hart, Collier and Cohen, 1990; Gussenhoven & Rietveld, 1992). With these two formal models at our disposal, the stage seems set for further explorations, notably of the relationship between form and function. The present study focuses on acoustic and perceptual correlates of one major functional contrast, viz. the opposition between declarativity (statement) and interrogativity (question), two functions featuring prominently in everyday communication.

Generally speaking, the statement mode is used for making announcements, relating events, stating conclusions, and so on. By contrast, the interrogative mode makes a direct appeal to a listener for a reply. While the statement mode usually has the most basic form of clause available in a language, interrogativity may be marked by special syntactic and/or lexical means, in particular by inversion of subject and finite verb or by the presence of a question word. These, however, are by no means the sole indicators of the contrast between declarativity and interrogativity. It is assumed that intonation, also, plays an important role (see 6.2). In our research we target the phonetics of Dutch statements and three types of questions:

A. Botinis (ed.), Intonation, 119-143.

S *Statements*, with normal word order between subject and finite, and lacking question words, i.e., with zero interrogativity markers.

W *Wh-questions,* beginning with a question word (*who, where, when, what, why*, etc.), and often followed by inversion of subject and finite, i.e., with two interrogativity markers.

Y *Yes/no-questions*, syntactically marked by inversion of subject and finite only, i.e., with one interrogativity marker.

D *Declarative questions*, with the same lexical items and word order as the corresponding statement, again lacking any interrogativity marker.

Lexico-syntactic marking of interrogativity is strongest in W, weaker in Y, and absent in D (as well as in S, of course). All three question types are claimed to be prosodically marked as well, but not necessarily to the same degree. For instance, Lindblom's (1990) hyper and hypo (H&H) theory of speech production would predict that prosodic interrogativity cues will be weaker (i.e. hypo- or underarticulated) when the listener has enough information elsewhere in the signal to reconstruct the speaker's intentions. Conversely, prosodic interrogativity marking will be more forceful (hyper- or overarticulated) as there are fewer or no lexico-syntactic question markers in the utterance. On the basis of such functional considerations we expect phonetic/prosodic interrogativity marking to be stronger in inverse proportion to the (number of) lexico-syntactic markers, i.e., in the order D>Y>W>S.

6.1.2 Literature on Question Intonation, with Emphasis on Dutch

As far as Dutch question intonation is concerned, there are early claims in the literature that the question contour is hammock-shaped, i.e., has a high beginning, a low stretch in between and an equally high ending (van Es, 1932; Daan, 1938). It has also been claimed that Dutch questions are realized in a higher register[1] (van Alphen, 1914; van Es, 1932). So far, however, no experiments have been carried out to check such claims. In the two formal accounts of Dutch intonation mentioned above the function interrogativity is not explicitly dealt with; however, both models feature formal units which may serve as a final interrogative rise. That the element of high(er) pitch in Dutch questions is perceptually relevant was demonstrated by Gooskens and van Heuven (1995). Also, their results indicate that Dutch listeners, when deciding whether an otherwise ambiguous utterance is a statement or a question, favour questions from which the canonical F_0 downtrend has been removed.

Taking the above theoretical claims into account, we formulate the hypotheses that, in comparison with statements, Dutch questions are characterized by higher pitch in at least the following ways: (i) questions will have final rises, (ii) questions will be generally realized on a higher register, (iii) questions will show less global downtrend of F_0 or even an upward trend, and finally (iv) the prosodic interrogativity-marking properties will be stronger in the order D>Y>W>S (see 6.1.1).

6.1.3 Structure of the Paper

In the following pages we will report three experiments. The first is a production experiment, designed to test the hypotheses derived above on the basis of a carefully controlled corpus of play-acted speech utterances spoken by five male and five female Dutch speakers. The results of this experiment provide strong support for our hypotheses. In fact, the experimental data allow us to draw up clearly differentiated prosodic profiles for each of the four sentence types under analysis, i.e., for statements and for each of the question types W, Y and D. Moreover, the results suggest that the differentiation among the sentence types is not solely due to the presence versus absence of the terminal question-marking pitch rise. Acoustic differentiation between the various types seems to take place much earlier in the sentence. This article therefore includes two subsequent perceptual experiments in which we aimed to determine in some detail at which point(s) in the time course of the utterance the human listener hears whether the speaker intends to communicate a statement or a (specific type of) question. The results show that statements can be told apart from declarative questions well before the onset of the terminal question-marking rise (experiment IIB); however, no such early differentiation is observed in the time course of statement and W-questions on the one hand, and between Y and D-questions on the other (experiment IIA). A post-hoc stimulus analysis was run to determine the most likely acoustic correlates of the early differentiation between statement and (D-) question.

6.2 Production Experiment

6.2.1 Materials and Measurements

A corpus of 800 target utterances was made up from two two-accented core statements (S), each of which could take the shape of a wh-question (W, marked by a wh-word and inversion), a yes/no question (Y, marked by inversion) and a declarative question (D, lacking both wh-word and inversion)[2]. Five male and five female native speakers of Dutch twice read out the eight target utterances, *in pausa* as well as in pairs (i.e. preceded or followed by an utterance providing a context). Recordings were made onto digital audio tape (DAT) in a sound-proofed studio (48.1 kHz, 16 bit) using a Sennheiser MKH-416 condenser microphone. Subjects were asked to speak the sentences as if they were actors in a radio play. For details on materials and procedures see Haan, van Heuven, van Bezooijen and Pacilly (1997a, 1997b); for the full set of stimuli see appendix 1A-B.

F_0 was extracted by the method of subharmonic summation (Hermes, 1988), followed by curve smoothing over a 50-ms time window. F_0-values were expressed in Equivalent Rectangular Bandwidths (ERB), which is currently held to be the psychophysically most relevant F_0-scale in intonation languages (Hermes & van Gestel, 1991; Ladd & Terken, 1995)[3]. Measurements included the F_0-values of the onsets, the offsets (both with and without the final rise, if present), the minima and the maxima (discounting a final rise), as shown in Figure 6.1.

The presence of a terminal rise was determined interactively in the visual display of the smoothed F_0-curve. Its onset was placed at the local minimum F_0 within the 500-ms time interval before the pause marking the end of the utterance. Figure 6.1 also shows upper and lower trend lines, which were fitted to the F_0-curve as follows:

- A linear F_0-with-time regression line ('all points', cf. Lieberman, Katz, Jongman, Zimmerman & Miller, 1985) was drawn through the utterance (minus the terminal rise); the F_0-measurement points were (automaticaly) divided into an upper and lower half, i.e., those points lying above the all-points regression line, and those lying below it, respectively.
- Next, an upper regression line was fitted through the upper F_0-points, and a lower regression line through the lower points.

The slope coefficients of the lines approximate the steepness of what is called declination lines in 't Hart *et al.* (1990). F_0 at the temporal midpoint of the upper and lower trend lines allows comparison of height and width of register across utterances (for a validation of this claim see 6.2.3).

The data were then subjected to analyses of variance with Sentence type or Question type as a three or four-level factor: (S,) W, Y, D.

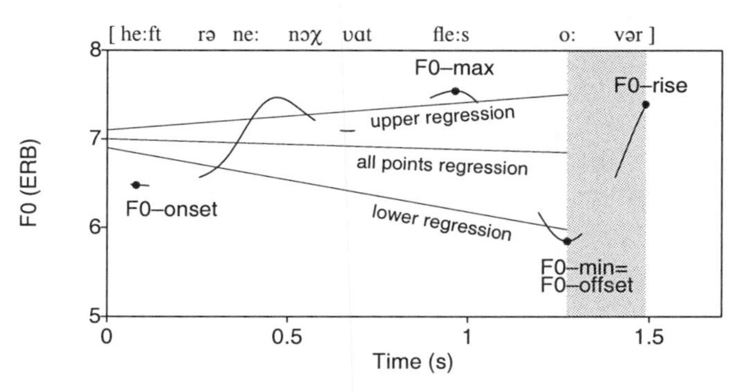

Figure 6.1. Raw parameters and regression lines for an isolated Y-question spoken by a female *Heeft Renée nog wat vlees over?* 'Does Renée have any meat left?'. The terminal rise extends over the grey area.

6.2.2 Results

Final rises, traditionally the canonical question markers, were considered first. Figure 6.2 presents the number of final rises found for each of the four sentence types, broken down by Sex of speaker.

Final rises never occur in statements. Rises are nearly always present in D-questions, and less often in Y and W-questions in the order W<Y<D. Moreover, women clearly use more final rises than men; the difference is especially marked in W-questions. The greater incidence of final rises with women has been reported earlier in the literature on American English (Edelsky, 1979).

Figure 6.3 presents the onset and offset F_0, and implicitly excursion size, (in ERB) of those final rises that were present, broken down by Question type and by Sex of speaker. Clearly, the excursion size of the rise is the same for the three question types, $F(2,502)=2.1$ ($p=0.122$, ins.). However, the onset and offset F_0 does differ, with higher values in the predicted D>Y>W order, $F(2,502)=27.6$ ($p<0.001$) for F_0-onset and $F(2,502)=21,5$ ($p<0.001$) for F_0-offset; all contrasts were significant (Newman-Keuls procedure with $\alpha<0.05$) except the one between W and D for offset- F_0. Note that the excursion size of the female rises is roughly twice as large as that of males, $F(1,503)=52.1$ ($p<0.001$). Since F_0 is already scaled so as to optimally reflect auditory distance, this finding indicates that women mark their question rises perceptually more saliently than men.

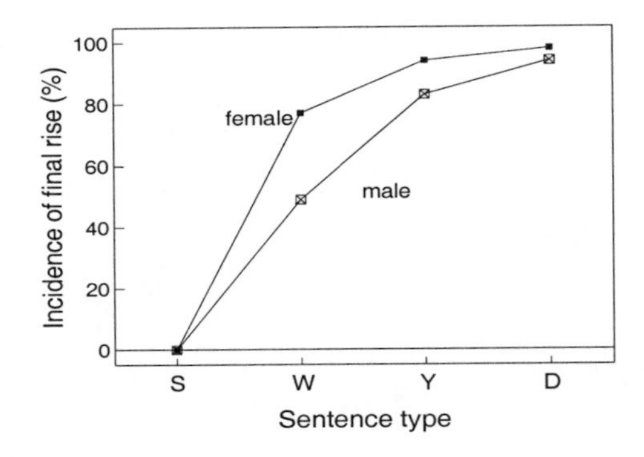

Figure 6.2. Incidence of terminal rises for four sentence types broken down by Sex of speaker.

Sentence type also showed strong effects on F_0-minima, $F(3,796)=53.8$ ($p<0.001$, all contrasts significant), maxima, $F(3,796)= 26.7$ ($p< 0.001$; without final rise; contrast between Y and D insignificant) and offsets, $F(3,796)=149.6$ ($p<0.001$; without final rises; all contrasts significant). F_0-values were highest, again, in D, lowest in S and intermediate in W and Y, respectively (D>Y>W> S). However, F_0-values of the utterance onsets surprisingly deviated from this pattern: while W and Y were characterized by high onsets, the onsets of S and D were both low and did not differ significantly from each other, $F(3,796)= 142.9$ ($p<0.001$). Figure 6.4 sums up the results.

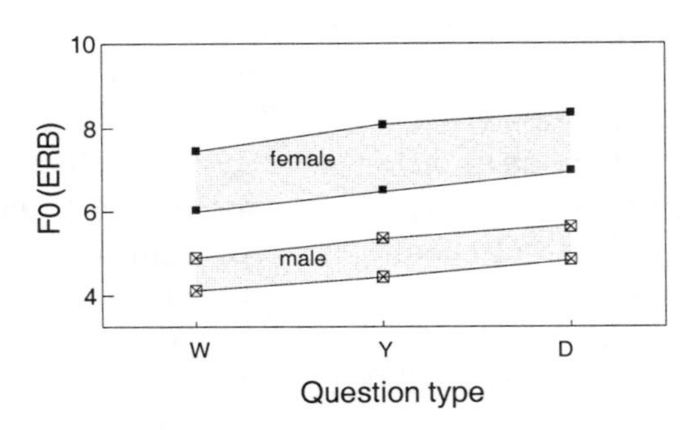

Figure 6.3. Onset and offset F_0 of question-marking final pitch rise (ERB) for three question types broken down by Sex of speaker. Utterances without a final rise were excluded from the analysis.

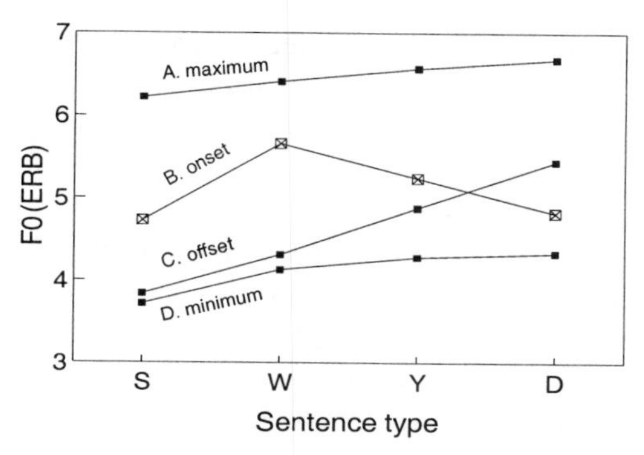

Figure 6.4. Mean F_0 (in ERB) at F_0-maximum (line A, discounting the final rise), sentence onset (line B), sentence offset (line C, last measurement before onset of final rise), and at F_0-minimum (line D).

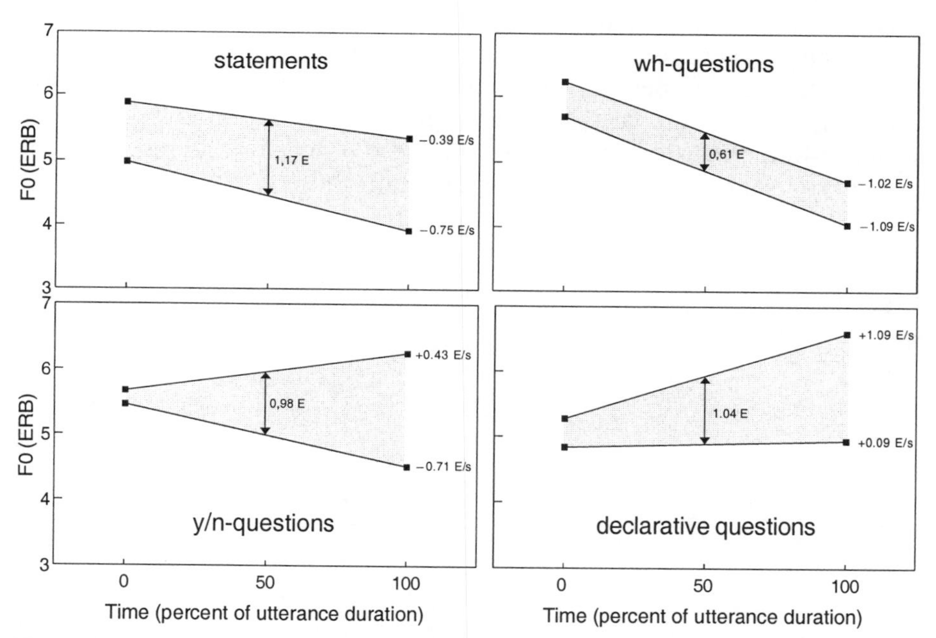

Figure 6.5. Onset and terminal frequencies (in ERB) of upper and lower regression lines and their respective slope coefficients (in ERB/s) for four sentence types; vertical arrows indicate mean register width at the temporal midpoint.

Figure 6.5 shows the effects of Sentence type on the course of the upper and lower trend lines (and by implication on relative register width. Typically, S and W have an overall downward trend in both their upper and lower lines. By contrast, Y combines a declining lower line with an inclining upper line and D even displays uptrends in both lines. All contrasts between the slopes of the *upper* lines were significantly different in the order D>Y>S>W, $F(3,796)=160.5$ ($p<0.001$).

As to the prototypical register width of each sentence type, in W this is narrowed by almost 50%; register widths of S, Y and D do not significantly differ from each other, $F(3,796)=138,8$ ($p<0.001$).

6.2.3 Validation of Automatic Trend Line Estimation Procedure

To determine the validity of the automatic regression parameter extraction, the F_0-curves of all 800 utterances were also segmented and labelled by hand. In each utterance the (potentially) accent-lending pitch configuration on the subject and on the object were stylised as a pointed hat, i.e., as a rise-fall sequence (where a rise or fall could be absent). In the case of W-questions an additional accent-configuration was measured on the sentence-initial question word itself. Figure 6.6 displays the results of the manual stylisation of rises and falls on potential accents.

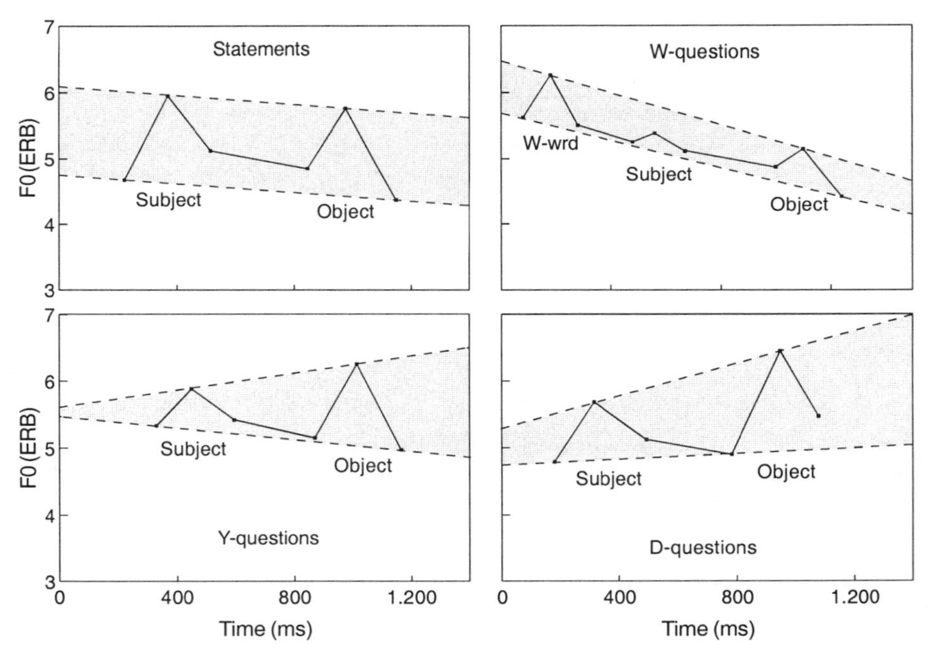

Figure 6.6. F_0-pivot points (ERB) as a function of time (ms) for four sentence types, as obtained from manual stylisation. Declination lines were fitted by hand.

The results show, quite clearly, that the automatic regression parameters adequately capture the essentials of the intonation patterns of the four sentence types. In fact, the details of the rises and falls in the stylised contours largely explain why the upper and lower trend lines show up the way they do. The accents on subject and object in statements are roughly equally large with normal declination between them. The accent (peak F_0) on the subject is reduced in Y-questions, and even more so in D-questions, and the accent on the object is increased in size roughly in inverse proportion to the reduction of the subject. This, obviously explains the diverging trend lines for Y and D-questions, as well as the tendency for the upper trend line to incline. We conclude, therefore, that the global differences between the four sentence types, as were apparent in the upper and lower trend lines in Figure 6.5, are largely determined by the height of local accents on subject and object. W-questions, moreover, typically start on a high F_0 (where it is not always clear whether there is an accent on the question word, or whether this is 'just' a high onset). The accents on the subject and object in W-questions are generally small. These local phenomena, then, explain why the register of W-questions is narrow, and downtrend steep. Note, finally, that some independent contribution of global characteristics of the four sentence types remains. For instance, the lower declination line, as visible between the accents on the subject and on the object, differs from one question type to the next: the trend is downward for S and Y, and even more so for W, but gently rising for D.

6.2.4 Conclusions and Discussion

Clearly, the functional contrast between Dutch declarativity and interrogativity has acoustic correlates that can be adequately captured in terms of low vs. high pitch. Thus, apart from ending predominantly in terminal rises, Dutch questions are also realized in a higher as well as a narrower register. Furthermore, in two of the three question types the overall trend of the upper regression lines is upward rather than downward. In general, the strength of these pitch-level properties was inversely related to the presence of lexico-syntactic markings of interrogativity; this was reflected in the frequent occurrence of the rank order D>Y>W.

However, even though the three question types share major properties as a category, it should also be emphasized that they have distinct pitch profiles of their own. Thus, if D has the highest offset (both with and without final rise), the highest local minima and maxima and the steepest global upward trend, its onset is unexpectedly low. In fact, it might be argued that D is actually an upward realization of S. Lexically and syntactically, the two are identical; also, their onset pitch levels are greatly similar, as are their accentuation patterns since both lack the reduction of the first accent frequently found in W and Y (see below). Thus it would seem that the functional distinction declarative~interrogative is actually signalled by the strongly diverging F_0-trends: downward for declarative, upward for interrogative. This would be in line with interpretations of D claiming that it is biased in that it solicits confirmation rather than information. Seen in this light, D would actually be a statement the truth of which has only to be confirmed by the hearer; the request for confirmation is then expressed by means of a steeply upward intonation.

If prototypical W starts from the highest pitch and is realized in the highest and narrowest register, its very steep downward slope nevertheless comes as a surprise. Again, this might be caused by the accentuation pattern which, typically, renders the (utterance-initial) wh-word prominent, mostly at the expense of one or even both of the non-initial accents. This, of course, may steepen the downward slope of the top line.

Between D and W as extremes, Y holds an intermediate position. Like D, its upper and lower trend lines typically diverge but Y is like S and W in that its lower trend line takes a downward course. The positive tilt of the upper line springs from the contrast between the first accent which is often reduced, and the second accent, which usually has quite a large excursion.

6.3 Perception Experiments

6.3.1 Introduction

The previous experiment showed that Dutch statements and questions are acoustically distinct in the absence versus presence, respectively, of a question-marking terminal pitch rise. Within the category of questions, D-questions were always marked by a terminal rise, Y-questions in about 85%, and W-questions in no more than 60% of the cases. Moreover, the global characteristics of the F_0-curve were systematically different for each of the four sentence (sub)types, to the effect that F_0-downtrend was present in W-questions and statements, but absent, or even inverted to an uptrend, in Y and D-questions, respectively. Finally, there were local differences among the various sentence types in the size of the accent-marking pitch movements on the subject and object constituents. Taken together, these findings indicate that the non-final portion of the utterance contains potentially useful information that allows the listener to identify the type of sentence he is hearing soon after the sentence onset. The present experiment addresses the following three questions:

1. Are listeners able to identify the correct sentence type from four possible sentence types (Y-question, W-question, D-question and statement) before the final part of the sentence?
2. How does the discrimination between the four sentence types develop as the utterance progresses?
3. What specific acoustic cues enable early identification of sentence type?

We ran two gating experiments in which sentence-initial fragments of increasing size were made audible in successive passes; listeners were asked to identify the (most likely) sentence type from these fragments. Experiment IIA tested the identification of all four sentence types; experiment IIB narrowed down the research to just the identification of statements versus declarative questions, as these are lexico-syntactically identical and differ in their prosodic characteristics only. A post hoc stimulus analysis was carried out to isolate – through regression analysis – the most likely acoustic cues that the listeners used when performing their task.

6.3.2 Experiment IIA

6.3.2.1 Method

From the database described in 6.2.1 only sentences of four (out of ten) speakers (two male, two female) were selected that had been spoken in isolation (rather than in two-sentence paragraphs), yielding four tokens of two lexically different sentences in four different versions (yes/no question, wh-question, declarative question, and statement). The speakers were selected on the basis of separate Linear Discriminant Analyses ran in order to automatically classify the utterances in terms of our four sentence types (van Heuven, Haan & Pacilly, 1997). The two male and female speakers who had kept the sentence types most successfully distinct, were included in the gating experiments.

In order to further reduce the work load on our listeners, only the sentences based on *Marina wil haar mandoline verkopen* were included in the listening materials. Since this sentence contains almost exclusively voiced (and sonorant) segments, it was deemed that it would be easier for the listeners to perceive the details of the pitch curve. Moreover, only the first reading of each sentence was used[4].

The four versions of the sentence used in this experiment were (for glosses see appendix 1A):

Y *Wil Marina haar mandoline verkopen?*

W *Waar wil Marina haar mandoline verkopen?*

D *Marina wil haar mandoline verkopen?*

S Marina wil haar mandoline verkopen.

Since most sentence types differ in the first two or three words, the audible portion of the utterance always started at the beginning of the third word. Four fragments of increasing size were cut from the recordings (using a high resolution waveform editor on a Silicon Graphics Iris Indigo computer), as follows:

1. *haar mando* no accent, no final rise
2. *haar mandoLIne* final accent included (small caps)
3. *haar mandoLIne verko* no final rise
4. *haar mandoLIne verkopen(?)* final rise included (if present)

The material was presented over headphones to 18 native Dutch listeners. Subjects were instructed to decide, for each fragment they heard, the most likely sentence that the fragment was a portion of, with forced choice from four alternatives (S, W, Y, D). For each fragment, the subjects' answer sheets listed the four alternative sentences in full print.

6.3.2.2 Results

Figure 6.7 presents the listeners' responses to each of the four stimulus sentence types (panel A: Y-questions; panel B: W-questions; panel C: D-questions; panel D: Statements) broken down by gate number and response alternative. In each panel, the correct responses are indicated by the thick line labelled with a bold-faced letter; the other (thin) lines with normal-face labels represent the confusions.

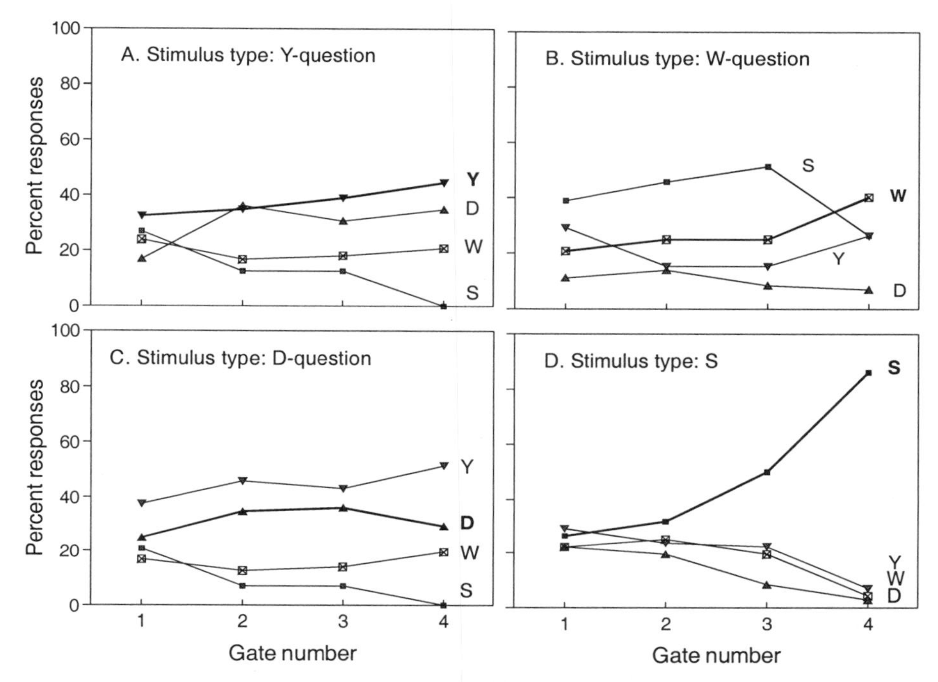

Figure 6.7. Percent responses in each of four sentence type alternatives, as a function of gate length (number). Each panel present the data of a different stimulus type. The thick lines, labeled in bolf face, indicate the correct alternative (chance = 25%).

Percent correct increases with gate number, except for D-questions, which were heavily confused with Y-questions. A confusion analysis (not presented) showed that Y and D-questions were confused on the one hand, as were S and W on the other, during the first three gates. Therefore, Figure 6.8a presents the data once more, but now Y and D-question responses have been lumped together, as have W and S-responses. Figure 6.8a shows that from gate 2 onwards, identification of the correct group {Y+D} versus {W+S}) is above chance level. This figure is deceptive to some extent in that statements can be told apart from all question types at gate 4, where a terminal rise is a sure sign of interrogativity; therefore Figure 6.8b presents the data lumped together in yet another fashion: all question types {Y+D+W} versus statements (S).

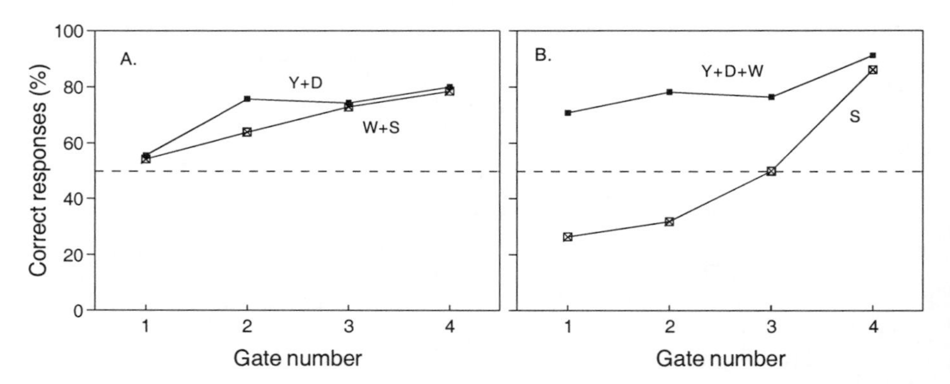

Figure 6.8. Percent correct identification of sentence type as a function of gate length (number). Panel A: Sentences have been lumped together two-by-two (Y+D-questions versus W-questions+statements); panel B: results have been regrouped into all question types versus statements.

6.3.2.3 Discussion and conclusion

In 6.3.2.1 we asked whether subjects are able to identify the correct sentence type before the end of the sentence. Our results show that listeners were unable to recognise the correct sentence type above chance level at gate 1. From gate 2 onwards, i.e., when a pitch accent has been made audible, listeners identify two broad sentence types. Apparently Y and D-questions share certain characteristics that cause our listeners to lump these two together, as happens in the case of W-questions and statements. Only at gate 4, when the terminal rise is made audible, do listeners properly differentiate W-questions from statements. Even here, however, the identification is not perfect: some 10% error remains. The reason for this is that a number of W-questions did not end in a final rise; the global characteristics of W-questions resembles that of statements very much, so that some ambiguity remains.

If the entire sentence had been made audible, i.e., including the sentence-initial question word, the ambiguity could have been solved lexically. In experiment IIB, therefore, we limit the number of stimulus sentence types to just two: statement versus D-question. We know from experiment I (6.2.3) that statements are never marked by a final rise, and D-questions always end in a rise. The prosodic ambiguity inherent to W-questions is thus eliminated. Moreover, D-questions and statements contain the same words in the same order, so that the stimuli may be gated from the beginning of the sentence. This will allow us to trace the contribution of global prosody to the identification of sentence type in better detail than before.

6.3.3 Experiment IIB

6.3.3.1 Method

This experiment was set up to also investigate the influence of the first part of the sentence on sentence type recognition. Only D-questions and statements were compared; since these two sentence types have identical syntactic and lexical information, the utterances could be truncated into six different gates, as follows (small caps denoting accented syllables):

1. *maRIna* first accent included
2. *maRIna wil*
3. *maRIna wil haar mando*
4. *maRIna wil haar mandoLIne* second accent included
5. *maRIna wil haar mandoLIne verko*
6. *maRIna wil haar mandoLIne verkopen(?)* terminal rise included

Note that the sentence-initial proper name *maRIna* was always accented, its pitch movement aligned with the lexically stressed second syllable.

These stimuli were presented to the same group of 18 listeners and under the same conditions as in experiment IIA. Sentences were presented in random order, but shorter gates always preceded longer gates derived from the same sentence.

6.3.3.2 Results and conclusion

Figure 6.9 presents the percentages of correct responses broken down by stimulus sentence type and gate number. Figure 6.9 shows that, when confronted with a statement, listeners chose the correct sentence type from the first gate onwards well above chance level. For the first three gates, percent correct responses remained around 70. From gate 4 to 6 there is a steady increase of correct responses, culminating in a near-perfect identification of the statement type by the end of the sentence. For the D-questions, however, listeners did not respond correctly above chance until gate 2. At the first gate there is no clear preference for either one of the response types. This may be a case of response bias towards statements, since these are more frequent in everyday speech than (D-) questions. So listeners will identify a sentence as a statement unless there is compelling prosodic evidence to the contrary. Apparently, such evidence comes available at gate 2. In the stimulus analysis below we will examine various parameters in the acoustical make-up of the 16 stimuli in the present experiment to see if, indeed, these allow us to understand why our listeners behaved as they did.

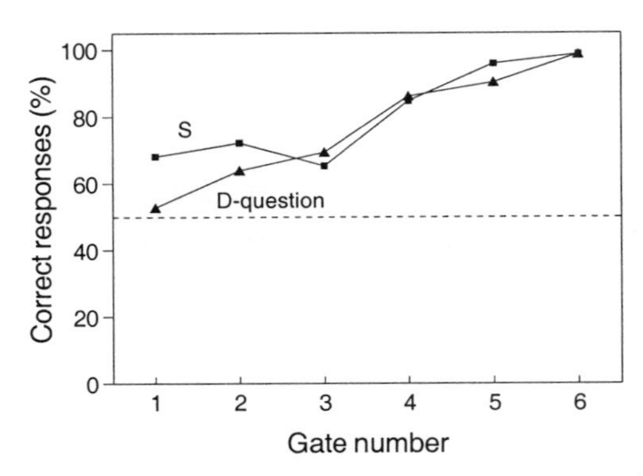

Figure 6.9. Percent correctly identified sentence type as a function of Gate length, broken down by stimulus Sentence type (chance = 50%).

6. 3.4 Stimulus Analysis

For experiment IIA, an explanation is needed for the early recognition of W and D-questions on the one hand versus W-questions and statements on the other. Possible early cues seem to be (i) declination slope and (ii) excursion size of the second accent. In experiment IIB, an additional, even earlier, potential cue was found, viz. (iii) the excursion of the first accent (on the subject constituent *maRIna*)[5]. Any remaining perceptual ambiguity should disappear once the terminal question-marking rise is included in the audible stimulus. The following acoustic parameters were selected from the measurements performed in experiment I:

- *Declination rate* (in ERB/s), separately for the upper and lower trend lines that had been fitted through the F_0-curves, after exclusion of final rises.
- *Excursion size* (in ERB) of the second accent in the sentences. It was defined as the distance between the peak F_0 associated with the accent and the preceding or following local minimum, whichever was the lowest.

Figure 6.10a presents *declination rates* for the four sentence types. Y-questions and D-questions show a very slight declination and an inclination, respectively. W-questions and statements, on the other hand, both show considerable downtrend. Figure 6.10b shows the Y and D-question responses taken together, as a function of declination rate at gate 2 ($r=0.61$, $p<0.025$). A similar figure for the other combined response category, i.e. W-questions+statements, would show the mirror image, of course. We conclude, then, that with a declining F_0-pattern responses will typically be W or S; subjects will give Y and D responses if F_0 runs level or inclines.

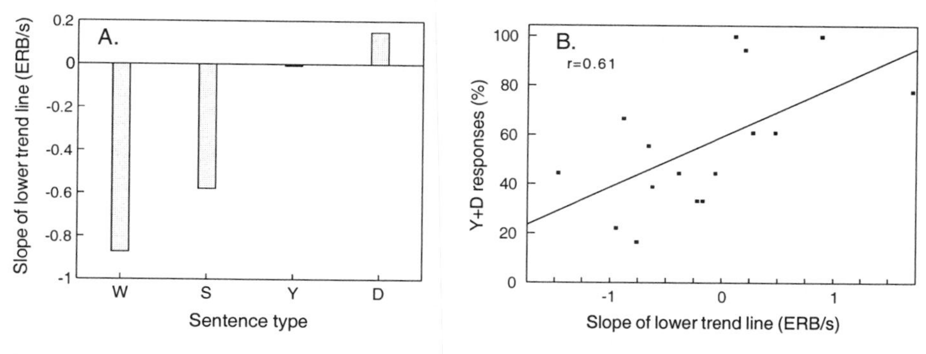

Figure 6.10. Panel A: mean declination rate (ERB/s) of lower trend line for four sentence types. Panel B: percent Y+D questions combined as a function of declination rate of lower trend line.

Mean *excursion size* for the second accent is given for each of the four sentence types in Figure 6.11a. W-questions show the smallest excursion of all.

Figure 6.11b shows Y+D-question responses combined as a function of excursion size at gate 2: clearly, increasing the excursion size of the accent on the object leads to an increasing number of Y+D-question responses ($r=0.87$, $p<0.005$).

If at the final gate a *final rise* is present, listeners nearly always gave a question response (99%). If in that case the final rise is absent, listeners still give an occasional question response (12%), presumably on the basis of other cues.

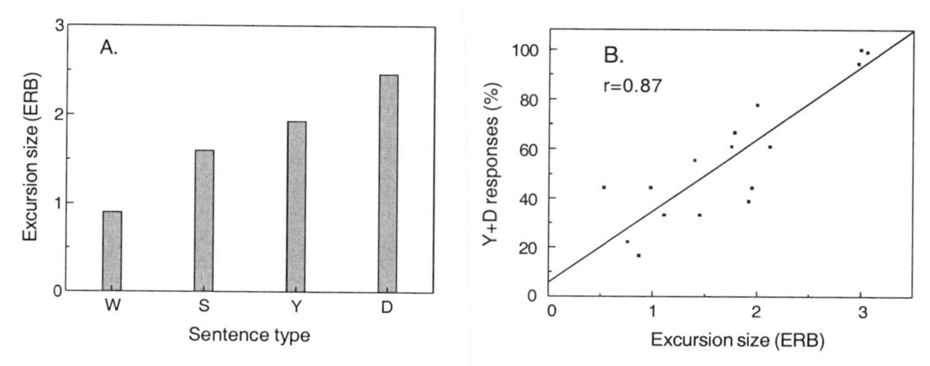

Figure 6.11. Panel A: mean excursion size of second accent (in ERB) broken down by Sentence type. Panel B: Percent Y+D questions combined as a function of excursion size (in ERB).

For experiment IIB, a third parameter was measured, which was never made audible to the listeners in experiment IIA:

- *Excursion size of the first accent.* Since the basic stimulus in experiment IIB consisted of entire sentences, the excursions of the first accent were measured. This was done as described above for the second accent.

The mean *declination rates* for statements and declarative questions were presented in Figure 6.10 above. Percent D-responses as a function of declination rate for gate 4 is given in Figure 6.12a ($r=0.84$, $p<0.005$). The result seems almost categorical, in the sense that a negative declination rate yields hardly any D-responses and an inclination yields many. Gate 4 (*marINa wil haar mandoLIne*) is the first gate at which the correlation between F_0-slope and D-responses is significant. This finding would explain why S and D were not convincingly differentiated until gate 4.

Excursion size of the first accent (on *maRIna*) is given for each of the four speakers in Figure 6.11b. Unfortunately, excursion size of the accent on the subject constituent does not afford any clear differentiation between S and D-utterances. Males have small excursions on the subject in D-questions and larger ones in statements. This behaviour conforms to the general trend observed in Figures 6.4 and 6.5. However, the effect is absent or even reversed for the two female speakers. This intricate interaction seems to explain why differentiation between S and D is only little above change (never more than 70% correct) during the first three gates in experiment IIB.

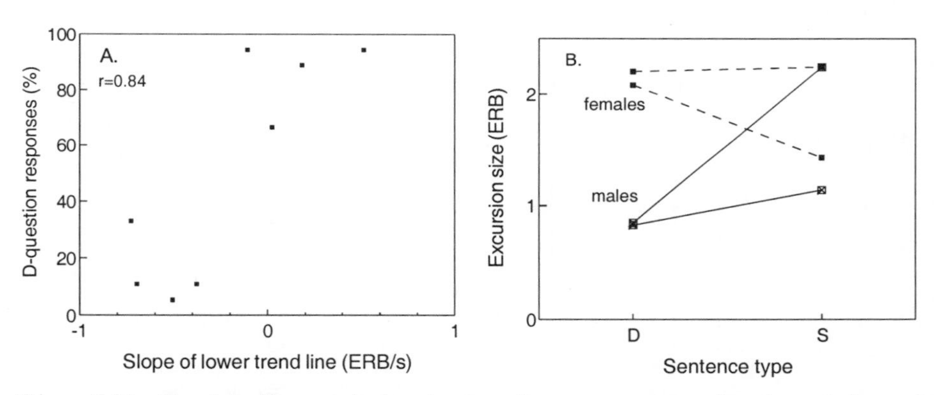

Figure 6.12. Panel A: Percent declarative question reponses as a function of slope of lower F_0-trend line (in ERB/s). Panel B: Excursion size of first accent (in ERB) for individual male and female speakers, broken down by Sentence type.

6.3.5 Conclusion

From these perception experiments, it appears that listeners can make an accurate distinction between Y and D-questions on the one hand and W-questions and statements on the other, as soon as the second accent of the sentence is heard. By then the slope of the lower declination line and excursion size of the second accent (i.e., the accent on the grammatical object of the sentence) provide enough information about the intended sentence type. As expected, the presence of a sentence-final rise is a very reliable cue in the recognition of questions. One cue which was thought to have an effect on listeners' responses, namely excursion size of the first accent, proved insignificant.

6.4 Overall Conclusions and Discussion

We set out to compare and contrast intonational characteristics of two major communicative functions, declarativity and interrogativity. The results of experiment I show that, as a category, interrogative utterances have a number of acoustic properties in common which clearly distinguish them from declarative utterances. Within the category of interrogative utterances, however, the three Dutch question types investigated cannot be simply lumped together. Pragmatic factors affect the magnitude of the pitch accents, thereby causing them to deviate from categorical means to an extent which cannot be ignored.

We hypothesized on functional grounds that the interrogative nature of an utterance will be more clearly marked as such by prosodic means as the number of lexico-syntactic questions markers is smaller. Support for this hypothesis was found in a rather straightforward fashion in (i) the number and (ii) the onset and terminal F_0 (but not the excursion size) of the sentence-final question-marking pitch rises typically found after questions. No such final rises were found after statements, but their incidence as well as their onset and terminal F_0 increased for W, Y and D-questions, in the predicted order. However, the interrogative status of an utterance was found not only in the terminal rise. Rather, questions of various types are differentiated from statements, as well as from each other, by more subtle characteristics of the pitch pattern that extend over the utterance, as follows, where the features of S serve as the reference condition:

S. Parallel downward upper and lower declination lines; pitch accents on subject and object have standard sizes.

D. Parallel declination lines with reduced distance between upper and lower line, i.e., with smaller excursions on the pitch accents. The upper declination line starts at a higher F_0 than in S, and the downward trend of both declination lines is somewhat steeper.

Y. Diverging declination lines: lower declination has downtrend but upper declination has uptrend. The final pitch accent (on the grammatical object) has a higher peak F_0 and larger excursion size (than the standard accent as in S).

D. As Y, but all characteristics deviate more strongly from S: lower declination is flat to slightly rising, upper trend line rises strongly. The final accent is much larger than the standard (and larger than that in Y as well).

The predicted order S<W<Y<D is systematically observed for S, Y and D. It is reflected in the slopes of both lower and upper trend lines, which are less negative (i.e., less downtilted) or even positive (uptilted) in the predicted order. However, it is not easy to evaluate the behaviour of W-questions; these are differentiated from the other sentence types by a different set of parameters, notably by strongly reduced pitch accents, high onset and a stronger downtrend even than S. The stronger downtrend groups W together with S, whilst the higher onset as well as the reduced pitch span between upper and lower trend lines (such that the lower trend line is shifted to an overall higher pitch) class W with the questions[6]. Given these 'mixed signals', it seems reasonable to conclude that W is the question type that is closest to statements. If this line of reasoning is accepted the functional hypothesis S<W<Y<D is fully confirmed.

Apart from the greater incidence of terminal rises after questions, there are no sex-specific differences in the marking of sentence type, neither between statements and questions in general, nor among the three question types internally. We do note, however, that the excursion size of the terminal rise is roughly twice as large for females than for males. Here is does not matter whether the excursion size is expressed linearly (in Hz), logarithmically (in semitones) or psychophysically (in ERB-units), the women invariably have the larger excursion sizes. It is unclear at this juncture if this should be taken as evidence that women mark the question status of an utterance more clearly than men do. Recent analyses (not presented here) of the accent-lending pitch movements produced in experiment I, as well as in additional spontaneously produced Dutch speech materials (van Heuven, Haan & Pacilly, 1998; van Donzel, Koopmans-van Beinum & Pols, 1998) show that excursion sizes in female speech are about double the size of male excursions. Obviously, if all pitch movements produced by women are larger than those of men, one cannot claim that women make a *special* effort to mark *interrogativity*, although one can still claim that women are intent on speaking more clearly than men *in general*.

Our acoustic measurements show that the difference between the various sentence types is signalled both by local pitch phenomena, such as presence, peak F_0 and excursion size of specific pitch movements, and by such global characteristics as the degree of tilt of the declination lines. Unfortunately, the local and global pitch measures are statistically correlated so that it is impossible, a priori, to make claims about causality. Nevertheless, it seems highly implausible that our speakers should aim to produce an inclining upper trend line in questions. Rather, we believe, speakers specifically focus on the object (or just the sentence-final constituent) in Y and, even more so, in D-questions. As a result of this, the excursion size and/or peak F_0 of the accent on the object is larger, which in turn causes the upper trend line to incline. However, such a line of reasoning cannot easily explain why the *lower* trend line should have a different slope depending on the type of sentence being communicated. Therefore we would like to claim that interrogativity is independently signalled by both local and global means: locally by the magnitude of the accents as well as by the presence and height of the terminal rise, and globally by the tilt of the lower trend line.

The viability of the above account is supported by the results of our perception experiments. First al all, S and W sentences were strongly confused as long as the terminal rise could not be heard in the stimulus; similarly, our listeners were largely unable to tell Y and D questions apart. This confirms, independently, that the interrogative status of W-question is less clearly marked by prosodic means than that of Y and D. Second, the results demonstrate that interrogativity is clearly signalled at a point in the utterance well before the onset of the sentence-final pitch rise (which is invariably picked up as a interrogativity marker); listeners know that the utterance is a question (either Y or D) as soon as the accent on the object is heard. It is impossible to say, on the basis of the present experiments, which property of the pitch pattern triggers the early perception of interrogativity: is it the peak F_0 of the final accent, or its excursion size, or the tilt of the lower declination line? Our gating experiments were not suited to answer this type of question. One way to answer this question is to run another gating experiment, with more fine-grained steps such that an extra gate is presented in which the entire stimulus can be heard up to the onset of the final accent. If the slope of the lower trend line rather than the accent on the object is the relevant property, the sudden increase in correct identifications should be located at this gate, rather than at a longer gate that also includes the final accent. A more principled way to go about establishing the relative importance of local and global features of the intonation patterns as interrogativity cues is, of course, to systematically vary these aspects through speech (re-)synthesis, as was done in Gooskens and van Heuven's (1995) pilot experiment, and combine the advantages of systematic, orthogonal manipulation of acoustic properties with those of gating.

Notes

1. Register is defined as an area within a given speaker's overall pitch range, enclosed by the highest and lowest frequency within which tones of a particular utterance are realized (cf. Clements, 1990).
2. The materials collected in this experiment were designed to serve a wider purpose than just a characterization of four sentence types. Therefore the design included a number of factors that will not, or only tangentially play a role in the present paper. For instance, we will not deal in any detail with the effect of sex of speaker on the prosodic marking of sentence type (but cf. van Heuven & Haan, 1998) nor will we be concerned with the effect of preceding or following context on the accentuation, and prosody in general, of our 800 target utterances.
3. We used the various algorithms as implemented in the PRAAT speech processing package developed by Boersma (1998) at the University of Amsterdam.
4. In two cases a slight disfluency could be heard in an utterance; here the second reading of the sentence was used instead.
5. This restricted set of acoustic variables remained after studying the correlations between a much larger variety of potential cues and the identification scores.
6. These differences, i.e., excluding the information in sentence-final rises, are in fact systematic enough on a token-by-token basis to afford a relatively successful automatic identification of sentence type. van Heuven, Haan and Pacilly (1997): statements were correctly identified in 82% of the cases, and the supercategory of questions in 90%. Within the category of questions, the subtypes were less successfully identified but still well above chance (between 53 and 75% overall, and between 58 and 80% for those utterances that were not realised with a terminal rise).

Acknowledgements

This research was funded in part by the Netherlands organisation for Scientific Research (NWO) under project nr. 200-50-073 (principal investigators R. van Bezooijen & V.J. van Heuven).

References

Boersma, P. 1998. *Praat: A system for doing phonetics by computer.* {http://fonsg3.hum.fon.uva.nl/praat/praat.html}.

Bolinger, D.L. 1982. Nondeclaratives from an intonational standpoint. In Schneider, Tuite and Chametzky (eds.), 1-22.

Brown, G., K. Currie and J. Kenworthy. 1980. *Questions of Intonation*. London: Croon Helm.

Clements, G.N. 1990. The status of register in intonation theory: comments on papers by Ladd and by Inkelas and Leben. In Kingston and Beckman (eds.), 58-71.

Cooper, W.E. and J.M. Sorensen. 1981. *Fundamental Frequency in Sentence Production*. New York: Springer-Verlag.

Cutler, A. and D.R. Ladd (eds.). 1983. *Prosody: Models and Measurements*. Berlin: Springer-Verlag.

Daan, J. 1938. Dialect and pitch-pattern of the sentence. *Proc. 3rd ICPhS* (Ghent, Belgium), 473-481.

de Hoop, H. and M. den Dikken (eds.). 1997. *Linguistics in the Netherlands 1997*. Amsterdam: John Benjamins.

Docherty, G.J. and D.R. Ladd (eds.). 1992. *Papers in Laboratory Phonology II*. Cambridge University Press.

Eady, S.J. and W.E. Cooper. 1986. Speech intonation and focus location in matched statements and questions. *J. Acoust. Soc. Am.* 80, 402-415.

Geluijkens, R. 1986. *Questioning Intonation*. Antwerp Papers in Linguistics 48.

Gooskens, C. and V.J van Heuven. 1995. Declination in Dutch and Danish: Global versus local pitch movements in the perceptual characterisation of sentence types. *Proc. 13th ICPhS* (Stockholm, Sweden), vol. 2, 374-377.

Gussenhoven, C. and T. Rietveld 1992. A target-interpolation model for the intonation of Dutch. *Proc. ICSLP* '96 (Banff, Canada), vol. 2, 1235-1238.

Haan, J., V.J. van Heuven, J.J.A. Pacilly and R. van Bezooijen. 1997a. Intonational characteristics of declarativity and interrogativity in Dutch: A comparison. *Proc. ESCA Workshop on Intonation* (Athens, Greece), 173-176.

Haan, J., V.J. van Heuven, J.J.A. Pacilly and R. van Bezooijen.1997b. An Anatomy of Dutch question intonation. In de Hoop and den Dikken (eds.), 97-108.

Hadding-Koch, K. 1961. *Acoustic-Phonetic Studies on the Intonation of Southern Swedish*. Lund: Gleerup.

Hadding-Koch, K. and M. Studdert-Kennedy. 1964. An Experimental Study of Some Intonation Contours. *Phonetica* 11, 175-185.

Hardcastle, W.J. and A. Marchal (eds.). 1990. *Speech Production and Speech Modelling*. Dordrecht: Kluwer Academic Publishers.

Hart, J. 't, R. Collier and A. Cohen. 1990. *A Perceptual Study of Intonation*. Cambridge University Press.

Hermann, E. 1942. *Probleme der Frage*. Nachrichten Akademie von Wissenschaft, Göttingen.

Hermes, D.J. 1988. Measurement of pitch by subharmonic summation. *J. Acoust. Soc. Am.* 83, 257-264.

Hermes, D.J. and J.C. van Gestel. 1991. The Frequency Scale of Speech Intonation. *J. Acoust. Soc. Am.* 90, 97-102.

Inkelas, S. and W. Leben. 1990. Where phonology and phonetics intersect: The case of Hausa intonation. In Kingston and Beckman (eds.), 35-57.

Kingston, J. and M.E. Beckman (eds.). 1990. *Papers in Laboratory Phonology I*. Cambridge University Press.

Ladd, D.R. and J.M.B. Terken. 1995. Modelling intra- and inter-speaker pitch range variation. *Proc. 13th* ICPhS (Stockholm, Sweden), vol. 2, 386-389.

Lieberman, Ph. 1967. *Intonation, Perception and Language*. Cambridge, Mass.: MIT Press.

Lieberman, Ph., W. Katz, A. Jongman, R. Zimmerman and M. Miller. 1985. Measures of the sentence intonation of read and spontaneous speech in American English. *J. Acoust. Soc. Am.* 77, 649-659.

Lindblom, B. 1990. Explaining phonetic variation: A sketch of the H&H theory. In Hardcastle and Marchal (eds.), 403-439.

Lindsey, G.A. 1985. *Intonation and Interrogation*. PhD dissertation, UCLA.

Schneider, R., K. Tuite and R. Chametzky (eds.). 1982. *Papers from the Parasession on Nondeclaratives*. Chicago Linguistic Society.

Thorsen (Grønnum), N. 1980. A study of the perception of sentence intonation – Evidence from Danish. *J. Acoust. Soc. Am.* 67, 1014-1030.

Vaissière, J. 1983. Language-independent prosodic features. In Cutler and Ladd (eds.), 53-66.

van Alphen, J. 1914. De Vraagzin. *De Nieuwe Taalgids* 8, 88-95.

van den Berg, R., C. Gussenhoven and T. Rietveld. 1992. Downstep in Dutch: Implications for a model. In Docherty and Ladd (eds.), 335-358.

van Donzel, M., F.J. Koopmans-van Beinum and L.C.W. Pols. 1998. Speaker strategies in the use of prosodic means in spontaneous discourse in Dutch. *Proc. ESCA Workshop on Sound Patterns of Spontaneous Speech* (La Baume-les-Aix, France), 135-138.

van Es, G.A. 1932. Syntaxis en dialectstudie IV, intonatie en syntaxis. *Onze Taaltuin*, 122-128.

van Heuven, V.J., J. Haan and J.J.A. Pacilly. 1997. Automatic recognition of sentence type from prosody in Dutch. *Proc. EUROSPEECH* '97 (Rhodes, Greece) 3, 1483-86.

van Heuven, V.J., J. Haan and J.J.A. Pacilly. 1998. Global and local characteristics of Dutch questions in play-acted and spontaneous speech. *Proc. ESCA Workshop on sound patterns of spontaneous speech* (La Baume-les-Aix, France), 139-142.

Appendix 1A. Overview of stimulus sentences experiment I.

Type	Target sentence	Context sentence
S	*Renée heeft nog wat vlees over.* 'Renée has still meat left'	*Onze poes moet nog wat eten hebben.* 'Our cat must some food have'
Y	*Heeft Renée nog wat vlees over?* 'Has Renée still some meat left?'	*Wil de poes nog wat eten hebben?* *'wants the cat still some food have?'*
W	*Wat heeft Renée nog voor vlees over?* 'What has Renée still of meat left?'	
D	*Renée heeft nog vlees over?*	
S	*Marina wil haar mandoline verkopen.* 'Marina wants her mandolin sell'	*Er is donderdag weer een rommelmarkt.* 'There is Thursday again a jumble sale'
Y	*Wil Marina haar mandoline verkopen?* 'Wants Marina her mandolin sell?'	*Is er donderdag weer een rommelmarkt?* 'Is there Thursday again a jumble sale?'
W	*Waar wil Marien zijn mandoline verkopen?* 'Where wants Marien his mandolin sell?'	
D	*Marina wil haar mandoline verkopen?*	

Appendix 1B. Schematic representation of design stimuli experiment I. Each cell frequency must be multiplied by 4 (*Renée ~ Marina* sentence, 2 repetitions).

Type of target sentence	Type of sentence context					
	Statement		Y-question			
	1st pos	2nd pos	1st pos	2nd pos	None	Total
Statement	10	10	10	10	10	50
W-question	10	10	10	10	10	50
Y-question	10	10	10	10	10	50
D-question	10	10	10	10	10	50
Total	40	40	40	40	40	200

7

Pitch Movements and Information Structure in Spontaneous Dutch Discourse

MONIQUE VAN DONZEL AND FLORIEN KOOPMANS-VAN BEINUM

7.1 Introduction

The study reported on in this paper is part of a larger project on the acoustic determinants of focusing in discourse (van Donzel & Koopmans-van Beinum, 1995, 1996, 1997a, 1997b; van Donzel, 1997, 1999; Koopmans-van Beinum & van Donzel, 1996). It is generally assumed that speakers make use of F_0-variations to signal various types of information in their speech. For instance, information that is new in the discourse (new to the listener), and thus has to be put in focus, will generally be pronounced with a pitch accent. Information that has been mentioned previously will generally not be realized with an accent (cf. Nooteboom & Terken, 1982; Nooteboom & Kruyt, 1987). In this way, the speaker may give 'instructions' to the listener as to where he/she should pay attention, in other words, which parts of the discourse are important.

Furthermore, speakers may use F_0-variations to chunck their discourse into smaller parts, and to indicate what the relation is between those parts. For this purpose, boundary tones are used: falling tones generally indicate finality, whereas rising or level tones indicate non-finality or continuity (Blaauw, 1995, and Swerts, 1994 for Dutch; Brown, Currie & Kenworthy, 1980 for Edinburgh Scottish English). A high boundary tone may, however, also coincide with a pitch accent, resulting in a word being prominent, for instance in (monosyllabic) words with final lexical stress carrying a pitch accent on that stressed syllable, and occurring at the end of a phrase, but not at the end of a sentence. In such a case, the pitch accent and the boundary tone can not convincingly be separated on the basis of the speech signal. We will have intonation experts indicate pitch accents and

A. Botinis (ed.), Intonation, 145-161.

boundary tones, on the basis of auditory cues. In case pitch accents and boundary tones coincide, they are able to indicate whether both movements occur on the same syllable, and judge the movements as both accent lending and boundary marking.

Variations in F_0 need not necessarily be the only prosodic means available to the speaker to mark information structure. Other means include, among others, pausing, variations in intensity, and in vowel quality. In this paper we will focus on the F_0-variations specifically.

An earlier study on Dutch (Blaauw, 1995) found that in spontaneous speech pitch accents are mainly realized as so-called 'pointed hats' rather than 'flat hats'. In that study, the occurrence of pitch accents was however not in detail related to information structure in terms of evoked and new information. Another study (Streefkerk & Pols, 1996), using parts of the read versions of the material used in the present experiment (see below) found that almost all perceptually prominent items were realized with a pitch accent. Again, this was not related to the information structure of the sentence. Furthermore, these studies made use of either read material, or so-called 'instruction monologues'. Even if the instruction monologues are produced spontaneously, results obtained for this type of speech are not necessarily true for a more unrestricted type, where speakers are for instance telling a story and are not restrained as such. On a scale expressing the different types of spontaneous speech, with 'careful' on the one side and 'sloppy' on the other, the instruction monologues would be on the 'careful' side, whereas retelling a story would be on the 'sloppy' side. Still, both are instances of spontaneous speech.

In the present paper we want to focus on the question of how discourse structure is prosodically marked by speakers in spontaneous discourse, and how listeners use the cues to detect discourse structure in spontaneously spoken discourse.

7.2 Methods

7.2.1 Speakers, Discourse Analysis, and Listeners

Four male and four female native speakers of Dutch were selected as speakers. They were all students or staff members of the Institute of Phonetic Sciences. The speakers were asked to read aloud a short story in Dutch ('A Triumph' by S. Carmiggelt, 1966). After a short break they were asked to retell the same story in their own words, with as many details as possible. During the retelling of the story a listener was present to create a more natural story-telling situation. This procedure resulted in eight spontaneously retold versions of the same story (hereafter 'retold version').

All recordings were made in an anechoic room on DAT-tape. The retold versions were stored as digitised audio files (sample rate 48 kHz, 16-bit precision).

Verbatim transcriptions of each of the eight retold versions were made by the first author. These transcriptions were subsequently analysed by the first author for discourse structure, using the purely text-based framework mentioned above. This was done in collaboration with a panel of discourse analysts. On a global level, a distinction was made in paragraphs, sentences, and clauses. On a more detailed level, words or word groups were labelled according to their information status in new, inferable, or given (evoked) information. Furthermore, discourse markers and modifiers were labelled, as well as auxiliary and main verbs. Concepts are comparable to noun phrases or word groups: a noun accompanied by determiners, adverbs, and/or adjectives. For instance, 'an enormous herd of wild swines' would count as one concept.

From a *computational* point of view, some independent frameworks have been developed to study the relation between discourse structure and prosodic aspects in spontaneous speech, for example the discourse labelling manual by Nakatani, Grosz, Ahn and Hirschberg (1995), used in experimental studies by Grosz and Hirschberg (1992), Nakatani, Hirschberg and Grosz (1995) for American English, among many others. These studies are primarily concerned with the overall global structure of whole discourse. See also the work by Carletta, Isard, Doherty-Sneddon, Isard, Kowtko and Anderson (1997), Hearst (1997), and Passoneau and Litman (1997) for recent studies using a computational approach towards discourse segmentation. In the present study we want to concentrate not only on the structure of the discourse in terms of 'chunking', but also on the internal focal structure of a discourse in terms of 'given' and 'new' information. For this purpose, an independent framework was developed for discourse analysis, in which the internal focal structure of a discourse is based on *pragmatic* features of discourse structure (cf. Mann & Thompson, 1988; Chafe, 1987, 1994; Prince, 1981, 1992). This model of is able to label 'global' structures of discourse, such as different types of boundaries, as well as the information status of the various concepts of which the discourse is build up. For further details on the framework and the labelling procedure, see van Donzel and Koopmans-van Beinum (1995) and particularly van Donzel (1999, chapter 2).

In order to obtain perceptual judgements, twelve listeners, all students, participated in a listening test. They were asked to mark the structure of each of the retold versions, on the basis of the speech signal rather than on the basis of the text. Other means available to the listener in determining the structure of a spoken message include lexical, morphological, syntactical, and semantic aspects. A text-bias in the listeners' judgements can be avoided by using delexicalised speech, for instance by using methods such as the ones proposed by Pagel, Carbonell and Laprie (1996) and by Strom and Widera (1996). In the present study, this was not an available option, for several reasons. First of all, listeners needed to select smaller pieces of discourse and playback a number of times, keeping track of a delexicalised spoken version would be hardly possible at all. Secondly, in order to have reliable listening judgements, each spoken version had to be evaluated by all listeners. Listeners found this task hard to do, evaluation of the delexicalised version would have taken much longer, and would have been even more difficult to perform.

The listeners had to indicate non-final, sentence final, and paragraph final boundaries, using conventional punctuation (',' for non-final, '.' for final, '//' for discourse unit-final). They also had to underline which words or word groups ('concepts') they perceived as being emphasized by the speaker. The verbatim transcriptions without any punctuation were used as an answer sheet. See van Donzel (1997, 1999) for a detailed description of the experiment.

7.2.2 Pitch Measurements, Stylisation, and Labelling

The digitised audio files were analysed using the speech editing program 'Praat' (Boersma & Weenink, 1996). Pitch extraction was done automatically (time step 0.01 sec., range 75-600 Hz, silence threshold 0.03). Stylisation of the pitch contours was done by hand, using both visual and auditory cues. The contours were stylised by interpolating the labelled pitch points (cf. 't Hart, Collier & Cohen, 1990) in such a way that the manipulated contour sounded equivalent to the original, to make sure that 'micro intonation' ('t Hart *et al.*, 1990) was not included in the stylisation, and perceptually essential pitch movements were not omitted. This stylisation was done also to eliminate octave jumps and to 'smooth' the contour, and does not pretend to actually be a 'close copy stylisation' in the sense that 't Hart *et al.* (1990) would require.

Finally, a group of eight Dutch intonation experts was asked to evaluate the retold versions for the location of pitch accents at the syllable level and boundary marking pitch movements (for the boundaries, the experts also had to mark whether the boundary was high/non-low or low/non-high). The eight retold versions were randomly ordered and distributed over the experts in such a way that each version was evaluated by three different experts. In order for a word to count as being accented, 2 out of 3 experts had to agree (based on the syllable scores). The same criterion was set for boundaries, the type of boundary ('high' or 'low') being determined by the majority. In case a boundary was marked by 2 experts, but no decision could be made on the exact type of boundary (1 'high' and 1 'low'), it was labelled as 'ambiguous'. By using the intonation expert procedure, we are sure to include all the relevant pitch accents and boundary tones. Automatic detection of these movements will inevitably contain mistakes, since not all peaks in the intonation contour are necessarily pitch accents. Furthermore, we are able to separate the pitch accents and boundary tones coinciding on one syllable, experts were asked to indicate in such cases that both movements occurred, hereby splitting the accent from the boundary tone. Six of the eight experts had a traditional IPO background. The two others were from the autosegmental background. This difference in background presents no problem for the task, since in both approaches, the decision of whether a pitch accent (the general term) is present or not is independent. The major difference between the two approaches is the issue whether a pitch accent is realized by means of a (combination of) movements (IPO approach) or as a tonal target (autosegmental approach). This issue will not be dealt with in this study.

The pitch accents, as obtained from the expert judgements, consisted mostly of an accent lending rise followed by a fall ('pointed hat'-pattern). F_0-maxima of the pitch accent peaks were measured using both visual and auditory cues. For each pitch accent, the information status of the lexical word associated with the peak was indicated. For each high or low boundary tone, the type of discourse boundary associated with it was indicated. Thus, a boundary was either clause-internal, clause-final, sentence-final, or paragraph-final. These specific measures could obviously only be taken after the experts had located the pitch accents and the boundary tones.

In the remainder of this paper, the term 'accents' concerns the pitch as determined by the intonation experts, whereas 'prominence' concerns the judgements given by the naïve listeners.

7.2.3 Hypotheses

On the basis of the findings in the literature, and the text-based discourse analysis (see below), we expect new and inferable information to be marked by a pitch accent, while given information will not be marked. Strictly speaking, inferable information is given information, since it can be recovered from the preceding context. However, this type of information is not 'given enough' to be pronominalised, and thus the full noun has to be used to avoid ambiguity: it is 'discourse-new & hearer-old' (cf. Prince, 1992). Therefore, we expect it to be accented. Modifiers (clause internal

adverbial expressions of time or place) and discourse markers (clause initial adverbial expressions of time or place) will be marked by a pitch accent, since these add important information to the discourse, or mark an important turning point (linguistically) in the discourse respectively (cf. the results by O'Shaughnessy and Allen, 1983). Verbs are not accounted for separately in the taxonomy proposed by Prince (1981, 1992). They are seen as belonging to an NP, for instance in 'writing a book'. This implies that they are expected not be accented and neither be perceived as prominent. We expect high boundary tones, associated with non-finality, to be realized on phrase and/or clause boundaries, and low tones, associated with finality, on sentence and paragraph boundaries.
In summary, the more detailed questions to be answered in this paper are:

1. How are pitch accents and boundary tones, as realized by the speakers, related to the information structure?
2. Do all speakers use these pitch movements in the same way?
3. How do these F_0-variations relate to listeners' judgements of perceived prominence and perceived discourse boundaries?

7.3 Results

7.3.1 Speaker Characteristics

Table 7.1 shows how the speakers can be characterized in terms of intonational aspects. For each speaker the mean F_0-values in Hz and semitones (ST), standard deviations, and the range relative to the mean F_0-value (maximum F_0 - mean F_0) in semitones are presented. The value of zero ST corresponds to 100 Hz. Since not all speakers produced texts of equal length, the total number of words and of discourse boundaries per speaker are given as well. These data will later be used to weigh the number of pitch accents and boundary tones per speaker.

A look at the data for mean F_0-values shows a clear distinction between female and male speakers (speakers 1, 3, 5, and 7 vs. speakers 2, 4, 6, and 8). Mean F_0-values in semitones are significantly different for male and female speakers ($F(1,6)=28.149$, $p=.002$), as could of course be expected. Note that the mean F_0-value (in ST) for speaker 3 is almost twice as low as for the other 3 female speakers, the difference in mean F_0 (in ST) for female speakers is significant ($t=5.774$, $df=3$, $p=.010$).

As for the range used by the speakers, we see that most speakers have a range of about 10 ST above the mean F_0-value. Speakers 1 and 3, however, have a range of 14 and 13 ST respectively, which is larger compared to other female speakers. This effect is significant ($t=6.715$, df=3, p=.007). For the male speakers, we see that speaker 8 is lower than the other male speakers, this effect is also significant (t=10.832, df=3, p=.002). Ranges do not differ significantly for the male and the female speakers (F(1,6)=0.684, p=.440).

Table 7.1. Mean F_0-values (in Hz and ST) and standard deviations (in ST), range relative to mean F_0 (in ST), as well as absolute total number of words and discourse boundaries, broken down by speaker (Female or Male).

Sex	F	M	F	M	F	M	F	M
Speaker	1	2	3	4	5	6	7	8
Mean F_0 (Hz)	195	88	137	97	192	117	212	87
Mean F_0 (ST)	11	-2	5	0	11	2	13	-2
sd (ST)	3	2	2	3	2	2	2	1
Range (ST)	14	10	13	9	8	11	8	7
# Words	537	459	582	504	491	361	417	511
# Boundaries	73	71	76	70	64	46	61	67

Table 7.2. Mean overall F_0-value (ST above 100 Hz) per speaker, and mean F_0-values and standard deviations (ST) for the different types of pitch movement (Peaks, High boundary tones H%, Low boundary tones L%, and ambiguous boundary tones B), as well as Δ F_0 relative to the mean, broken down by speaker. In case of empty categories, '--' is used (cf. speaker 2).

Sp	Mean F_0 (ST)	Peak	Δ F_0	H%	Δ F_0	L%	Δ F_0	B	Δ F_0
1	11	14 (3.3)	3	17 (3.6)	6	7 (3.2)	-4	11 (6.7)	0
2	-2	-1 (2.7)	1	0 (3.3)	2	-- (--)	--	-4 (1.1)	-2
3	5	7 (2.8)	2	9 (3.8)	4	2 (2.3)	-3	4 (1.3)	1
4	0	3 (2.8)	3	3 (3.4)	3	-2 (2.1)	-2	-2 (1.9)	-2
5	11	13 (2.3)	2	14 (3.3)	3	8 (0.1)	-3	10 (1.5)	-1
6	2	5 (2.7)	3	7 (4.0)	5	0 (1.0)	-2	2 (0.7)	0
7	13	15 (1.8)	2	15 (2.6)	2	10 (2.0)	-3	13 (0.4)	0
8	-2	-.4 (1.9)	2	0 (21.)	2	-4 (0.0)	-2	-2 (2.0)	0

The next step is to look at the mean F_0-values at which the pitch accents and boundary tones are realized. These values are best interpreted relative to the mean F_0-value for each speaker. These data are shown in Table 7.2.

As can be seen, the various types of pitch movements (peaks, high, low, and ambiguous boundary tones) are realized at different frequencies. This effect appeared to be significant (F(3,1419)=16.56, p=.000). Post hoc tests (Tukey multiple comparison) revealed that all differences are significant, except between low and ambiguous boundary tones.

The peak values (ΔF_0) for most speakers are about 2-3 ST higher than the average F_0. The difference is significant (t=-8.401, df=7, p=.000). The variation in the standard deviations for peak values indicates that the excursion size per speaker is quite different.

7.3.2 Distribution of Pitch Accents and Boundary Tones

Secondly, we want to know how pitch accents and boundary tones are realized by the speakers in relation to the information status of the discourse. The location and presence of *pitch accents* will be related to the local level of information structure (i.e. focal structure), whereas the location of *boundary tones* will be related to the global level of discourse structure (i.e. discourse structure). Table 7.3 presents the total number of pitch accents and the total number of concepts perceived as prominent, per speaker. Mean number of accents and mean number of prominent concepts per phrase are given as well. The total number of phrases is equal to the total number of boundaries in Table 7.4 (see below). Table 7.4 presents the total number of boundary tones and the total number of perceived discourse boundaries, per speaker. Since the absolute numbers are not suitable to compare between speakers, we also included the ratio of pitch movements per speaker, relative to the total number of words for 'Accent' and to the total number of discourse boundaries for 'Boundaries'.

7.3.2.1 Pitch Accents and Perceived Prominence

The data in Table 7.3 show that ratios for Accents range between .20 (speaker 8) and .32 (speakers 3 and 7). This means that, relative to the length of the discourse, speakers produced between 20% and 32% of all words with a pitch accent. The ratios for concepts perceived as prominent are more comparable for speakers, they all range between .30 and .36. Thus, overall, between 30% and 36% of the discourse may be perceived as prominent. The difference between realized pitch movements and perceived prominence is remarkable: 20-30% of the discourse is realized with a pitch accent, while 30-36% is perceived as prominent. The mean number of accents per phrase is around 2 for all speakers, which means that in each phrase 2 pitch accents are realized. The number of prominent concepts per phrase is somewhat higher, around 2.5 for all speakers. These numbers reflect the overall percentages accented and prominent as they are matched with the mean number of words per phrase. As for the relation between accentuation and prominence, if the ratios were equal, it could be assumed that perceived prominence is triggered by the presence of a pitch accent. Our data, however, indicate that apparently pitch accents are not the only cue to perceived prominence for the listeners, at least in our retold stories. Pausing may have had an influence as well (see below).

Table 7.3. Total number of pitch accents, and total number of concepts perceived as prominent, as well as ratios relative to the total number of words and mean number of accents (accents/phrase) and prominent concepts per phrase (prom./phrase), broken down for speaker.

Speaker	1	2	3	4	5	6	7	8
Words	537	459	582	504	491	361	417	511
Phrases	73	71	76	70	64	46	61	67
words/phrase	7.3	6.4	7.6	7.2	7.6	7.8	6.8	7.6
Accents	158	124	191	113	107	84	134	103
Ratio	.29	.27	.32	.22	.22	.23	.32	.20
accents/phrase	2.1	1.7	2.5	1.6	1.6	1.8	2.1	1.5
Prominence	195	143	185	162	150	114	139	163
Ratio	.36	.31	.31	.32	.30	.31	.33	.31
prom./phrase	2.6	2.0	2.4	2.3	2.3	2.4	2.2	2.4

7.3.2.2 Realized and Perceived Boundaries

Table 7.4 shows that ratios for *realized* discourse boundaries vary considerably per speaker. Speaker 6, for instance, realized more boundary marking pitch movements than would be expected on the basis of the discourse analysis, whereas speaker 2 realized only half of the expected number. The other speakers are somewhere between these extremes; speaker differences are thus very large. Thus, the use of boundary marking pitch movements as a tool in the structuring of discourse is applied in different ways by the speakers. Ratios for *perceived* discourse boundaries show a different picture: for most speakers, more boundaries are perceived than are expected (ratios above 1, up to 1.3). For speakers 5, 7, and 8, however, only about 80% of the expected boundaries are perceived. Speaker differences are smaller than for the realized boundaries. This means that listeners are apparently not influenced by the exact realization in terms of pitch movements in their perception of discourse boundaries, and that other cues may have played a role as well, for instance pausing. Pausing may be used as a cue by the speaker to indicate the listener that something important is coming up, thereby indicating 'pay attention to what I am about to utter'.

Table 7.4. Total number of boundary tones (H%, L%, and B) and number of perceived discourse boundaries, as well as ratios relative to total number of structural discourse boundaries, broken down per speaker.

Speaker	1	2	3	4	5	6	7	8
Boundaries	73	71	76	70	64	46	61	67
Realised	66	37	50	54	59	47	53	43
ratio	.90	.52	.65	.77	.92	1.02	.86	.64
Perceived	74	72	86	85	53	60	48	57
ratio	1.0	1.0	1.1	1.2	.83	1.3	.79	.85

7.3.3 Accentuation, Boundary Marking, and Information Structure

As indicated in the introduction, we expect new information to be accented more often than for instance given information. This will be tested by looking at the way the pitch accents are distributed over the various types of information structure. Table 7.5 presents the data for pitch accents and boundary tones separately.

Table 7.5. Distribution of ratio pitch accents and boundary tones for various categories of information structure (New, Inferable, Evoked, Modifier, Discourse Marker, and Verb) and discourse boundaries (I=internal, C=clause, S=sentence, P=paragraph), broken down by speaker. Values are expressed as ratios; ratios add up to 1 per type of pitch movement per speaker.

	Pitch accent						Boundary tone			
Speaker	New	Inf	Evk	Mod	Dm	Verb	I	C	S	P
1	.20	.22	.03	.19	.07	.29	.21	.35	.29	.15
2	.30	.22	0	.16	.07	.25	.05	.43	.38	.14
3	.23	.24	.03	.27	.05	.18	.08	.40	.36	.16
4	.27	.25	.01	.12	.07	.28	.15	.39	.30	.17
5	.22	.20	0	.19	.06	.33	.17	.41	.25	.17
6	.33	.24	.01	.17	.04	.21	.13	.43	.25	.19
7	.23	.20	.01	.20	.04	.32	.17	.28	.40	.15
8	.33	.15	.02	.15	.06	.29	.05	.28	.42	.25

The data show that for all speakers, about half of all measured accent peaks are realized on ‘new’ or ‘inferable’ information, while virtually no peaks are realized on ‘evoked’ information or on ‘discourse markers’. ‘Modifiers’ and ‘verbs’ also receive some pitch accents. This is in accordance with our hypotheses and earlier findings on perceived prominence (van Donzel & Koopmans-van Beinum, 1995; van Donzel,

1997). As for the boundary tones, we see that most boundary tones are realized on clause boundaries, less on sentence boundaries. This is true, except for speakers 7 and 8, where more boundary tones are realized on sentence boundaries than on clause boundaries. Internal boundaries and paragraph boundaries received even less boundary tones.

In general, speakers seem to behave in a similar way in their use of pitch accents and boundary tones to signal discourse structure. The relative number of realized pitch accents per speaker is more stable than the relative number of realized boundary tones. The listener apparently used pitch accents as a primary cue in the perception of prominence, and some other cues, since the ratios for *perceived* prominence are higher than for *realized* pitch accents. We expect this additional cue to be pausing; further testing is necessary. In the perception of discourse boundaries, listeners definitely did not use pitch movements only, since up to twice as many boundaries were perceived as pitch movements were realized. Additional measurements (van Donzel, 1999) revealed that pausing had a much bigger effect on boundary perception than pitch movement. This is in accordance with results from De Pijper and Sanderman (1994).

The data from Table 7.5 give information only about how the *realized* pitch accents and boundary tones are distributed over information categories. The question now is: what percentage of *all* 'new', 'inferable', etc., words are accented? And, what percentage of *all* discourse boundaries is also perceived as such? And, what is the overlap between pitch accented and prominent words? Since speaker differences did not appear to be significant at this level, speakers are treated as one group. Table 7.6 gives an overview of the mean number of concepts for the essential information categories, as well as the mean percentage of concepts *realized* with a pitch accent and *perceived* as prominent, and *prominent pitch accents*, across speakers. 'Evoked' information is not presented since the percentage of evoked information realized with a pitch accent is negligible.

These data show that the percentages accented and prominent per information type are roughly in accordance with what we expected on the basis of the literature and point into the same direction of what we found in earlier experiments on perceived prominence (van Donzel & Koopmans-van Beinum, 1995; van Donzel, 1997). Brown (1983), using the same taxonomy by Prince (1981) on structured dialogues, found that 87% of all new information, 79% of all inferable information, and 4% of all evoked information was accented in her material; there were no categories for discourse markers and for modifiers in her analysis. Percentages are somewhat higher in our data, this may be explained by the difference in material.

Table 7.6. Mean number of concepts within the essential information categories on the basis of the discourse analysis (Mean N), mean percentage of concepts realized with a pitch accent (% Accented), mean percentage perceived as prominent (% Prominent), and mean percentage pitch accents also perceived as prominent (% Prominent accent) across speakers.

Information	Mean N	Accented (%)	Prominent (%)	Prominent accent (%)
New	34	104	121	98
Inferable	32	86	96	89
Modifier	51	41	50	82
Discourse marker	35	18	28	67
Verb	72	45	53	82

Some additional remarks are in order. The percentages for 'new' exceed 100% in both cases. This means that within each concept, which may consist of more than one word, at least one (and sometimes a second one) is accented and perceived as prominent. These results also suggest that the concept is not some kind of 'domain' in spontaneous speech, in which only one accent or perceptually salient item is allowed. The data furthermore reveal that between 67% and 98% of all pitch accented words were also perceived as prominent. This indicates that not all pitch accents, as determined by the intonation experts, were perceptually relevant.

The data also show that there is some kind of ordering in the types of information: 'new' information is realized and perceived more often as important than 'inferable' information, followed by 'verbs', 'modifiers', and 'discourse markers'. This ordering is identical for accent placement and prominence perception, as well as for prominent pitch accents. 'Verbs' are more important than was predicted in the model by Prince (1981, 1992). Traditionally, they are not taken into account in her taxonomy of information structure. These data show that they should, at least in spontaneous speech. Verbs are realized and perceived as important in about 50%, which is not negligible. 'Modifiers' are somewhat lower in the ordering than expected in the hypotheses. They add extra information to the discourse, but that is not crucial to the content, and are apparently not equal to 'real' new information, both from production and from perception point of view. 'Discourse markers' are not as important as expected, and play a relatively small prosodic role. One explanation could be that the clear linguistic form and function of discourse markers is taken as a sufficient cue by the speakers in the production of the discourse, and therefore not marked acoustically more prominent, and thus not perceived as such. This could mean that the speaker assumes that the listener does not need the 'extra' prosodic information to recognize and/or process discourse markers.

Table 7.7 presents the number of perceived boundaries (non-final, sentence final, and paragraph final) in relation to the structural discourse boundaries, across listeners and speakers. These are not given in percentages, since the perception of discourse boundaries is not a binary choice as the perception of prominence is (a word is prominent or not prominent); boundaries are seldom perceived as for instance only 'non-final', but very often as 'non-final' by some listeners and as 'sentence final' by others.

Table 7.7. Mean number of discourse boundaries (clause, sentence, and paragraph) as well as mean number of perceived boundaries (non-final, sentence final, and paragraph final), across listeners.

Boundary type	Mean N	Non-final	Sentence final	Paragraph final
Clause	37	22	6	0
Sentence	18	11	10	1
Paragraph	11	8	11	6

These data show that clause boundaries are mainly perceived as 'non-final', and rarely as 'sentence final'. They are never perceived as 'paragraph final'. Sentence boundaries, on the other hand, are equally often perceived as 'non-final' and as 'sentence final', and almost never as 'paragraph final'. Paragraph boundaries, finally, are mostly perceived as 'sentence final', less as 'non-final', and even less as 'paragraph final'. This suggests that listeners do not have a clear idea of how 'paragraph final' boundaries sound, also since this type of finality is perceived only occasionally. This also may suggest that the notion of 'paragraph' is based on written rather than on spoken text. Apparently, listeners make a distinction between either 'non-final' or 'final', and perceive boundaries accordingly. Still, discourse boundaries are overall mostly perceived as 'non-final'.

7.4 Conclusions and Future Research

The questions we wanted to answer in this paper were the following.

1. How are pitch accents and boundary tones, as realized by the speakers, related to the information structure?
2. Do all speakers use these pitch movements in the same way?
3. How do these F_0-variations relate to listeners' judgements of perceived prominence and perceived discourse boundaries?

Summarizing the results presented above, we may come to the following conclusions. The realization of pitch accents and boundary tones is closely related to the perception of prominence and perceived discourse boundaries (phrasing). When naïve listeners are asked to judge 'emphasis' (neutral term) in spontaneous discourse, they use F_0-variations as a primary cue. However, more concepts were perceived as prominent than were realized with a pitch accent, meaning that listeners also used other prosodic cues in the perception of prominence, probably pausing: an acoustic pause placed before a highly informative word may be an instruction from the speaker to the listener that something important is coming up. The relation between accentuation and information type is as predicted: the tendency to accent information that is new to the discourse, while given information is not accented. Many more boundaries were perceived than predicted on the basis of the discourse analysis, also more boundaries were perceived than boundary tones were realized. This indicates that listeners did use boundary marking pitch movements as cues to boundary perception, but that other cues, i.e. pausing, are important as well. Furthermore, discourse boundaries are primarily realized with high boundary tones, even on locations where a low one was expected on the basis of the prosody-independent discourse analysis. This is also picked up by the listeners: discourse boundaries are mainly perceived as 'non-final'.

The answers to the questions can now be formulated in the following way:

1. The relation between pitch accents and information structure is roughly as expected. There is a hierarchy in the accentability of focal structure, ranging from much accented to little accented: new information > inferable information > verbs > modifiers > discourse markers. Realization in terms of target F_0-values are not dependent on hierarchy in types of information. Discourse boundaries are mostly realized with high boundary tones, even where low ones are expected. Variation between the speakers in the ratio of realized

boundary marking pitch movement is high. Again, realization in terms of F_0-targets are not dependent on the type of boundary that is marked, i.e. a H% is realized in the same way, whether it is placed on a clause, sentence, or paragraph boundary.

2. Even though the speakers produced texts of different lengths and content, they used the same means to signal information structure. They did, however, not use the same means in the same way for all structures. Pitch accents are rather constant across speakers, both in terms of acoustic realization and distribution over information types, but differences are most apparent for the boundary tones, that is in the distribution of boundaries marked by a pitch movements relative to the total number. Acoustic realizations, again, are rather constant.
3. Listeners are little influenced by the difference in acoustic realization by the speaker, and are very flexible: the perception of prominence and discourse boundaries across speakers is comparable. Apparently, speakers can allow themselves a lot of liberty in the realization of spontaneous discourse, without disturbing the listener. The acoustic means available to the speaker may be used in various ways, and this has no perceptual effect for the listener in detecting the structure of the message.

Acknowledgements

The authors would like to thank Louis ten Bosch, Johanneke Caspers, Carlos Gussenhoven, Dik Hermes, Vincent van Heuven, Toni Rietveld, Marc Swerts, and Jacques Terken for their participation as intonation experts. We would also like to thank Louis Pols for careful reading of and for giving comments on earlier versions of this paper. Furthermore, the comments by Gérard Bailly, Gösta Bruce, and an anonymous reviewer are also greatly acknowledged.

References

Blaauw, E. 1995. *On the Perceptual Classification of Spontaneous and Read Speech.* PhD dissertation, University of Utrecht.

Boersma, P. and D. Weenink. 1996. *PRAAT: A system for doing phonetics by computer (*version 3.4). {http://fonsg3.let.uva.nl/paul/praat.html}.

Brown, G. 1983. Prosodic structure and the given/new distinction. In Cutler and Ladd (eds.), 67-77.

Brown, G., K. Currie and J. Kenworthy. 1980. *Questions of intonation.* London: Croom Helm.

Carletta, J., A. Isard, S.D. Isard, J.C. Kowtko, G. Doherty-Sneddon and A.H. Anderson. 1997. The reliability of a dialogue structure coding scheme. *Computational Linguistics* 23, 13-32.

Carmiggelt, S. 1966. *Fluiten in het Donker*. Amsterdam: ABC Boeken.

Chafe, W.L. 1987. Cognitive constraints on information flow. In Tomlin (ed.), 21-51.

Chafe, W.L. 1994. *Discourse, Consciousness, and Time. The Flow and Displacement of Conscious Experience in Speaking and Writing*. The University of Chicago Press.

Cole, P. (ed.). 1981. *Radical Pragmatics*. New York: Academic Press.

Cutler, A. and D.R. Ladd (eds.). 1983. *Prosody: Models and Measurements*. Berlin: Springer-Verlag.

De Pijper, J.R. and A.A. Sanderman. 1994. On the perceptual strength of prosodic boundaries and its relation to suprasegmental cues. *J. Acoust. Soc. Am.* 96, 2037-47.

Grosz, B.J. and J. Hirschberg. 1992. Some intonational characteristics of discourse structure. *Proc. ICSLP* '92 (Banff, Canada), vol. 1, 429-432.

Hart, J., 't, R. Collier and A. Cohen. 1990. *A Perceptual Study of Intonation*. Cambridge University Press.

Hearst, M.A. 1997. TextTiling: Segmenting text into multi-paragraph subtopic passages. *Computational Linguistics* 23, 33-64.

Koopmans-van Beinum, F.J. and M. van Donzel. 1996. Discourse structure and its influence on local speech rate. *Proc. ICSLP* '96 (Philadelphia, USA), vol. 3, 1724-27.

Mann, W.C. and S.A. Thompson. 1988. Rhetorical Structure Theory: toward a functional theory of text organization. *Text* 8, 243-281.

Mann W.C. and S.A. Thompson (eds.). 1992. *Discourse Description*. Amsterdam & Philadelphia: John Benjamins.

Nakatani, C.H., B.J. Grosz, D.D. Ahn and J. Hirschberg. 1995. Instructions for annotating discourse. *Technical Report Number* TR-21-95. Center for Research in Computing Technology, Harvard University.

Nakatani, C.H., J. Hirschberg and B.J. Grosz. 1995. Discourse structure in spoken language: studies on speech corpora. *AAAI Spring Symposium on Empirical Methods in Discourse Interpretation and Generation* (Stanford, USA).

Nooteboom, S.G. and J.G. Kruyt. 1987. Accents, focus distribution, and the perceived distribution of given and new information: An experiment. *J. Acoust. Soc. Am.* 82, 1512-1524.

Nooteboom, S.G. and J.M.B. Terken. 1982. What makes speakers omit pitch accents? An experiment. *Phonetica* 39, 317-336.

O'Shaugnessy, D. and J. Allen. 1983. Linguistic modality effects on fundamental frequency in speech. *J. Acoust. Soc. Am.* 74, 1155-1171.

Pagel, V., N. Carbonell and Y. Laprie. 1996. A new method for speech delexicalisation, and its application to the perception of French prosody. *Proc. ICSLP* '96 (Philadelphia, USA), vol. 2, 821-824.

Passoneau, R.J. and D.J. Litman. 1997. Discourse segmentation by human and automated means. *Computational Linguistics* 23, 103-140.

Prince, E.F. 1981. Toward a taxonomy of given-new information. In Cole (ed.), 223-255.

Prince, E.F. 1992. The ZPG letter: Subjects, definiteness, and information status. In Mann and Thompson (eds.), 295-325.

Streefkerk, B.M. and L.C.W. Pols. 1996. Prominent accent and pitch movement. *Proc. of the Institute of Phonetic Sciences* 20, 111-119. Amsterdam.

Strom, V. and C. Widera. 1996. What's in the 'pure' prosody? *Proc. ICSLP* '96 (Philadelphia, USA), vol. 3, 1497-1500.

Swerts, M. 1994. *Prosodic Features of Discourse Units*. PhD dissertation, Eindhoven University of Technology.

Tomlin, R.S. (ed.). 1987. *Coherence and Grounding in Discourse*. Amsterdam & Philadelphia: John Benjamins.

van Donzel, M. 1997. Perception of discourse boundaries and prominence in spontaneous Dutch speech. *Working Papers* 46, 5-23. Dept of Linguistics and Phonetics, Lund University.

van Donzel, M. 1999. *Prosodic Aspects of Information Structure in Discourse*. PhD dissertation, University of Amsterdam (LOT series 23).

van Donzel, M. and F.J. Koopmans-van Beinum. 1995. Evaluation of discourse structure on the basis of written vs. spoken material. *Proc. 13th ICPhS* (Stockholm, Sweden), vol. 3, 258-261.

van Donzel, M. and F.J. Koopmans-van Beinum. 1996. Pausing strategies in discourse in Dutch. *Proc. ICSLP* '96 (Philadelphia, USA), vol. 2, 1029-1032.

van Donzel, M. and F.J. Koopmans-van Beinum. 1997a. Pitch accents, boundary tones, and information structure in spontaneous discourse in Dutch. *Proc. ESCA Workshop on Intonation* (Athens, Greece), 313-316.

van Donzel, M. and F.J. Koopmans-van Beinum. 1997b. Evaluation of prosodic characteristics in retold stories in Dutch using semantic scales. *Proc. EUROSPEECH* '97 (Rhodes, Greece), vol. 1, 211-214.

8

Discourse Constraints on F_0 Peak Timing in English

ANNE WICHMANN, JILL HOUSE AND TONI RIETVELD

8.1 Introduction: Studies of F0 Alignment

A number of recent studies have investigated the alignment of fundamental frequency (F_0) contours to the segmental string. Understanding how alignment works will (i) provide a key to establishing boundaries between phonologically distinct intonational categories; (ii) identify contextual constraints on the realisation in time of such categories; and (iii) contribute to prosodic modelling for more natural-sounding synthetic speech. Languages for which detailed studies have been reported include American English, British English, German, Swedish, Dutch, Mexican Spanish and Greek[1].

A survey of some recent alignment literature is to be found in Ladd (1996). Some experimental studies, including Kohler (1987, 1990), Pierrehumbert and Steele (1989), Arvaniti and Ladd (1995), are primarily concerned with the identification of tonal categories and/or the location of boundaries between them. These are not further discussed in the present paper, which is concerned with within-category constraints on alignment. An underlying assumption has to be that we can reliably identify instances of a particular phonological F_0 category, such as a specified pitch accent, and then examine its realisation in context. We also assume that we can characterise the shape of these pitch accents in terms of their salient turning points (H, L), that we can identify and measure these points on an F_0 display, and relate them successfully to units on the segmental string. Many studies have investigated the alignment of tonal peaks – the H* tone in the autosegmental-metrical framework – which is presumed to be associated to some tone-bearing unit, typically identified as the accented syllable. We

A. Botinis (ed.), Intonation, 163-182.

make the same assumptions below when reporting on the effects of discourse context on peak alignment.

8.2 Known Constraints on F0 Timing

Results of earlier studies are not always easy to compare, but many alignment constraints appear to be language-independent. Bruce (1990, 107) summarises known constraints under four categories: intrinsic properties of the pitch accent itself (e.g. mono- or bitonal composition); prosodic context; segmental context (e.g. vowel length); and speaking rate. Prosodic context was further categorised as follows:

a. *Boundaries* (word, phrase, utterance, etc.)

b. *Rhythmical organisation* (rhythmical grouping, e.g. stress clash)

c. *Focus* (prefocal, focal, postfocal position)

d. *Tonal environment* (tonal interaction within and between successive pitch accents, e.g. tonal crowding)

e. *Pitch range* (local or global, e.g. differences in degree of overall emphasis due to degree of involvement)

f. *Global intonation* (e.g. absence/presence of downdrift due to interrogative/declarative structure)

Bruce does not explicitly include any higher-level prosodic organisation of utterances ('discourse intonation'), which we would argue plays a role. Some level of paragraph or topic organisation was identified by Silverman (1987) as the highest of three tiers which jointly determine the shape of F_0 contours. A consistent finding which emerges from experimental evidence, based on laboratory speech, is that F_0 timing correlates with the segmental string in at least two different ways:

1. *Segmental effects*. When the prosodic context of the accented syllable is held constant, we observe straightforward adjustments to F_0 alignment depending on intrinsic properties and durations of the segments within the syllable.

2. *Prosodic effects*. The local prosodic context of the accented syllable is determined by its position in relation to larger constituents, such as the stress foot, the word or the phrase, and will have predictable, robust effects on F_0 alignment.

The key findings under these headings are outlined below.

8.2.1 Segmental Effects

The effects reported below mainly relate to studies where the pitch accent under scrutiny is realised on a monosyllable, in nuclear position, with the organisation of segments within syllabic constituents (onset, rhyme) systematically varied.

- *Rhyme duration*: in monosyllabic realisations of pitch accents, the rhyme of the accented syllable carries the relevant F_0 contour. The duration of voicing in the syllable, particularly of the sonorant rhyme, will determine how much space is available to carry the F_0 movement. In H*-type pitch accents, the longer the sonorant rhyme, the later the peak is likely to be (Steele, 1986; House, 1989; van Santen & Hirschberg, 1994; Rietveld & Gussenhoven, 1995).

- *Intrinsic differences in vowel duration*: make a major contribution to rhyme duration, so F_0 peaks are typically later in long vowels than in short. Shortening of the rhyme by voiceless coda consonants, or by an increase in speaking rate, causes a straightforward temporal adjustment of the F_0 peak position. (Steele, 1986; House, 1989; Silverman & Pierrehumbert, 1990; Caspers & van Heuven, 1993; Rietveld & Gussenhoven, 1995).

- *Onset duration*: overall duration of onset consonants is one of the factors determining F_0 peak position in the syllable (van Santen & Hirschberg, 1994; Rietveld & Gussenhoven, 1995). Onset consonant manner is less important, though van Santen and Hirschberg (1994) have found that allocating the final sonorant in an onset cluster to the domain of the 's-rhyme' (sonorant rhyme) improves their F_0 alignment algorithm. The algorithm used by van Santen and Hirschberg (1994) to model nuclear monosyllables in American English calculates peak position by using a weighted combination of onset duration and s-rhyme duration, plus a constant. They claim that onset duration has the largest effects on the timing of the first part of the F_0 contour, s-rhyme duration on the later parts.

- *Microprosody*: vowel-intrinsic F_0 and consonantal perturbations appear to be salient in the frequency domain, though Silverman (1987) demonstrated that listeners were skilled at factoring out these effects when interpreting intonation. Their contribution to time-alignment is harder to assess; in a visual analysis of an F_0 signal, consonantal perturbations may obscure the location of nearby F_0 peaks (House, 1989). Microprosodic effects are not considered further here.

8.2.2 Prosodic Effects

The evidence reported here derives from studies where the local prosodic context, particularly the immediate right context, has been systematically varied, with some control over the segmental material inside the accented syllable. Both nuclear and pre-nuclear accents have been investigated.

- *Structure of stress-group* (foot): when the pitch accent is associated with a syllable at the head of a polysyllabic foot, the alignment of the F_0 peak shows a strong rightward shift; in some cases the peak may be aligned with the unstressed syllable following the stressed one (Steele, 1986; House, 1989; Silverman & Pierrehumbert, 1990; Prieto, van Santen & Hirschberg, 1995).
- *Nuclear vs. pre-nuclear accents*: F_0 peaks in accents in pre-nuclear position are consistently later than in nuclear position (Silverman, 1987; Silverman & Pierrehumbert, 1990; Prieto *et al.*, 1995).
- The effects of both foot structure and of nuclear vs. pre-nuclear position can be interpreted in terms of the *strength of the prosodic boundary* in the immediate right context: the stronger the boundary, the more the F_0 peak is pushed leftwards (Silverman & Pierrehumbert, 1990; Prieto *et al.*, 1995). The weakest boundary is a syllable boundary which is word-internal and foot-internal; strongest is the boundary of an intonational phrase (IP). An intermediate boundary is that of the accent group. A pre-nuclear accented syllable immediately followed by another accented syllable will show a leftward shift in F_0 peak, apparently to avoid 'stress clash', a phenomenon discussed in terms of tonal repulsion or gestural overlap (Silverman & Pierrehumbert, 1990).

8.2.3 Discussion

We thus find quite different constraints on F_0 alignment resulting from segmental and prosodic factors. It has often been observed that syllables occurring immediately before a prosodic boundary are subject to pre-boundary lengthening. Nonetheless the lengthened rhymes in these syllables have the opposite effect on F_0 peak alignment to rhymes which are relatively long because of an intrinsically long vowel. In a controlled prosodic context, the longer the vowel, the later the peak. In a prosodically lengthened syllable, the longer the rhyme (including the stretched vowel), the earlier the peak. In some circumstances peak alignment is pushed rightwards outside the accented syllable itself, suggesting that it may be better to associate pitch accents with the larger constituent of which the stressed syllable is part, such as the foot, or accent group. The proposed association with the foot is not new – Pierrehumbert and Beckman (1988) describe accent as "a foot-level property that is attracted to the head syllable" – but most experimental studies have assumed the stressed syllable to be the tone-bearing unit. Current work by Möbius (1994, 1995) and van Santen and Möbius (1997) accepts the relevance of the larger accent group domain for modelling pitch accent curves.

8.3 The Discourse Issue

So far this paper has mainly been concerned with the constraints of low-level structures (segment, word, foot, intonational phrase) on intonation. It is however well known that intonation is also affected by a higher level 'discourse' structure. The information contained in a text is not simply expressed by a sequence of sentences, but by sentences grouped together around a topic or sub-topic to make up a meaningful unit, often referred to as a discourse unit. In written texts the boundaries of such topical units are often highlighted by typographic means such as paragraph divisions, headings, sub-headings etc.

In speech, both the internal coherence of a discourse unit and the demarcation of its boundaries can be indicated prosodically. The prosodic features most commonly associated with the transitions between discourse units are low boundary tones, long pause and high pitch reset. These observations were made for example in conversational data by Brown and Yule (1983) who use the term 'paratone' to indicate a prosodic domain comparable to a paragraph in writing.

8.3.1 F0 Peak Timing and 'finality'

Swerts (1994; cf. also Swerts, Bouwhuis & Collier, 1994), in a study of the prosodic correlates of finality, observed that in synthesised utterances the timing of the F_0 peak (H*) on the nuclear accent affects the degree of perceived finality. A single utterance in Dutch (*een gele driehoek*: 'a yellow triangle') was synthesised with two different contours from the Dutch (IPO[2]) system, the 'flat hat' and the sequence of two 'pointed hats', and with two different timings: an early fall beginning 20 ms before vowel onset, and a late fall 80 ms after vowel onset. The results of the perception experiment showed that, regardless of the contour, the early falls constituted stronger finality cues than the late falls. This is consistent, as Swerts points out, with the autosegmental model of Dutch intonation proposed by Gussenhoven (1988) which has two accent-lending falls: downstepped (early) and non-downstepped (late), as opposed to the IPO system which only has one ('t Hart *et al.*, 1990). The meaning ascribed by

Swerts to the two falls, (degrees of 'finality'), supports the intuitions of Rietveld and Gussenhoven in their perception of the two falls: *'The downstepped contour sounds more as if it were meant as a definitive contribution to the discourse, and does not seem intended to draw the listener into a further discussion or evaluation of that contribution'* (1995, 377). This perception is what one would expect in an interactive context, although it is an inferred meaning, rather than a 'literal' meaning. In a different context, such as in a monologue, or reading aloud, the effect of different degrees of finality would be perceived not as interaction management but as a reflection of the structure of the text.

Swerts's study was based on one isolated utterance, and claims that the observed effects are necessarily related to discourse structure are therefore too strong. However, the perception of different degrees of finality in the performance of a single utterance suggests that there is at least a potential for this to be exploited for discoursal effect. A similar observation has been made in relation to the height of nuclear falls (Wichmann, 1991b), namely that the starting point of a fall affects the degree of perceived finality: the lower the starting point, the greater the perceived finality. There is clearly a relationship between late and early falls and high and low falls, but this has yet to be explored fully. Categorical or gradient distinctions have been claimed for both sets of patterns. This is still a matter for debate.

Most experimental studies of peak timing have focused on the shape of accents in nuclear position, though a few more recently have explicitly compared these with pre-nuclear, or phrase-medial, accents. A consistent observation, confirmed by these studies, is that the peaks in pre-nuclear accents tend to occur later in the accented syllable (if not outside it altogether) compared with those in nuclear accents. The conclusion drawn is not that nuclear and pre-nuclear accents differ intrinsically in their properties, but that the difference in alignment is due to context: it can be explained by the strong effects of prosodic lengthening reported in nuclear position, together with pressure from domain-edge tones which need to be realised. Our studies show that the notion of context may have to be extended to include discourse factors.

Inevitably, in controlled data for laboratory experiments, the variation in discourse structure is often somewhat limited. Implicit in the above studies is an assumption that pre-nuclear accents must be a homogeneous class - the argument was whether they differed from nuclei, not whether they differed among themselves. But in natural speech, pre-nuclear accents may be phrase-initial or phrase-medial, and there is ample evidence from the many studies of accent scaling that position within the phrase has an important effect in the F_0 domain. However the intonational phrasing is phonologised (not at issue here), the first accent in a phrase, described as the intonational 'onset' in the British tradition, but abbreviated henceforth to IO, to avoid confusion with syllable onsets, is typically observed to have a higher F_0 than subsequent ones in a defined domain. When text material is organised into spoken 'paragraphs', according to discourse topic, this effect is enhanced: the start of a new topic appears to be signalled by extra high pitch on the first IO. The question is whether there are also consistent generalisations to be made about the timing of F_0 contours within non-final accents according to the position of the accent in the phrase and in the discourse.

8.3.2 F0 Peak Timing and 'initiality'

Wichmann (1991a) investigated the intonational correlates of topic shift in a complete 5-minute broadcast news summary, text B04 in the Spoken English Corpus (henceforth SEC) (Knowles, Taylor & Williams, 1996, 83-85). The results showed that, in addition to the expected high pitch reset at the beginning of each news item there was also a clear link between topic-initial IOs and a very late alignment of the extra-high F_0 peak, which occurred at the end or even beyond the accented syllable. This observation led to the hypothesis that the links between late alignment and topic initiality were predictable.

In order to test this hypothesis we have carried out two studies: the first is a re-analysis of the natural data (news summary) described above, taking into account segmental and low-level prosodic constraints on peak timing. The second study tests the same observations in an experimentally controlled context.

8.4 Looking for Evidence in Natural Data for Reported Constraints on Peak Timing

8.4.1 Procedure

8.4.1.1 Phonological Analysis

The text used was that referred to in Wichmann (1991a), B04 in the SEC. It was chosen for its relatively simple topic structure (a sequence of news items). Certain textual elements, however, have discourse organising and deictic properties. In the past this has been described most frequently for spoken conversation, and is most readily applied to what are known as 'discourse markers' (Schiffrin, 1987) such as 'anyway', 'as I was saying' or 'by the way'. Some of these elements have a global signposting function, signalling macro-organisation of discourse (cf. Aijmer, 1996) and can point backward, forward or both. They often accompany coherence breaks in discourse, such as topic shifts, and are to be seen as 'external to the proposition proper' (Aijmer, 1996, 207). This also applies to three complete sentences in the broadcast text, which were therefore excluded from the analysis.

> *Now it's one o'clock and this is Peter Bragg in London with the Radio three news summary.*
>
> *And finally the weather forecast.*
>
> *And that's the end of the news and weather with the time just after half past one.*

In the prosodic transcription of the SEC no formal distinction is made between nuclear and non-nuclear accents, and each accented syllable, including IOs, is assigned a tonetic stress mark indicating a tone (fall, fall-rise etc.). There are two levels of intonation boundary, major and minor. The IO was taken to be the first accented syllable in a major tone-group (roughly equivalent to a sentence). The IOs concerned here were for the most part marked with a high level tone, but also included the fall, fall-rise and rise. The level, fall and fall-rise tones were assumed to be varieties of H* pitch accents. Those marked with a rise were excluded from the analysis for two reasons: firstly a rising tone has a late F_0 peak by definition, and secondly, it is assumed to be in a phonologically different category (L*H). Some examples of the relevant text fragments are as follows:

> *The BOARD of ....*
> *The WESTland chairman ....*
> *It became CLEAR afterwards ....*
> *The TRADE and industry secretary ....*
> *The conFEderation of British industry ....*
> *The PRIME Minister ....*

8.4.1.2 Measurement Procedure

F_0 analysis was carried out using the CSL (Kay Elemetrics Corp.) workstation. The corpus text was digitised at 10 kHz and pitch was extracted using the autocorrelation method. The segmentation and analysis was carried out jointly by two of the authors. Both waveform and broadband spectrogram were used in order to identify as nearly as possible the relevant segment boundaries. In the case of vowel-vowel transitions, auditory analysis provided additional information.

We considered a number of options on how to align pitch events with events in the tone bearing units. Possible pitch events are onset, maximum or minimum (for H*L and L*H respectively) and offset of F_0 movements, while possible events in the tone bearing units are the P-centre, and onset, offset, and duration of the stress group, metrical foot, or syllable. We opted for the position of the F_0 maximum in relation to the syllable. The P centre was disregarded since it was found to be a poor predictor of peak timing in a previous study (Rietveld & Gussenhoven, 1995). We expected the foot or stress group to be too large a unit, since metrical effects on the timing of the component syllables might obscure the timing effects which are the object of this study. The vowel was a possible candidate, as it had been the focus of previous studies of segmental timing effects (cf. Rietveld & Gussenhoven, 1995). However, pilot studies carried out by the authors revealed that the timing of F_0 peaks in relation to the vowel was far less sensitive to discourse structure than their timing in relation to the syllable. The final choice of the syllable as the relevant tone-bearing unit also reflects the association of F_0 targets with the syllable in the autosegmental approach to pitch contours ('starred tone segments'), and default settings in many speech synthesis systems.

The timing of the F_0 peak was calculated as a percentage of the total duration of the accented syllable. Syllables were measured as consistently as possible, respecting the principle of maximum onset (in the case of the geminate nasals in '*Prime Minister*' the midpoint was chosen as the syllable boundary). Some contours did not have an obvious peak but flattened out to a plateau. The F_0 value at the beginning of the plateau was taken as the target value (see Figures 8.2 and 8.3).

8.4.1.3 Weighting Procedure

Naturally-occurring speech is of course not controlled for segmental and prosodic environment. Claims for a discoursal effect on peak timing cannot be made without taking into account other timing constraints. Each accented syllable was therefore examined for its segmental makeup and prosodic environment. All factors, other than tempo, identified by past studies as having an influence on the timing of the peak were taken into account. Those exerting a *leftward pull* were taken to be:

- short vowel
- each consonant in the syllable onset
- presence of voiced segments in syllable onset
- stress clash (upcoming accent)
- upcoming word boundary
- upcoming intonation phrase boundary

those exerting a *rightward push* were:

- long vowel or diphthong
- sonorant syllable coda

Since each syllable contains a vowel, the combination of a rightward push for long vowels and a leftward pull for short vowels results in the absence of a neutral position.

The relatively small number of observations (11 topic-initial and 27 non-topic-initial sentences) placed restrictions on the analysis. The numerous factors which are known to influence F_0 peak alignment justify, in principle, a corresponding number of independent variables in the analysis of variance. With the available data this would yield a large number of (quasi-)empty cells. We therefore constructed a new variable: "Accumulated Timing Constraint" (*ATC*), on which the influencing factors were mapped.

The features listed above were used to arrive at a crude weighting for each syllable. Since past studies have not been able to quantify the degree to which timing constraints operate, we followed the practice of Rietveld and Gussenhoven (1995) and assigned the same unitary value to each feature. Each leftward pulling factor was assigned a unitary negative value, and for each rightward pushing factor a positive value was assigned. The final weighting was the sum of these values (e.g. three leftward pulling factors and one rightward gives a final weighting of -3 + 1 = -2).

The resulting values were subsequently dichotomised, distinguishing syllables with a tendency to leftward ($\leq$ -2, classed as '0'), and rightward alignment (> -2, classed as '1') respectively. The position of the F_0 peak in each IO was calculated as a percentage of the total duration of the syllable.

8.4.2 Results

8.4.2.1 Evidence of Local Timing Constraints

The mean peak position for the factor ATC (Table 8.1) shows the expected effect of known timing constraints, namely that in syllables with a leftward ATC the peak occurs earlier (73% into the syllable) than in those with a rightward ATC (98% into the syllable). This confirms that the timing effects observed in the past under experimental conditions also operate in unconstrained texts. The results also justify the following investigation, using the same binary classification for the timing constraints, into possible additional effects of discourse structure.

Table 8.1. Mean F_0 peak position (expressed as % of the syllable duration) in syllables with a leftward or rightward timing constraint (8,1).

ATC	Mean Peak Position
0	73
1	98

8.4.2.2 Evidence of a Discourse ('topic') Constraint

When the timing of the F_0 peak within the syllable is examined according to whether the sentence is topic-initial or non-initial (Table 8.2), we see that, regardless of the ATC, the peak in the +topic condition is consistently later than in the -topic condition, even to the extent of occurring outside the accented syllable itself. For syllables with an ATC of 0, the mean peak position is 67% if topic non-initial, and 88% if topic initial. With an ATC of 1 the mean is 89% if topic non-initial and 119% if topic initial. A value greater than 100% means that the peak occurs after the accented syllable.

Table 8.2. Mean F_0 peak position (expressed as % of the syllable duration) according to leftward and rightward timing constraints (0,1) and topic initiality (-topic, +topic).

ATC:	-topic	+topic
0	67	88
1	89	119

An analysis of variance was carried out on the relative position of F_0 peaks in the syllable. Two factors were included in the design: TOPIC (+/- top) and ATC (0,1). None of the factors reached significance at the 5% level. However, in view of the small power achieved with the available data (0.399 for the factor TOPIC, and 0.420 for ATC), the obtained F-ratios suggest the presence of the hypothesised effect of discourse structure on F_0 peak alignment: $F_{1,34} = 3.08$, p=0.088; the effect size was estimated by $\eta^2 = 0.083$. The F-ratio for ATC was $F_{1,34} = 3.28$, p= 0.079; with η^2 for effect size = 0.088.

In order to assess the extent to which it is possible to distinguish between topic initial and topic non-initial sentences on the basis of F_0 peak alignment only, we also carried out a discriminant analysis, with one variable: the relative position of the F_0 peak in the syllable. Obviously, this did not yield a significant result either ($\chi^2 = 2.92$, p= 0.088, df = 1) but the positions of the F_0 peaks on the resulting discriminant function (Figure 8.1) suggest again that F_0 peaks of topic-initial sentences tend to be right aligned with the syllable, whereas the non-initial peaks are more evenly distributed over the domain of the syllable.

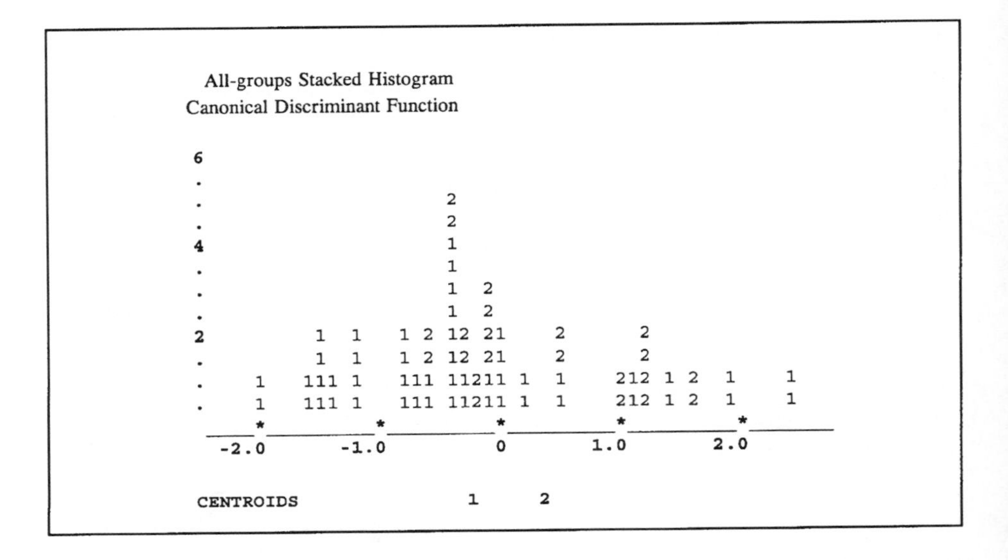

Figure 8.1. Discriminant function with positions of topic initial F_0 peaks ("2") and topic non-initial peaks ("1").

The results suggest that, for this speaker at least, the topic structure of the text has an effect on peak timing in the IO, exerting a strong rightward push even to the extent of causing the peak to occur beyond the accented syllable itself.

8.5 Experimental Evidence

The second study tested further the tendencies observed above for topic structure to exert an influence on peak timing, this time under experimental conditions. One of the difficulties of using natural data is that each IO falls on a different syllable/word. An experimental text was therefore designed in which a defined set of syllables/words were placed in different positions in the text. In this way the segmental make-up of the syllables under scrutiny was kept constant, obviating the need to take segmental timing effects into account.

8.5.1 Procedure

Three texts were constructed, based loosely on an existing Open University lecture (SEC, text D02) on the subject of "The Enlightenment in France". Each text was approximately 380 words in length, and contained 3 or 4 paragraphs. The texts were considerably more complex than the news broadcast described above, of an academic nature, and requiring some preparation before performance.

There were 10 readers (5 male, 5 female), all speakers of Southern British English (near-RP), of whom 9 were used for subsequent analysis. In order to elicit a reading style as close to a professional performance as possible, readers were instructed to read as if they were experts in their field, recording an Open University lecture. If they stumbled during the reading, they were asked to repeat the paragraph in which the mistake occurred.

The words which were used for the analysis are as follows (the relevant accented syllable is highlighted):

carTESian
COmmon
comPENdium
enlightenment

Each word occurred in the course of the three texts in three different positions, which we coded as +/- Sentence Initial (SI), and +/- Paragraph Initial (PI):

1. Sentence Final: -SI, -PI
2. Sentence Initial (but paragraph medial): +SI, -PI
3. Paragraph Initial: +SI, +PI

"Sentence Initial" here implies the first accented syllable in the sentence (IO) rather than absolute initial position. We hypothesised that readers would interpret "Paragraph Initial" position as essentially topic-initial, although we were aware that there is no precise correlation between paragraphs, an essentially typographic convention, and topics (Wichmann, 2000).

The four target words were incorporated into the text as follows:

Paragraph Initial: ***Cartesian*** *philosophy in France was to be challenged quite suddenly in the late seventeen twenties.*

Sentence Initial: *But* ***Cartesian*** *thought was incorporated in a system of Christian apologetics which most Catholic scholars accepted.*

Sentence Final: *During the first quarter of the eighteenth century, French natural philosophy was almost totally* ***Cartesian****.*

Paragraph Initial: *The* ***common*** *ground shared by the 'philosophes' was clearly being lost by the seventeen seventies.*

Sentence Initial: *The* ***common*** *ground they shared was, despite much diversity of thought, most obviously, their scepticism about the forms and rituals of traditional Christianity.*

Sentence Final: *It was continuing belief in providence which sustained Voltaire's deism, and gave him and other contemporary thinkers some important ground in* ***common****.*

Paragraph Initial: *The* ***compendium*** *of scientific information conceived by Diderot was known as the Great Encyclopaedia.*

Sentence Initial: *The* ***compendium*** *was written, like most encyclopaedias, to convey truths, but truths arrived at in the only way that the philosophers thought it possible to acquire certain knowledge - by reason and experience.*

Sentence Final: *But the encyclopaedia was more than a wide-ranging* ***compendium****.*

Paragraph Initial: *The* ***Enlightenment*** *and its ideas are nowhere more evident than in the great Encyclopaedia published in 28 huge folio volumes between 1751 and 1772.*

Sentence Initial: *The* ***Enlightenment*** *was now, although often thought of as a unified system of thought, in fact becoming increasingly diverse.*

Sentence Final: *The movement they created has become famous as the* ***Enlightenment****.*

If a reader repeated a paragraph and unwittingly produced a target utterance twice, we included only the first version. Some target utterances had to be excluded from analysis, either because an F_0 trace could not be extracted or because the reader failed to accent the appropriate syllable. Six data points were omitted in this way, leaving 102 for the final analysis.

The texts were recorded under anechoic conditions onto DAT tape using a Sony 1000 ES DAT recorder, and a B&K 2231 sound level meter fitted with a 4165 microphone cartridge. F_0 analysis was carried out using the LSI (Loughborough Sound Images) speech processing system. The signal was sampled at 10 kHz; the pitch analysis option 'cepstrum' was used with a frame advance of 5 ms. The same measurement procedures were adopted as in the earlier study of the news broadcast (section 8.4.1).

8.5.2 Results

Subjects provided a good reading performance in the style we had hoped to elicit, although one male subject's recording was excluded from analysis because in a number of target sentence-final utterances he used a rising tone (L*), categorically different from, and therefore not comparable to the other speakers. Although reading speeds varied considerably from subject to subject, the performances were highly proficient, clearly reflecting the structure and meaning of the text, and in a varied, expressive style.

The positions of the F_0 peaks were defined as their relative locations (expressed as a percentage) in or beyond the interval stretching from the start of the syllable until its end. An analysis of variance was carried out on these positions as a function of three fixed factors: speakers (9 levels), word (4 levels) and discourse position (3 levels); the factor 'speaker' was considered to be a fixed factor as the speakers had been selected by one of the authors on the basis of the criterion 'good and experienced reader'. Three main effects turned out to be significant at the 1%-level: speaker ($F_{9,41} = 4.60$, $p = 0.001$), word ($F_{3,41} = 28.50$, $p < 0.001$) and discourse position ($F_{2,41} = 88.58$, $p < 0.001$; two two-way interactions were significant: word × condition: ($F_{6,41} = 3.97$, $p = 0.003$) and speaker × condition ($F_{14,41} = 2.55$, $p = 0.010$). The relative positions of the pitch peak as a function of discourse position are: 116%, 105% and 62%, with the highest value for the paragraph initial position and the lowest for sentence final. Post-hoc comparisons (Tukey's HSD) showed significant differences ($p < 0.05$) between all discourse positions. As the interactions were ordinal, the order of the F_0 peaks as a function of discourse position did not change at the different levels of the factors 'word' and 'speaker'.

Discourse position also affected syllable duration (see Table 8.3). However, no difference in syllable duration was found between paragraph initial and sentence initial positions, while sentence final position showed the expected lengthening effect. There was a further effect of discourse position on the height of the pitch peak: $F_{2,41} = 133.51$, $p < 0.01$. Post-hoc comparisons (Tukey's HSD) showed significant differences between all discourse positions at the 5%-level.

Table 8.3. Mean values of F_0 peak position (expressed as % of syllable duration), F_0-peak (Hz) and Syllable durations (ms) as a function of discourse position (pooled data).

	Paragraph Initial	**Sentence Initial**	**Sentence Final**
F_0 peak position (%)	115.7	105.2	62.2
F_0 peak (Hz)	314	285	189
Syllable duration (ms)	206	207	217

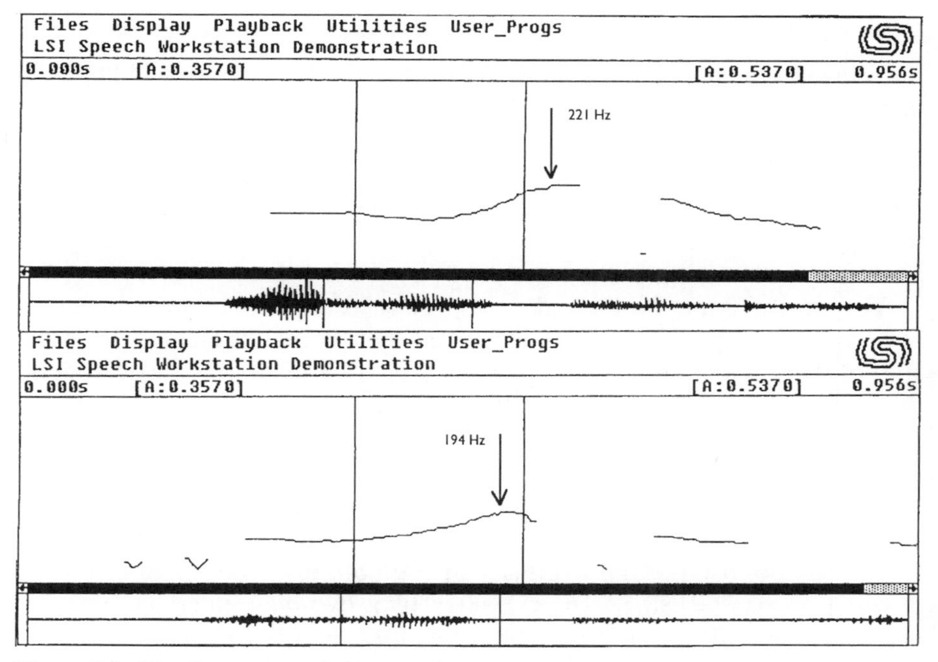

Figure 8.2. The F_0 contour of the target word 'enLIGHtenment' in (i) Paragraph Initial position (upper panel) and (ii) Sentence Initial position (lower panel). The cursors enclose the accented syllable. (Speaker 10).

8.5.3 Conclusion

The above results confirm the tendencies observed in natural, unconstrained data. In addition to any segmental or prosodic constraints on the timing of pitch peaks there is also a discourse effect. The experimental data shows first of all a very marked difference between the timing of sentence-final pitch accents and the timing of those in sentence-initial position. In addition we find that speakers distinguish between degrees of initiality. The F_0 peak at the beginning of a topic or paragraph occurs later (further to the right) in relation to the syllable as a whole than if the syllable is sentence-initial but not topic- or paragraph-initial.

8.6 Discussion

8.6.1 Segmental Effects and Prosodic Context

Results of the study described in section 8.4 provide first of all clear confirmation in uncontrolled natural data of those segmental and prosodic influences on peak timing observed in experimental studies, particularly those constraints assumed to exert a leftward pull. Further studies might usefully investigate the relative strength of effect of segmental and prosodic constraints, and indeed whether the rather crude unitary measure used here to assess the overall weighting should be refined.

8.6.2 Discourse Effects

The results of both studies suggest in addition that peak timing in relation to the associated accented syllable is also affected by discourse structure. Paragraph or topic initiality exerts a strong rightward push even to the extent of causing the F_0 peak to occur beyond the accented syllable itself. It is of course well known that topic-initial IOs are also markedly higher in pitch than those which are sentence-initial but topic-medial, and this is confirmed in our experimental study. Since the relative height of the target (measured in Hz) correlates highly in our text with discourse position (Kendall's tau(b) $= 0.81$, $p < 0.01$, $n = 102$), it is possible that the timing effect is only indirect. The delayed peak may simply be a function of increased height, on the assumption that the greater the step up in pitch, the longer it may take for a speaker to reach the target. However, the text analysed in section 4 above contains evidence that this is not necessarily the case. The very first utterance in the broadcast (omitted from the analysis since it is metatextual) begins with a very high IO in which the F_0 peak nonetheless occurs very early in the syllable. This speaker at least is therefore physically able to combine a high onset with an early peak.

In our experiment we found a number of cases where the readers did not show the clear effect we were looking for, but instead appeared to choose an alternative strategy, illustrated in Figure 8.3.

We identified the accent-related F_0 peak in terms of the earliest point at which it reached maximum F_0. This might, however, obscure the relevance of another aspect of the contour, namely the position of the start of the fall. In Figure 8.3 we see that the F_0 maximum is reached at a similar point in both the sentence-initial and paragraph-initial conditions, but that in the latter there is a longer plateau, which has the effect of delaying the point at which the contour begins to fall again. This may simply be an alternative way of achieving the effect of topic or paragraph initiality. An alternative interpretation might be that the underlying strategy is the same, namely to delay the falling contour, but that this is achieved in different ways: either the F_0 peak is delayed, thus inherently delaying the starting point of the fall, or only the fall itself is delayed, resulting in a plateau between the F_0 peak and the fall. Perception tests in which the timing of both peak and fall are manipulated independently might shed more light on this.

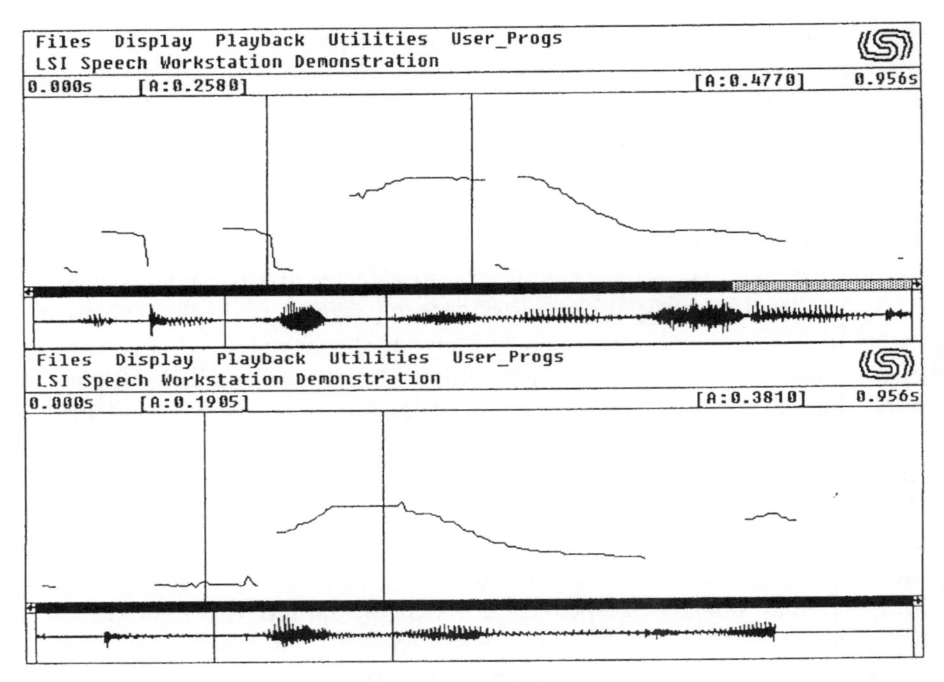

Figure 8.3. The F_0 contour of the target word "comPENdium" uttered in (i) paragraph-initial position (upper panel) and (ii) sentence-initial position (lower panel). The cursors enclose the accented syllable. (Speaker 5).

Our results reflect those reported earlier on the differences between nuclear and non-nuclear accent timing (Silverman & Pierrehumbert, 1990), and support the view that the timing of accents is context-dependent. This context-dependent variation is not limited, however, to a simple distinction between initial and final. It has already been observed that different timing of utterance-final falls can create different degrees of perceived finality. Our observations suggest that the greater perceived 'finality' of an early peak in a nucleus has its counterpart in the discoursally greater 'initiality' of a late peak in an intonational onset.

Notes

1. For example, American English: Steele (1986), Pierrehumbert & Steele (1989), Silverman & Pierrehumbert (1990), van Santen & Hirschberg (1994); British English: Silverman (1987), House (1989); German: Kohler (1987, 1990); Swedish: Bruce (1983, 1986, 1990); Dutch: Caspers & van Heuven (1993), Rietveld & Gussenhoven (1995); Mexican Spanish: Prieto *et al.* (1995); Greek: Arvaniti & Ladd (1995).
2. IPO, Center for Research on User-System Interaction, Eindhoven, The Netherlands.

References

Aijmer, K. 1996. *Conversational Routines in English*. London: Longman.

Arvaniti, A. and D.R. Ladd. 1995. Tonal alignment and the representation of accentual targets. *Proc. 13th ICPhS* (Stockholm, Sweden), vol. 4, 220-223.

Brown, G. and G. Yule. 1983. *Discourse Analysis*. Cambridge University Press.

Bruce, G. 1983. Accentuation and timing in Swedish. *Folia Linguistica* 17, 221-238.

Bruce, G. 1986. How floating is focal accent? *Nordic Prosody* IV, 41-49.

Bruce, G. 1990. Alignment and composition of tonal accents: comments on Silverman and Pierrehumbert's paper. In Kingston and Beckman (eds.), 107-114.

Caspers, J. and V. van Heuven. 1993. Effects of time pressure on the phonetic realization of the Dutch accent-lending pitch rise and fall. *Phonetica* 50, 161-171.

Gussenhoven, C. 1988. Adequacy in intonation analysis: the case of Dutch. In van der Hulst and Smith (eds.), 95-121.

Hart, J. 't, R. Collier, A. Cohen. 1990. *A Perceptual Study of Intonation*. Cambridge University Press.

House, J. 1989. Syllable structure constraints on F_0 timing. *LabPhon II*, Edinburgh (poster presentation, unpublished data).

Kingston, J. and M.E. Beckman (eds.). 1990. *Papers in Laboratory Phonology I*. Cambridge University Press.

Knowles, G., L. Taylor and B.J. Williams. 1996. *Spoken English Corpus*. London: Longman.

Kohler, K. 1987. Categorical pitch perception. *Proc. 11th ICPhS* (Tallinn, Estonia), vol. 5, 331-333.

Kohler, K. 1990. Macro and micro F_0 in the synthesis of intonation. In Kingston and Beckman (eds.), 115-138.

Ladd, D.R. 1996. *Intonational Phonology*. Cambridge University Press.

Möbius, B. 1994. A quantitative model of German intonation and its application to speech synthesis. *Proc 2nd ESCA/IEEE Workshop on Speech Synthesis* (New Paltz, USA), 139-142.

Möbius, B. 1995. Components of a quantitative model of German intonation. *Proc. 13th ICPhS* (Stockholm, Sweden), vol. 2, 108-115.

Pierrehumbert, J.B. and M.E. Beckman. 1988. *Japanese Tone Structure*. Cambridge, Mass.: MIT Press.

Pierrehumbert, J.B. and S. Steele. 1989. Categories of tonal alignment in English. *Phonetica* 46, 181-196.

Prieto, P., J. van Santen and J. Hirschberg. 1995. Tonal alignment patterns in Spanish. *Journal of Phonetics* 23, 429-451.

Rietveld, T. and C. Gussenhoven. 1995. Aligning pitch targets in speech synthesis: effects of syllable structure. *Journal of Phonetics* 23, 375-385.

Schiffrin, D. 1987. *Discourse Markers*. Cambridge University press.

Silverman, K. 1987. *The structure and Processing of Fundamental Frequency Contours*. PhD dissertation, University of Cambridge

Silverman, K. and J.P. Pierrehumbert. 1990. The timing of prenuclear high accents in English. In Kingston and Beckman (eds.), 72-106.

Steele, S. 1986. Nuclear accent F_0 peak location: effects of rate, vowel, and number of following syllables. *J. Acoust. Soc. Am.* 1, 80, s51.

Swerts, M. 1994. *Prosodic Features of Discourse* Units. PhD dissertation, Eindhoven University of Technology.

Swerts, M., D.G. Bouwhuis and Collier. 1994. Melodic cues to perceived 'finality' of utterances. *J. Acoust. Soc. Am.* 96, 2064-2075.

van der Hulst, H. and N. Smith. 1988. *Adavances in Nonlinear Phonology*. Dordrechet: Foris.

van Santen, J. and B. Möbius. 1997. Modelling pitch accent curves. *Proc. ESCA workshop on Intonation* (Athens, Greece), 321-324.

van Santen, J. and J. Hirschberg. 1994. Segmental effects on timing and height of pitch contours. *Proc. ICSLP* '94 (Yokohama, Japan), 719-722.

Wichmann, A. 1991a. *Beginnings, Middles and Ends: Intonation in Text and Discourse*. PhD dissertation, University of Lancaster.

Wichmann, A. 1991b Falls: variability and perceptual effects. *Proc. 12th ICPhS* (Aix-en-Provence, France), vol. 5, 194-197.

Wichmann, A. 1992. F_0 peak position as a cue to text structure, oral presentation. *Proc. 3th ICAME* (Nijmegen, unpublished data).

Wichmann, A. 2000. *Intonation in Text and Discourse*. London: Pearson Education.

Section IV

Intonation Modelling

9

Automatic Stylisation and Modelling of French and Italian Intonation

ESTELLE CAMPIONE, DANIEL J. HIRST AND JEAN VÉRONIS

9.1 Introduction

F_0 curves are often considered the combination of a macroprosodic component reflecting the speaker's choice of intonation pattern, and a microprosodic component (Di Cristo & Hirst, 1986) which is entirely dependent on the choice of phonemes in the utterance (lowering of F_0 for voiced obstruents, etc.). Numerous studies since the 1960's have attempted to factor out these two components and to extract automatically the relevant macroprosodic information from the speech signal. This extraction can be broken down into two stages:

- stylisation, i.e. the replacement of the F_0 curve by a simpler numerical function conserving the original macroprosodic information;
- symbolic coding, i.e. the representation by means of an alphabet of symbols, reducing the stylised curve to a sequence of discrete categories.

The first stage is often referred to as close-copy stylisation (De Pijper, 1979) which replaces the original F_0 curve by a stylised curve that is assumed to be perceptually identical to the original. The discrete categories of the second stage can be used to re-generate a curve which, unlike the perceptually identical version may be distinguishable from the original one but is nonetheless considered by listeners as linguistically equivalent. De Pijper (*op. cit.*) calls this "standardised perceptual equivalence".

A. Botinis (ed.), Intonation, 185-208.

Hirst, Di Cristo and Espesser (forthcoming) distinguish four distinct levels of representation for prosody. At the most abstract level they assume an underlying phonological representation and at the concrete level a physical (i.e. acoustic or physiological) representation. Between these two extremes they further distinguish a surface phonological representation constructed from discrete phonological categories and a phonetic representation expressed in terms of continuously variable values. In De Pijper's terms, utterances which have the same surface phonological representation might thus be assumed to be perceptually equivalent, whereas utterances with the same phonetic interpretation would be perceptually identical.

Generating F_0 curves from surface phonological representations obviously leaves quite a lot of unexplained variability in the data. This chapter explores the possibility of enriching surface phonological representations to the point where an utterance synthesised from a symbolic coding might be practically indistinguishable from the original utterance. Stylisation has been the object of a great number of studies and the technique has been mastered fairly satisfactorily. A number of systems of symbolic coding have also been proposed, but automatisation and reversibility (close copy) are far less advanced for symbolic coding than for stylisation.

A totally reversible system of analysis would obviously constitute an extremely valuable tool for the automatic coding of large speech corpora. Such tools would be useful both for the study of speech in general and for speech technology, in particular speech recognition and speech synthesis (Véronis, Di Cristo, Courtois & Lagrue, 1997; Di Cristo, Di Cristo & Véronis, 1997; Véronis, Di Cristo, Courtois & Chaumette, 1998). The rare prosodically labelled corpora which exist at present (e.g. Ostendorf, Price & Shattuck-Hufnagel, 1995) have required hand labelling by experts. Besides the laboriousness and difficulty of the task, the subjective nature of such labelling reduces the trustworthiness of the results or requires careful control by counter-experts thus increasing the cost yet more. Automatic techniques removing the need for manual intervention, or at least reducing it to a phase of checking and correcting, would obviously be extremely desirable. Unfortunately there are no automatic robust prosodic transcription systems available even for such widely used systems as ToBI (Silverman, Beckman, Pitrelli, Ostendorf, Wightman, Price, Pierrehumbert & Hirschberg, 1992), although work in this area is in progress (Black & Hunt, 1996; Dusterhoff & Black, 1997).

In this chapter we present six progressively more complex versions of an automatic transcription system based on the INTSINT model described in Hirst and Di Cristo (1998a). The six versions were tested on a corpus of read speech in French and the two most complex versions were subsequently also evaluated for a comparable corpus in Italian (representing readings from 20 different speakers for a total of about 90 minutes of speech). Although there is still some room for improvement, the best two versions provide a very close approximation to the original curves since the standard deviation of error (on a logarithmic scale) was reduced to less than a semi-tone both for the absolute pitch targets and for the relative pitch intervals.

9.2 Stylisation

In this section we give a brief overview of existing techniques including that which we adopted for this study.

The Instituut voor Perceptie Onderzoek (IPO) originally undertook research on the stylisation of F_0 contours with the aim of developing an intonation model for speech synthesis in Dutch (Cohen and 't Hart, 1965). The IPO approach subsequently evolved into a general theory of intonation structure ('t Hart, Collier & Cohen, 1990), through a procedure of analysis by synthesis. The approach is based on the principle that the simplified F_0 curve must be melodically identical to the original curve ('t Hart & Collier, 1975). In the late 70s, De Pijper (1979) introduced the concept of close-copy stylisation where a subject is asked to compare a recording of an utterance with a synthesised version of the utterance in which the F_0 curve has been replaced by a sequence of straight lines (on a logarithmic scale). A close copy is obtained when subjects are unable to distinguish the synthesised version from the original. For synthesis the piece-wise linear approximations obtained by close-copy stylisation were converted to standard pitch movements chosen from a (language-specific) inventory of "perceptually relevant pitch movements". The resulting output, while usually quite easily distinguishable from the original was nonetheless claimed to be as acceptable as the original and hence "perceptually equivalent" ('t Hart *et al.*, 1990). This approach, originally developed for the intonation of Dutch was subsequently applied to other languages: English (De Pijper, 1983; Willems, Collier & de Pijper, 1988), German (Adriaens, 1991), Russian (Odé, 1989), French (Beaugendre, 1994) and Indonesian (Odé & van Heuven, 1994).

More recently, Taylor (1993, 1994) proposed a model analysing an F_0 curve as a linear sequence of three primitive elements: Rise, Fall and Connection which can then be related to a phonological representation. The Rise and Fall are interpreted as piecewise parabolas and are hence equivalent to the quadratic spline presented below. The Connection element is interpreted as a linear transition.

D'Alessandro and Mertens (1995) present a technique for automatic stylisation based on a model of tonal perception which assumes (following House, 1990) that the syllable is the basic perceptual unit for speech. Syllabic pitch-contours are categorised as dynamic or static depending on the presence of a perceptible (and possibly complex) pitch movement. The F_0 contour is transformed into a sequence of tonal segments which are either static or dynamic as a function of a glissando threshold which varies with the duration of the syllable.

The method of stylisation used in this study: MOMEL (MOdélisation de MELodie) was originally proposed by Hirst (1980, 1983) and automated by Hirst and Espessser (1993) (see description in Appendix I). Contrary to many methods of stylisation which use a sequence of straight line segments, MOMEL uses a quadratic spline function (sequence of parabolic segments) resulting in a continuous, smooth curve, without the angles which occur when using straight lines. Unvoiced segments are interpolated so that the resulting curve presents no discontinuities at all. These characteristics of the quadratic spline function are also shared by the more complex stylisation functions used by Fujisaki and colleagues (Fujisaki & Hirose, 1982) as the continuation of earlier work by Öhman (1967). This model is based on the hypothesis that continuously varying F_0 curves are the result of a sequence of discrete commands produced by the speaker. Fujisaki distinguishes phrase commands and accent commands modelled respectively as an impulse-like command and a step-command.

It has been argued ('t Hart, 1991) that stylisation by curvilinear functions is not perceptually distinguishable from that using straight-lines. We note however that:

- stylisation by quadratic splines produces a curve which is closer to the original F_0 curve and hence introduces less noise into quantitative studies – in particular in the evaluation of models as in this paper;
- stylisation by quadratic splines produces a macroprosodic contour which is practically identical to the F_0 curves produced on utterances consisting entirely of sonorant segments which are both continuous and smooth[1].

The quadratic spline functions used for synthesis can be defined by a sequence of target points corresponding to the significant changes of the F_0 curve (zero of the first derivative[2]). Appendix I summarises the algorithm.

9.3 Symbolic Coding

The most widely used system for the symbolic coding of intonation at present is ToBI (Silverman *et al.*, 1992) which has been used successfully by numerous researchers for American English, although it has been criticised for falling unsatisfactorily between a phonetic and a phonological system (see for example Nolan & Grabe, 1997). ToBI is a system which is based on an extensive phonological analysis of the intonation system of English, and its application to other languages or dialects, while theoretically possible, would necessitate a considerable amount of prior research to establish the inventory of intonation patterns of the language (Pierrehumbert, forthcoming). ToBI labelling also relies on linguistic judgements made by experts and is consequently difficult to carry out automatically, although attempts have been made to do this (Wightman & Ostendorf, 1992; Ostendorf & Ross, 1997). Finally, the regeneration of an F_0 curve from the ToBI coding is far from obvious.

In our study, we use a symbolic coding system INTSINT (INternational Transcription System for INTonation) proposed by Hirst and Di Cristo (1998a) and Hirst *et al.* (forthcoming) and which has been used for the manual transcription of intonation patterns in a number of languages (see different chapters in Hirst & Di Cristo, 1998b). Unlike ToBI, which encodes events of a linguistic nature, INTSINT aims to provide a purely formal encoding of the macroprosodic curve. Each target point of the stylised curve is coded by a symbol either as an absolute tone, defined globally with respect to the speakers pitch-range, or as a relative tone, defined locally with respect to the immediately neighbouring target-points. Relative tones can further be subdivided into iterative and non-iterative categories where it is assumed that iterative tones can be followed by the same tone whereas non-iterative tones cannot. Within each category, positive, negative and neutral values are defined, which thus give a total of nine possible tone categories. Of these, it is assumed that one logical possibility, that of an iterative neutral tone, does not in fact occur so that the following eight possibilities are actually used (Table 9.1).

Table 9.1. Orthographic and iconic symbols for the INTSINT coding system. The letters stand for Top, Mid, Bottom, Higher, Same, Lower, Upstepped and Downstepped respectively.

		Positive	***Neutral***	***Negative***
ABSOLUTE		T [⇑]	M [⇒]	B [⇓]
RELATIVE	*Non-Iterative*	H [↑]	S [→]	L [↓]
	Iterative	U [<]	•	D [>]

This system can be thought of as a surface phonological system, something along the lines of the International Phonetic Alphabet for the transcription of segmental phonology. It can thus be seen as a first degree of abstraction which can be used for the automatic extraction of "linguistic-like" representations from spoken corpora which could provide the data necessary for the development of more abstract phonological systems such as ToBI.

The automatic and reversible coding using INTSINT poses a number of problems, however. Nicolas and Hirst (1995) showed, for example, that for the coding of continuous texts, considerable improvement was obtained when the values of the extreme tones T and B were varied to take into account paragraph level effects. In this chapter we concentrate on the coding of the relative tones for which the two basic problems are:

- determining a threshold to separate absolute tones T, B from relative tones H, S, L, U, D;
- determining the optimal criteria for distinguishing iterative and non-iterative tones

The distinction between iterative and non-iterative tones can be based on two rather different criteria. The first distinction is purely configurational in that basically H and L are respectively peaks and valleys whereas D and U are plateaus in rising/falling sequences of tone. If this were the only distinction, the coding would be redundant and U and D could be treated as allophones of H and L. There is, however, also a second scalar distinction between the two categories, since in general H and L are assumed to correspond to larger frequency intervals than U and D.

In the rest of this paper we present results on the evaluation of six different implementations of the INTSINT model. In particular we separate out the configurational and the size criteria in order to evaluate the relative importance of each. We also investigate the possibility of extending the scalar values in order to come nearer to a close-copy stylisation[3].

9.4 Common Properties of the Six Implementations

As mentioned above the six implementations we tested are all based on the automatic stylisation technique MOMEL. In all the implementations, two absolute symbols T and B are used to code extreme pitch values. The symbol S is used to code target points which are not significantly different from the preceding point.

- target points higher than a threshold τ_T are coded T, those below a threshold τ_B are coded B;
- target points less than 2.5% (on a log scale) from the preceding point are coded S.

The thresholds τ_T and τ_B are chosen so that 5% of the target points are coded T and another 5% are coded B, assuming a normal distribution of the values of target-points on a logarithmic scale, which a prior analysis (see below) showed to be a satisfactory approximation.

The resynthesis of the target points takes as a starting value a value situated before the beginning of each utterance coded M with an F_0 value equal to that of the overall mean of the target points. Subsequent points are generated in the following way:

- points coded T and B are assigned the values F_T and F_B corresponding to the mean of the target points respectively above τ_T and below τ_B assuming a normal distribution.
- points coded S are assigned the same value as the preceding target point[4];
- the value of the other target points are calculated by linear regression from the value of the preceding target point using regression coefficients estimated separately for each symbol on the z-transformed values for all speakers pooled for each language.

It should be noted that while there is no specific "downdrift" or "declination" component in this model, the typical lowering of a sequence of successive H and L targets observed in natural speech is an automatic consequence of applying the same regression coefficients iteratively. Thus sequences such as [M T L H L H L H...] or [M T D D D ...] will produced target values which decline towards asymptotic values (cf. Liberman & Pierrehumbert, 1984; Hirst *et al.*, forthcoming).

9.5 Specific Characteristics of the Different Implementations

9.5.1 Version HL

In a purely configurational interpretation, the distinction between iterative and non-iterative tones is treated as simply allophonic. The first implementation we tested is a minimal one conflating the two categories and assuming only two symbols, H and L (in addition to T, B and S which as mentioned above are common to all implementations) (Figure 9.1):

- H : rising;
- L : falling.

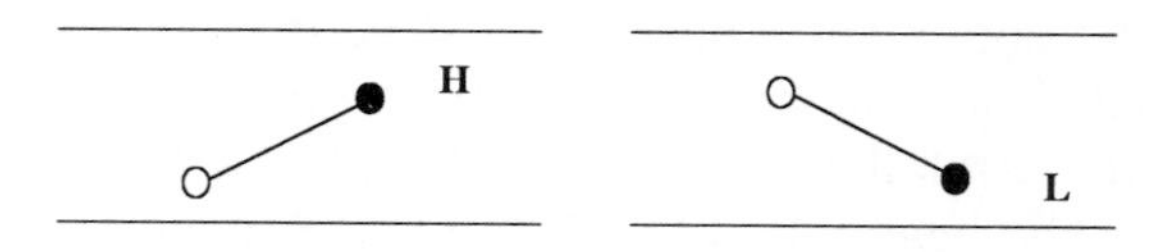

Figure 9.1. Examples of coding with version *HL*.

9.5.2 Version Config

The second implementation is designed to test the possibility that the difference between iterative and non-iterative values, while allophonic, nevertheless requires separate regression coefficients. Target points are consequently coded as H, L, U or D according to configuration (Figure 9.2):

- H : peak;
- U : raised plateau (upstepped);
- D : lowered plateau (downstepped);
- L : valley.

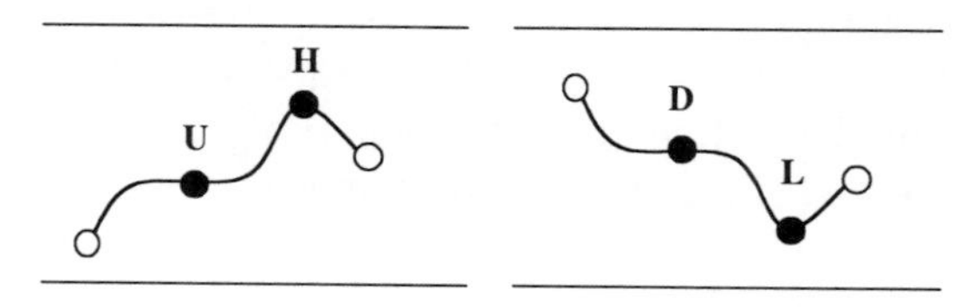

Figure 9.2. Examples of coding with version *Config*.

9.5.3 Version Mixed

This version is an implementation of the algorithm described in Hirst *et al.* (forthcoming). This version assumes that the distinction between iterative and non-iterative categories is based both on configuration and on size. Thus plateaus in rising or falling sequences will necessarily be coded U and D respectively as in the *Config* version, whereas peaks will be coded either H or U and valleys will be coded either L or D depending on the size of the pitch interval with respect to the preceding target (Figure 9.3).

- H: peak AND interval greater than a threshold α above the previous target point;
- U: raised plateau (upstepped) OR interval less than a threshold α above the previous target point;
- D: lower plateau (downstepped) OR interval less than a threshold α below the previous target point;
- L: valley AND interval greater than a threshold α below the previous target point.

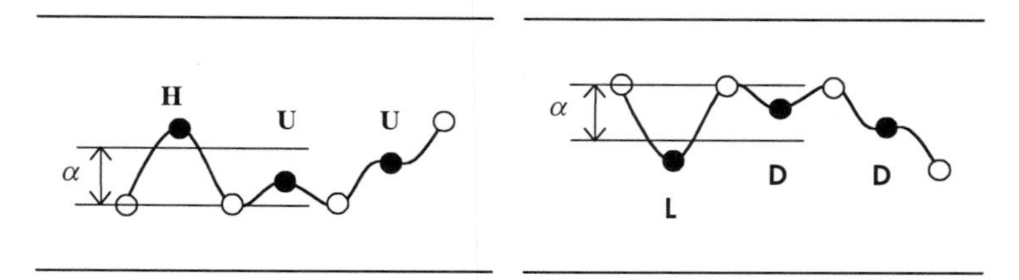

Figure 9.3. Examples of coding with version *Mixed.*

9.5.4 Version Ampli2

The implementation *Ampli2* assumes that the distinction between iterative and non-iterative tones is purely scalar, i.e. that H and L represent larger pitch intervals with respect to the preceding target point than do U and D respectively (Figure 9.4).

- H_1 : interval less than a threshold α above the previous target point;
- H_2 : interval greater than a threshold α above the previous target point;
- L_1 : interval less than a threshold α below the previous target point;
- L_2 : interval greater than a threshold α below the previous target point;

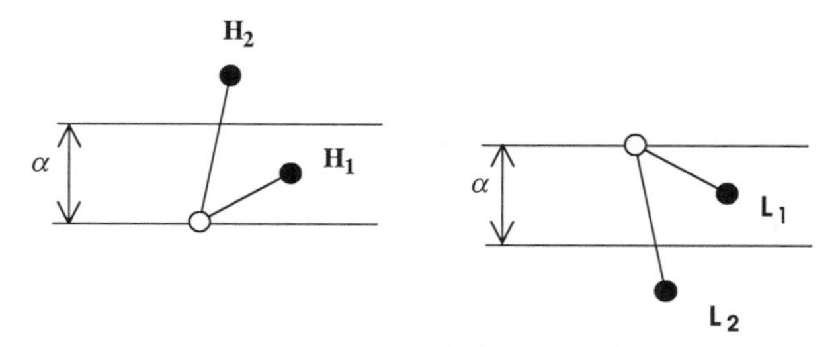

Figure 9.4. Examples of coding with version *Ampli2*.

9.5.5 Version Ampli3

The implementation *Ampli3* is based on the same principle as version *Ampli2* except that the scalar aspect is extended to distinguish three degrees of size defined by two thresholds α_1 and α_2 (Figure 9.5). This allows us to define six relative pitch levels:

- H_3 : interval greater than α_2 above the previous target point;
- H_2 : interval between α_1 and α_2 above the previous target point;
- H_1 : interval less than α_1 above the previous target point;
- L_1 : interval less than α_1 below the previous target point;
- L_2 : interval between α_1 and α_2 below the previous target point;
- L_3 : interval greater than α_2 below the previous target point.

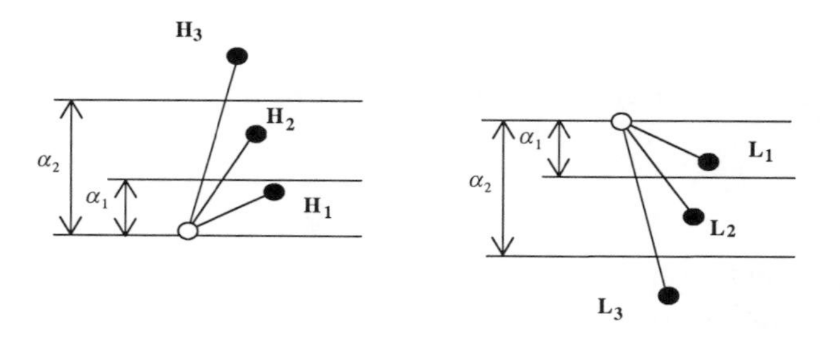

Figure 9.5. Examples of coding with version *Ampli3*.

9.5.6 Version Levels

The final implementation we tested explores a rather different principle, following earlier research by Rossi and Chafcouloff (1972). The central region between the two thresholds τ_T and τ_B is divided into three bands: G(rave), M(edium), A(cute), each corresponding to one third of the target points assuming a normal distribution. The coding of the target points takes into account both the direction with respect to the preceding target point and the band in which the targets are situated (Figure 9.6):

- H_A : higher than previous target and finishing in the acute band;
- H_M : higher than previous target and finishing in the medium band;
- H_G : higher than previous target and finishing in the grave band;
- L_A : lower than previous target and finishing in the acute band;
- L_M : lower than previous target and finishing in the medium band;
- L_G : lower than previous target and finishing in the grave band.

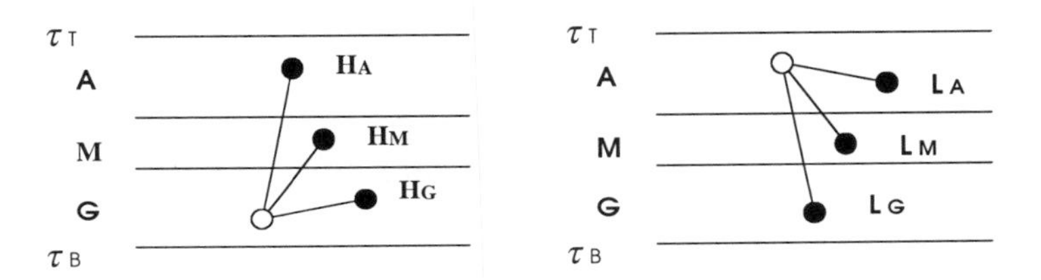

Figure 9.6. Examples of coding with version *Levels*.

9.6 Evaluation

9.6.1 Corpus

The corpus used for this study is part of the multilingual speech corpus EUROM1 which was defined and produced as a deliverable of the European ESPRIT project SAM (Multi-lingual Speech Input/Output Assessment, Methodology and Standardisation: Chan, Fourcin, Gibbon, Granström, Hucvale, Kokkinakis, Kvale, Lamel, Lindberg, Moreno, Mouropoulos, Senia, Trancoso, Veld & Zeiliger, 1995). We used the French and the Italian versions of the corpus, specifically the part consisting of 40 continuous passages read by ten speakers (5 male and 5 female) in each language. Each passage comprises 5 semantically linked sentences; the French and Italian versions are fairly free translations from the English version. The following is a sample passage in English, French and Italian:

> *Please take a request for an early-morning taxi. Mr Spencer of Chestnut Drive wishes to be at Heathrow terminal 4 by 6.15 a.m. His flight's not leaving till 7.50 but he has to arrange for excess baggage. Mark it as top priority and ensure punctuality. He expects you at 6.15sharp.*

Je voudrais commander un taxi pour demain matin de très bonne heure. C'est pour monsieur Durand, 4 rue des Châtaigniers. Il veut être à l'aéroport de Marignane, au départ des lignes intérieures, avant six heures et quart. Son avion décolle seulement à huit heures moins dix mais il a besoin de faire enregistrer un excédent de bagages. Il faudra absolument être à l'heure. Je compte sur vous. Il vous attendra en bas de chez lui à cinq heures et demie précises.

Vorrei prenotare un taxi per domani mattina presto per il dottor Rossi all'Hotel Sheraton. Deve essere all'aeropoorto per le sei e trenta. Puo arrivare per le sei? E' richiesta la massima puntualita'. Puo' lasciarmi la sua sigla per favore?

9.6.1.1 The French Corpus

The French corpus consists of forty passages of 5 sentences – a total of 200 different sentences. The passages are in four groups of 10, each of which is read by ten speakers so that:

- each of the ten speakers read one group of ten passages;
- the first two groups were read by three speakers;
- the second two groups were read by two speakers.

In all, 100 passages (500 sentences) were recorded, totalling 36 minutes 51 seconds of speech. Each sentence was read by either two or three speakers and the duration per speaker ranged from 3 minutes 28 seconds to 4 minutes 37 seconds.

9.6.1.2 The Italian Corpus

The Italian corpus also contains forty passages of five sentences, a total of 200 sentences. The passages are in eight groups of five and were recorded by ten speakers so that:

- each of the ten speakers recorded three groups of five passages;
- the first six groups were recorded by four speakers;
- the last two groups were recorded by three speakers.

In all, 150 passages (750 sentences) were recorded totalling 54 minutes 31 seconds of speech. Each sentence was recorded by either 3 or 4 speakers and the duration per speaker varied from 5 minutes 2 seconds to 7 minutes 11 seconds.

9.6.1.3 Stylisation and Correction

Once the entire corpus had been stylised using the MOMEL algorithm, manual corrections were made by experts using a minimalist strategy, making corrections only when otherwise there was an audible difference between the original recording and a recording produced from the stylised F_0 by means of PSOLA resynthesis (Hamon, Moulines & Charpentier, 1989). Apart from a slight distortion due to the lack of microprosodic effects in the stylised curves, the corrected versions were considered by the experts perceptually identical to the original recordings. Two different methods (qualitative and quantitative) were used to analyse the results (Campione & Véronis, 1998a).

9.6.1.4 Statistical Analysis

We first studied the characteristics of the distribution of target-points as well as the intervals between successive target-points for the French and Italian corpora (see a detailed analysis in Campione & Véronis, 1998b). The corpus analysed provided 6329 target points for French and 9804 for Italian. The fundamental frequency of the target points was converted to semi-tones (STs). The standard deviation varied from 2.79 to 4.18 STs for the French speakers and from 2.78 to 4.44 STs for the Italian speakers. The mean value of the initial targets of the passages for each speaker was very close to the overall mean for all targets for that speaker (mean difference -0.88 STs for French and 0.34 STs for Italian).

The shape of the distribution of target points for each speaker was approximately normal with neither mode nor discontinuity providing an objective criterion to set a threshold distinguishing T and B from the other tones. Similarly, the distribution of the successive pitch intervals (difference in STs between two successive targets) showed neither mode nor discontinuity (apart from that between rising and falling intervals) allowing such a classification.

While linear regressions between successive target points obviously showed no significant autocorrelation for the complete set of targets, a fairly strong correlation was observed when sequences of 2 points were classified as either rising or falling. Quadratic or exponential transformations did not significantly improve the correlation coefficient.

Finally it was observed that the size of the pitch interval was not significantly correlated with the temporal distance between the two corresponding targets.

The six implementations of the model were first evaluated on the French corpus. Subsequently, the best two versions were also tested for Italian.

9.6.2 Method and Result

The curves resynthesised from the six different versions of the automatic coding were compared with the stylised curves corrected by the experts. As has often been noted in the literature, measuring the similarity of two F_0 curves is not an easy task and ideally should make use of perceptual tests which are rarely practical for large corpora. We used the following measures:

- e_{abs} : standard deviation of error (SDE) of the fundamental frequency of the target points (in STs);
- e_{rel} : standard deviation of error (SDE) of the interval between two successive target points (in STs);
- p_{abs} : the percentage of target points less than 2 STs from the original target;
- p_{rel} : the percentage of intervals less than 2 STs from the original interval.

The results are summarised in (Figure 9.7) and (Figure 9.8). For French, the value of e_{abs} is minimal for the version *Levels* (0.96 STs), that of e_{rel} is minimal for the version *Ampli3* (1.09 STs). The percentage p_{abs} is maximal for the version *Levels* (96%) and the percentage p_{rel} is maximal for the version *Ampli3* (93.1%).

Similar results were obtained for the Italian corpus on the two versions *Levels* and *Ampli3* which were the only ones tested (Table 9.2). The fit for the data was globally slightly worse than for the French speakers. This was probably due to a greater variability of extreme values for the Italian speakers, particularly the female speakers (Campione, 1997).

Appendix II gives an example of regeneration of the same fragment using the six versions.

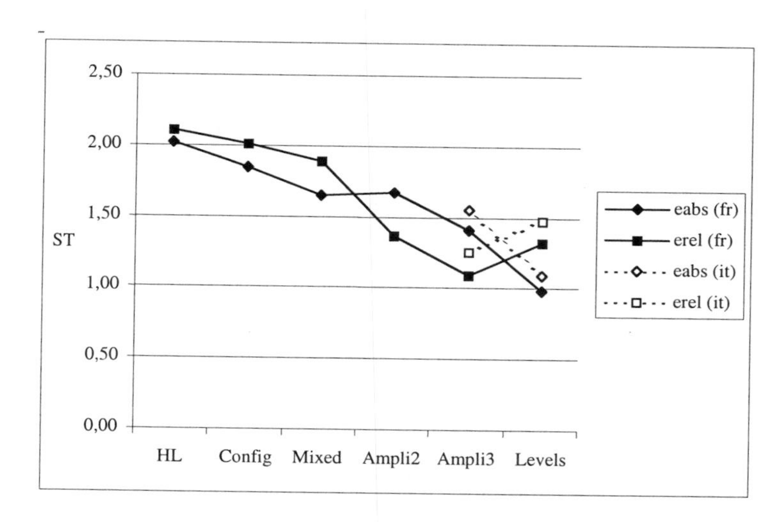

Figure 9.7. Standard deviation of error (SDE).

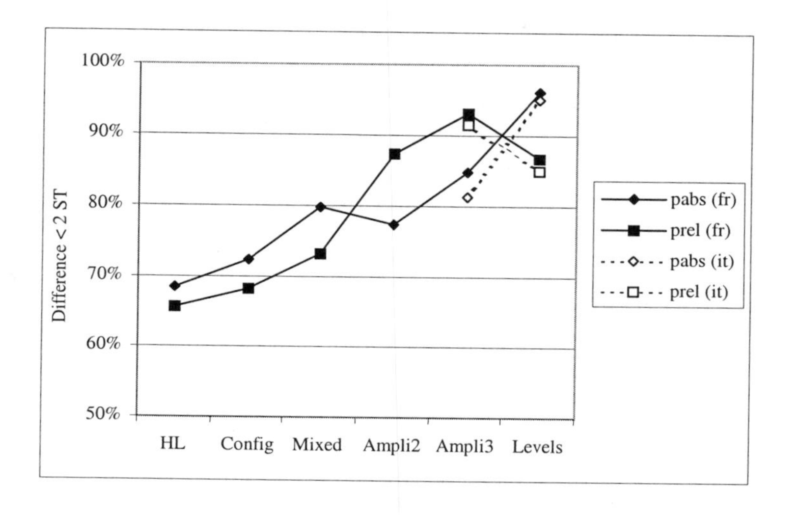

Figure 9.8. Percentage of intervals less then 2 STs from the original data.

Table 9.2. Comparison of result for the best two versions (*Ampli3 and Levels*) as applied to French and Italian.

	Ampli3		***Levels***	
	Fr	*It*	*Fr*	*It*
e_{abs} (ST)	1.41	1.55	0.98	1.09
e_{rel} (ST)	1.09	1.25	1.32	1.47
p_{abs} (%)	84.9	81.4	96.0	95.0
p_{rel} (%)	93.1	91.6	86.8	84.9

9.7 Discussion

As was to be expected, the precision of the coding generally increased with the number of symbols used. We note however the following points:

- The *HL* version is clearly insufficient for any practical use.
- For the same number of symbols, the *Mixed* version was superior to the *Config* version for both absolute and relative measures. In turn the *Ampli2* was superior to the *Mixed* version, but only for the relative measures, the absolute measures being slightly worse. However, it is likely that a better fit on relative movements is more desirable in perceptual terms than a perfect fit of absolute target values. A finer discrimination between these two versions (as well as between *Ampli3* and *Levels*) would need to take into account more results from subjective perception tests in order to determine whether relative intervals or absolute pitch levels are more crucial for the global perception of pitch curves. This result indicates that configuration alone is perhaps not sufficient to define the difference between iterative and non-iterative tones. On the other hand the fact that the *Ampli2* model performed better than the *Mixed* model for relative measures suggests that the distinction between small and larger intervals seems to be relevant not just for peaks and valleys but also for plateaus in rising and falling sequences. Further studies will be necessary in order to determine what linguistic or paralinguistic functions such a distinction embodies.
- The best two versions, *Ampli3* and *Levels*, for the same number of symbols have opposite results in terms of relative and absolute measures. *Ampli3* provides a better fit for relative pitch intervals whereas *Levels* provides a better fit for absolute target values. Similar

results were obtained for both French and Italian which suggests that this difference might be relatively language independent. It should be noted that while both *Ampli3* and *Levels* require six relative symbols for the symbolic coding, *Ampli3* used only six sets of regression coefficients whereas *Levels* used 12 since the band of the preceding target was treated as conditioning an allophonic realisation of the symbol. Thus separate coefficients were calculated for e.g. H_A when preceded by a target in the Acute, Medium or Grave band.

A number of other possibilities for improvement remain to be explored. The scalar factor introduced into the *Ampli2* and *Ampli3* models might be increased arbitrarily until some global criteria for fitting was satisfied. There is also, as we mentioned above, room for improvement in the fitting of the extreme values. It seems, nonetheless that the model as it stands is capable of providing a satisfactory model for the behaviour of F_0 target points in the type of data we have been looking at for French and Italian. Adding a scalar dimension to the symbolic coding provides a means of adapting the model with arbitrary precision to a given speaker's pronunciation of an utterance. This would allow us to derive both a perceptually equivalent coding (Basic INTSINT) and a close-copy coding (Scalar INTSINT) for the same utterance, which might prove to be a valuable distinction for the linguistic and paralinguistic interpretation of utterances. Finally a great deal of work remains to be done concerning the relationship between the "linguistic-like" representations which the algorithms described here allow us to derive automatically and the perceptual and linguistic or paralinguistic interpretation of such representations.

Notes

1. As pointed out by one anonymous reviewer, even sonorant consonants like /m/ may contain some deviations from the macroprosodic curve (often restricted though to one or two periods).
2. Or values very close to zero.
3. One anonymous reviewer has noted that this introduction of scalar values means that the resulting transcriptions are neither phonetic nor phonological but somewhere in between. We agree with this, but feel that the cut-off point between phonological and phonetic representations remains a question for empirical investigation. In our view scalar transcriptions of the type we present here could provide precisely the tool needed for work in this area.
4. Target-points coded S could equally well be assigned a slightly lower value than preceding ones (as in Hirst *et al.*, forthcoming) introducing a declination effect.

Acknowledgements

The authors would like to thank Robert Espesser for his technical assistance, Corine Astésano and Fabienne Courtois for their help with the correction of the MOMEL stylisation and Emmanuel Flachaire for his help with the data analysis. Our thoughts go in particular to the memory of Fabienne Courtois who, tragically, met a fatal accident while this work was in progress.

References

Adriaens, L.M.H. 1991. *Ein Modell Deutscher Intonation: Eine Experimentell-phonetische Untersuchung nach den Perzeptiv Relevanten Grunfrequenzünderrungen in Vorgelesenem Text.* PhD dissertation, Eindhoven University of technology.

Aranoff, M. and I. Sag (eds.). 1984. *Language Sound Structure.* Cambridge, Mass.: MIT Press.

Astésano, C., R. Espesser and D.J. Hirst. 1997. Stylisation automatique de la fréquence fondamentale : une évaluation multilingue. *Actes du 4ème Congrès Français d'Acoustique* (Marseille, France), 441-444.

Beaugendre, F. 1994. *Une Étude Perceptive de l'Intonation du Français.* Thèse d'Etat, Université de Paris XI.

Black, A., and A. Hunt. 1996. Generating F_0 contours from ToBI labels using a linear regression. *Proc. ICSLP* '96, (Philadelphia, USA), 1385-88.

Campione, E. 1997. *Stylisation et Codage Symbolique de l'Intonation.* Mémoire de DEA, Université de Provence.

Campione, E., and J. Véronis. 1998a. A multilingual prosodic database. *Proc. ICSLP* '98 (Sydney, Australia), vol. 7, 31-63.

Campione, E., and J. Véronis. 1998b. A statistical study of pitch target points in five languages. *Proc. ICSLP* '98 (Sydney, Australia), vol. 4, 1391-94

Chan, D., A. Fourcin, D. Gibbon, B. Granström, M. Hucvale, G. Kokkinakis, K. Kvale, L. Lamel, B. Lindberg, A. Moreno, J. Mouropoulos, F. Senia, I. Trancoso, C. Veld and J. Zeiliger. 1995. EUROM1 - A Spoken Language Resource for the EU. *Proc. EUROSPEECH* '95 (Madrid, Spain), vol. 1, 867-870.

Cohen, A., and J. 't Hart. 1965. Perceptual analysis of intonation pattern. *Actes du 5ème Congrès International d'Acoustique* (Liège, Belgium), 1-4.

Cutler, A. and D.R. Ladd (eds.). 1983. *Prosody: Models and Measurements.* Berlin: Springer-Verlag.

D'Alessandro, C. and P. Mertens. 1995. Automatic pitch contour stylisation using a model of tonal perception. *Computer Speech and Language* 9, 257-288.

De Pijper, J.R. 1979. Close-copy stylisation of British English intonation contour. *IPO Annual Progress Report* 14, 66-71.

De Pijper, J.R. 1983. *Modelling British English Intonation.* Dordrecht: Foris.

Di Cristo, A., Ph. Di Cristo and J. Véronis. 1997. A metrical model of rhythm and intonation for French text-to-speech synthesis. *Proc. ESCA Workshop on Intonation* (Athens, Greece), 83-86.

Di Cristo, A. and D.J. Hirst. 1986. Modelling French micromelody: analysis and synthesis. *Phonetica*, 43, 11-30.

Dusterhoff, K. and A. Black. 1997. Generating F_0 Contours for Speech Synthesis Using the Tilt Intonation Theory. *Proc. ESCA Workshop on Intonation* (Athens, Greece), 107-110.

Fujisaki, H. and K. Hirose. 1982. Modelling the dynamic characteristics of voice fundamental frequency with application to analysis and synthesis of intonation. *Proc. 13th ICPhS* (Stockholm, Sweden), 57-70.

Hamon, C., E. Moulines and F. Charpentier. 1989. A diphone system based on time-domain prosodic modifications of speech. *Proc. ICASSP* '89, 238-241.

Hart, J., 't. 1991. F_0 stylisation in speech: straight lines versus parabolas. *J. Acoust, Soc. Am.* 6, 3368-70.

Hart, J., 't, Collier, R. 1975. Integrating different levels of intonation analysis. *Journal of Phonetics* 3, 235-255.

Hart, J., 't, R. Collier and A. Cohen. 1990. *A Perceptual Study of Intonation.* Cambridge University Press.

Hirst, D.J. 1980. Un modèle de production de l'intonation. *Travaux de l'Institut de Phonétique d'Aix* 7, 297-315.

Hirst, D.J. 1983. Structures and categories in prosodic representations. In Cutler and Ladd (eds.), 93-109

Hirst D.J. and A. Di Cristo. 1998a. A survey of intonation systems. In Hirst and Di Cristo (eds.), 1-44.

Hirst D.J. and A. Di Cristo (eds.). 1998b. *Intonation Systems*. Cambridge University Press.

Hirst, D.J., A. Di Cristo and R. Espesser. Forthcoming. Levels of representation and levels of analysis for the description of intonation systems. In Horne (ed.).

Hirst, D.J., and R. Espesser. 1993. Automatic Modelling of Fundamental Frequency using a quadratic spline function. *Travaux de l'Institut de Phonétique d'Aix-en-Provence* 15, 75-85.

Horne, M. (ed.). Forthcoming. Prosody: *Theory and Experiment*. Dordrecht: Kluwer Academic Publishers.

House, D. 1990. *Tonal Perception in Speech*. Lund University Press.

Liberman, M.Y. and J.P. Pierrehumbert. 1984. Intonational invariance under changes in pitch-range and length. In Aranoff and Sag (eds.), 157-233.

Nicolas, P., and D.J. Hirst. 1995. Symbolic coding of higher level characteristics of fundamental frequency curves. *Proc. EUROSPEECH* '95 (Madrid, Spain).

Nolan, F. and E. Grabe. 1997. Can "ToBI" transcribe intonational variations in British English? *Proc. ESCA Workshop on Intonation* (Athens, Greece), 259-262.

Odé, C. 1989. *Russian Intonation: A Perceptual Description*. Amsterdam: Rodopi.

Odé, C. and V.J. van Heuven. 1994. *Experimental Studies of Indonesian Prosody*. Department of Languages and Oceania, University of Leiden.

Öhman, S.E.G. 1967. Word and sentence intonation: A quantitative model. *KTH-QPR* 2, 25-54.

Ostendorf, M.F., P.J. Price and S. Shattuck-Hufnagel. 1995. The Boston University Radio News Corpus. *Technical Report No. ECS-95-001*, Boston University.

Ostendorf, M.F and K. Ross. 1997. A multi-level model for recognition of intonation labels. In Sagisaka, Campbell and Higuchi (eds.), 291-308.

Pierrehumbert, J.B. Forthcoming. Tonal elements and their alignment. In Horne (ed.).

Rossi, M. and M. Chafcouloff. 1972. Les niveaux intonatifs. *Travaux de l'Institut de Phonétique d'Aix* 1,167-176.

Sagisaka, Y., N. Campbell and N. Higuchi (eds.). 1997. *Computing prosody*. New York: Springer-Verlag.

Silverman, K., M.E. Beckman, J. Pitrelli, M.F. Ostendorf, C.W. Wightman, P.J. Price, J.P. Pierrehumbert and J. Hirschberg. 1992. ToBI: a standard for labelling English prosody. *Proc. ICSLP* '92 (Banff, Canada), vol. 2, 867-870.

Strangert, E. and A. Aasa. 1996. Evaluation of Swedish prosody within the MULTEXT-SW project. *Proc. Swedish Phonetics Conference Fonetik* '96 (Stockholm, Sweden), *TMH-QPSR* 2, 37-40.

Taylor, P. 1993. Automatic Recognition of Intonation from F_0 contours using the Rise/Fall /Connection Model. *Proc. EUROSPEECH* '93 (Berlin, Germany), vol. 2, 789-792.

Taylor, P. 1994. The Rise/Fall/Connection Model of Intonation. *Speech Communication* 15, 169-186.

Véronis, J., Ph. Di Cristo, F. Courtois and C. Chaumette. 1998. A stochastic model of intonation for text-to-speech synthesis. *Speech Communication* 26, 233-244.

Véronis, J., Ph. Di Cristo, F. Courtois and B. Lagrue. 1997. A stochastic model of intonation for French text-to-speech synthesis. *Proc. EUROSPEECH* '97 (Rhodes, Greece), vol. 5, 2643-46.

Véronis, J., D.J. Hirst and N. Ide. 1994. NL and speech in the MULTEXT project. *AAAI* '94 Workshop *on Integration of Natural Language and Speech* (Seattle, USA), 72-78.

Wightman, C.W. and M.F. Ostendorf. 1992. Automatic recognition of intonational features. *Proceedings ICASSP '92*, vol. 1, 221-224.

Willems, N., R. Collier and J. De Pijper. 1988. A Synthesis scheme for British English Intonation. *J. Acoust, Soc. Am.* 1250-1260.

Appendix I: The MOMEL Algorithm Description [From Hirst and Espesser, 1993; Hirst *et al.*, forthcoming].

The algorithm comprises four stages. After a preliminary pre-processing of F_0 (stage 1), which eliminates aberrant values after voiceless sections, the central part of the algorithm (stage 2) consists in estimating target-candidates by a technique called *asymmetrical modal quadratic regression*. This stage works on the assumption that all relevant microprosodic effects consist in a lowering of the values of the underlying macroprosodic curve. The modal regression is applied within a moving window providing an optimal target for the F_0 values within the window. The next stage (stage 2) partitions the candidate targets and the final stage (stage 3) reduces the targets of each partition to a single candidate.

Stage 1: *Pre-processing of* F_0. All values more than a given ratio (typically 5%) higher than both their immediate neighbours are set to 0. Since unvoiced zones are coded as zero, this pre-processing has essentially the effect of eliminating one or two values (which are often dubious) at the onset of voicing.

Stage2: *Estimation of target-candidates.* This consists in three steps which are followed iteratively for each instant *x*.

- Within an analysis window of length A (typically 300ms) centred on *x*, values of F_0, (including values for unvoiced zones) are neutralised if they are outside of a range defined by two thresholds *hzmin* and *hzmax* and are subsequently treated as missing values. The threshold *hzmin* is a constant set to 50 Hz and the adaptive threshold *hzmax* is set to the mean of the top 5% of the F_0 values of the sequence multiplied by 1.3.
- A quadratic regression is applied within the window to all non-neutralised values. All values of F_0 which are more than a distance Δ below the value of F_0 estimated by the regression are neutralised (typical value of Δ fixed at 5%). This step is re-iterated until no new values are neutralised.
- For each instant *x*, a target point <*t,h*> is calculated from the regression equation:

  $y = a + bx + cx^2$

  where $t = - b/(2c)$ and $h = a + bt + ct^2$
- This target point corresponds to the extremum (maximum or minimum) of the corresponding parabola (Figure 9.9).

If *h* is less than *hzmin* or greater than *hzmax*, then *t* and *h* are treated as missing values.

Steps b) and c) are repeated for each instant *x*, resulting in one estimated target point <*t, h*> (or a missing value) for each original value of F_0 as in Figure 9.10 where the grey lines connect the value of *x* to the value of <*t, h*> for each window.

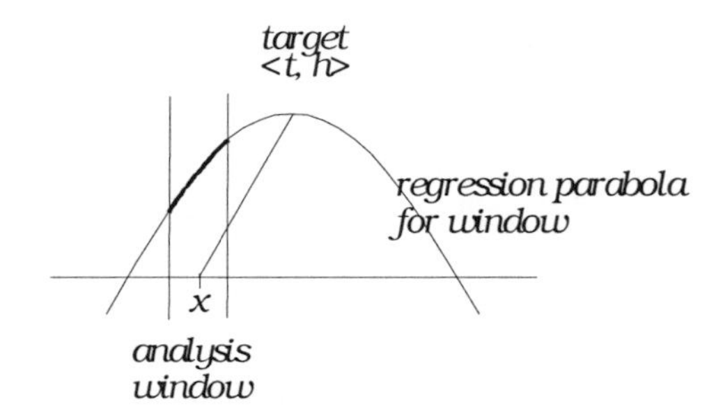

Figure 9.9. Calculation of a local target point.

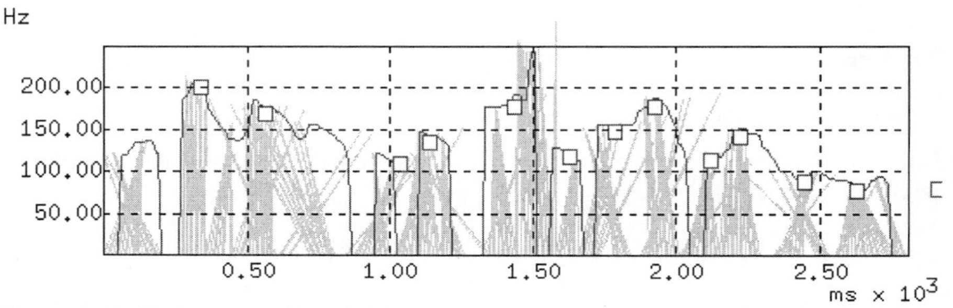

Figure 9.10. Estimation of candidate target point (grey lines) and final targets (white squares). The grey lines connect the centre of the moving window to the extremum of the parabola estimated for that window.

Stage 3: *Partitioning of target candidates.* The sequence of target candidates is partitioned by means of another moving window R (typically 200 ms) which is divided into two halves, left and right. The partition algorithm seeks values where there is a maximum difference between the targets in the left and right halves of the window. Specifically, a partition boundary is inserted when the difference between the average weighted values of t and h in the left and right halves of the window corresponds to a local maximum which is greater than a threshold (set to the mean distance between left and right halves for all windows).

Stage 4: *Reduction of candidates.* Within each segment of the partition, outlying candidates more than one standard deviation from the mean values for the segment) are eliminated. The mean value of the remaining targets in each segment is then calculated as the final estimate of t and h for that segment (Figure 9.11).

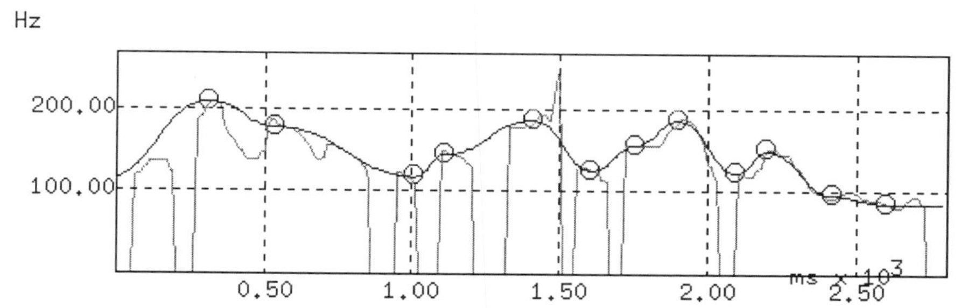

Figure 9.11. F_0 and quadratic spline curve obtained from MOMEL algorithm.

Estimation of parameters and evaluation

The three parameters used by the algorithm (the analysis window [*A*], the distance threshold [*D*] and the reduction window [*R*]), were estimated from a small corpus consisting of two sentences containing all the stops and fricatives of French and pronounced by 10 subjects (5 male and 5 female). For the different values of the parameters a subjective evaluation was carried out consisting of a visual and auditory comparison between the original signal and a signal generated by PSOLA resynthesis from the stylised curve. This was completed by an objective analysis consisting of a mean distance between the original and the stylised curve. The optimal values were as follows: $A = 300$ ms ; $D = 5\%$; $R = 200$ ms.

These optimised parameters were subsequently used for the stylisation of other corpora (Hirst and Espesser, 1993; Hirst *et al.*, forthcoming). The results showed that the percentage error (missing or erroneous targets), while slightly higher than for the first corpus was at a relatively reasonable level of 5%. The errors were moreover systematically of two or three different types, in particular missing targets in transitions from voiced to voiceless segments of speech, which suggests that an improved algorithm could probably eliminate the majority of them.

The stylisation technique has also been applied to a number of other languages (English, German, Arabic, Spanish, French, Italian, Swedish) in the framework of the European MULTEXT project (Véronis, Hirst & Ide, 1994), with comparable results. Astésano, Espesser and Hirst (1997) describe an evaluation of the algorithm for five languages (English, French, German, Spanish, Swedish) for a duration of 3 hours 45 minutes of speech (see also Strangert & Aasa, 1996 for Swedish). It was also evaluated in depth on French and Italian by Campione (1997). All these studies conclude that MOMEL is an efficient and robust technique for the representation of relevant information of F_0 curves, at least for the languages studied so far with a fairly low error rate of around 5%.

Appendix II. Examples of regeneration with the various versions.

Figure 9.12 shows the stylised curve regenerated with the six different versions of the model, along with the original (dotted line). The fragment is “Pourriez-vous m'indiquer à quelle heure est la correspondance à Valence? Si je dois partir avant midi de Marseille, j'aimerais savoir s'il y aura un wagon restaurant.” (female speaker). (Could you tell me at what time there is a connection at Valence? If I have to leave Marseilles before midday I’d like to know if there is a dining-car).

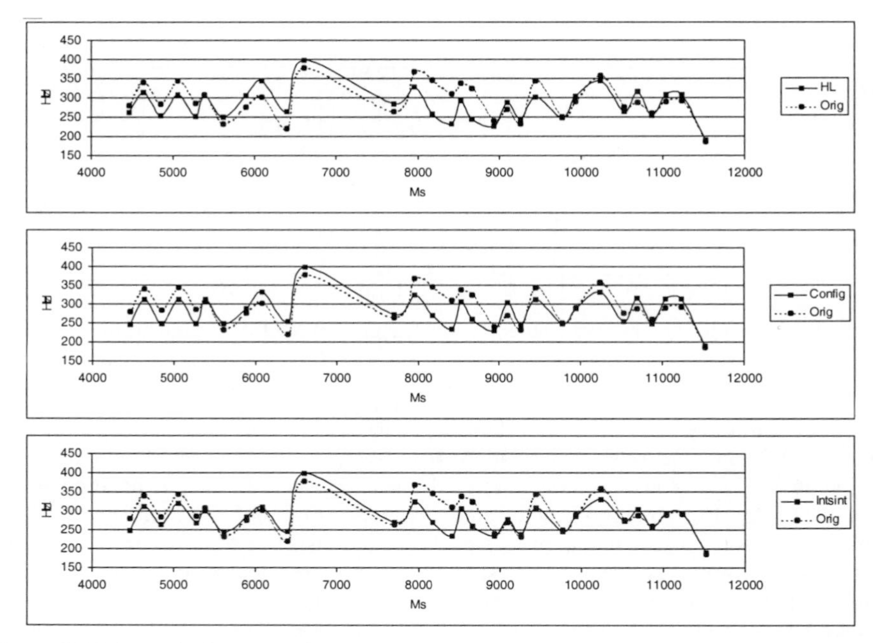

Figure 9.12 (part a). Regeneration with models *HL, Config, Mixed.*

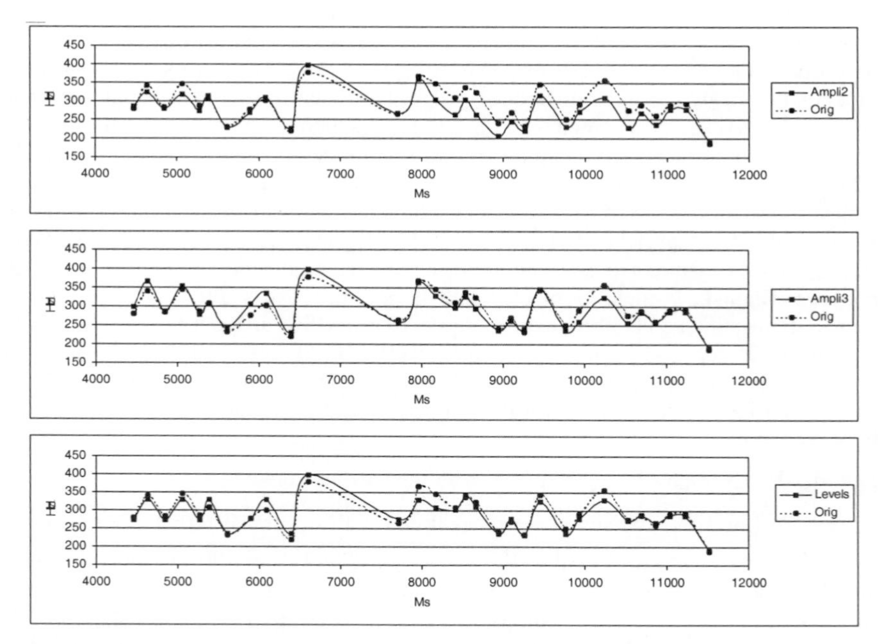

Figure 9.12 (part b). Regeneration with models *Ampli2, Ampli3* and *Levels*.

10

A Phonological Model of French Intonation

SUN-AH JUN AND CÉCILE FOUGERON

10.1 Introduction

French intonation has been characterized as having a sequence of rising pitch movements. Recently, this intonation pattern has been phonologically analysed by several researchers such as Hirst and Di Cristo (1984, 1996), Mertens (1987, 1993), Di Cristo and Hirst (1993a, 1993b, 1996), Post (1993), Jun and Fougeron (1995), and Hirst, Di Cristo, and Espesser (forthcoming), among others. These authors agree that a tone is associated with a stressed syllable, and that stress is rhythmic or postlexical. They also agree that an utterance is hierarchically organized into different prosodic levels, though some of these levels are not referred to in the same terminology. These models diverge principally in the levels of phrasing and their tonal representation. They also disagree in the notion of accent in French and in the degree of abstractness in the tonal representation, which is due to the conceptual differences linked to the application of the model: the focus is either on acoustic representation of models relevant for speech synthesis and recognition, or on abstract representation for phonological description, or on both levels.

In this paper, we propose a development of our previous model (Jun & Fougeron, 1995; Fougeron & Jun, 1998) based on the tonal pattern of focus in both declarative and interrogative sentences, and other syntactic/semantic constructions and intonational clichés. The organization of this paper is as follows. In section 2, we will elaborate our previous model: the underlying tonal pattern and the surface realizations of two prosodic units, an Accentual Phrase and an Intonation Phrase. We will compare our Accentual Phrase with the existing prosodic units proposed by

A. Botinis (ed.), Intonation, 209-242.

other intonational, especially phonological, models. In section 3, we will report on a phonetic experiment on focus intonation. We will suggest an analysis of how to represent the tonal shape of the focused phrase with our phonological model, introducing a third prosodic unit, an Intermediate Phrase. In Section 4, we will support the analysis based on other syntactic/semantic constructions and intonational clichés, and discuss an alternative analysis. All data discussed in this paper are from Parisian French.

10.2 Description of Jun & Fougeron's Early Model

Following the framework developed in Pierrehumbert (1980), Beckman and Pierrehumbert (1986), and Pierrehumbert and Beckman (1988), we assume that an intonational tune is composed of a sequence of underlying H and L tones, and that each tone is linked to a syllable which is either metrically strong or marks the boundary of a prosodic unit. It is assumed that not all syllables are specified as having a tone, and tonally unspecified syllables get their surface tone by interpolating between two adjacent tonal targets. We also assume that an intonationally defined prosodic unit is hierarchically organized, and obeys the Strict Layer Hypothesis (Selkirk, 1984, 1986; Nespor & Vogel, 1986). That is, a prosodic unit of a given level of the hierarchy is composed of one or more units of the immediately lower prosodic unit, and each unit is exhaustively contained in the superordinate unit of which it is a part. Under these assumptions, we proposed in our earlier paper (Jun & Fougeron, 1995) two intonational units in French: an Accentual Phrase and an Intonation Phrase. In the following subsections, we will describe the tonal patterns of each phrase and compare our Accentual Phrase with similar units proposed by others.

10.2.1 The Accentual Phrase

10.2.1.1 The Underlying Tonal Pattern

The Accentual Phrase (henceforth, AP) is the lowest tonal unit in French, and it can contain more than one word. The AP is the domain of primary and secondary stress, and has an underlying tonal pattern of /LHLH/. The final rise (LH) of AP has a demarcative function, similar to the AP in Korean (Jun, 1993, 1998). In order to reflect the difference between the two H tones of the AP (initial H and final H), we represent the underlying tonal pattern of AP as /LHiLH*/ (Fougeron & Jun, 1998).

The phrase final H tone (=H*) is a demarcative pitch accent associated with the primary stressed syllable of an AP, which is the phrase final full vowel ("accent primaire" or "accent final"). We assume that French has accentable syllables (i.e., the final full syllable of lexical words) and that this pitch accent is realized on one of these syllables. In other words, every word final full vowel can be the target of the primary accent (or pitch accent), but not every word final full vowel is realized with an accent. This view of accent being the property of the phrase or the group, not of the word, has been proposed by many previous researchers (e.g., Grammont, 1914; Delattre, 1939; Marouzeau 1956 (cited in Di Cristo, 1998)). Since an AP is delimited by this phrasal accent (H*), AP is quite similar to Delattre's "groupe accentué", "groupe rythmique", or "groupe de sens". The diacritic '*' is used to show that the tone is associated to a stressed syllable.

The phrase initial H tone (Hi) (also known as "accent initial", "accent secondaire", or "ictus mélodique") is optional in an AP. Its occurrence is influenced by several factors, mostly rhythm, style, and speaker[1]. The location of the Hi is also variable. It is realized on the initial stressed syllable of the first lexical word within an AP. But in surface it is generally said to fall on one of the first two syllables of the first lexical word within a phrase (Fónagy,1980; Lucci, 1983; Pasdeloup, 1990; Vaissière, 1974, 1997). To investigate the location of initial accent within a word, we ran an experiment where we varied the number of syllables within a word from two to eight (seven or eight 2~7-syllable words and three 8-syllable words): e.g. *médite, méditer, méditation, méditérranée, méditérranéen, méditérranéiser, méditérranéisation*; *invite, inviter, individu, individualisme, individualité, individualisation.* The target word (total 46 words) was embedded in the subject position of a carrier sentence, "X est un mot utilisé par les français" ('X is a word used by French people'), where the secondary (initial) stress is most likely to occur (Pasdeloup, 1990). Each sentence was repeated four times by three speakers. Results are shown in Figure 10.1. The X-axis refers to the number of syllables in a word and the Y-axis refers to the frequency of initial peak occurrences in percent. Data show that the initial peak, Hi, was realized either on the first syllable (when the word has two or three syllables) or on the second syllable (when the word has more than three syllables[2]). For a few tokens, an initial peak was realized over the first two or three syllables as a plateau (e. g., 'plateau s1-s2' in Figure 10.1 refers to the cases where the initial peak covers the first and second syllables).

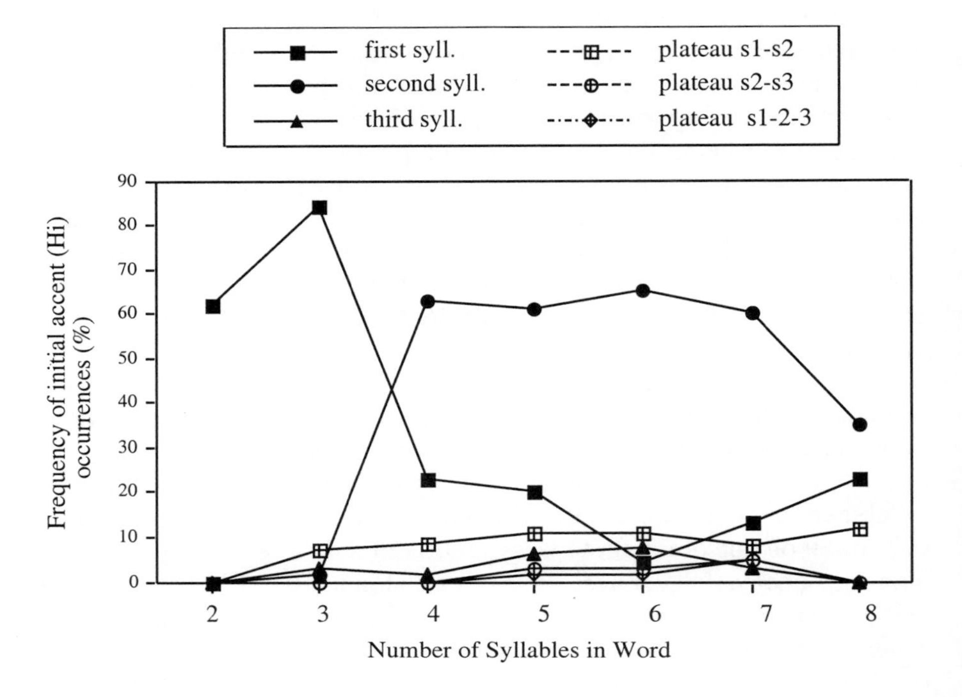

Figure 10.1. Location of initial accent (Hi) within a 2~8 syllable-word which forms an AP in sentence initial position: the frequency (in %) of realization of Hi on the 1st, 2nd, 3rd syllable, or as a plateau high realized over the initial two or three syllables (e.g., "plateau s1-s2": Hi over the 1st and 2nd syllables).

In a different corpus including several examples of an AP beginning with a function word (e.g., in the story "La bise et le soleil" 'The North Wind and the Sun'), we noticed that the realization of Hi is sensitive to the presence of a function word. When an AP begins with one or more function words, especially when the function words are monosyllables, Hi tends to be realized after all the function words[3]. For example, Hi was placed on *faire* in *à le lui faire ôter* 'in making him take off (his coat)' or on *bout* in *et au bout d'un moment* 'and after a moment'. We also noticed that in cases where the function word is a disyllable (e.g., *était*) or where there are more than one function word, Hi is sometimes realized on a function word. For example, Hi was placed on *était* in *qu'il était le plus fort* 'that he was the stronger' or on *sont* in *ils sont tombés d'accord* 'They made an agreement'.

The sensitivity of initial accent to lexical information in French has also been mentioned by others. Hirst and Di Cristo (1996) shows that the initial accent is linked to the beginning of the lexical word. They found a positive, though small, correlation (r^2=.26) between the number of unaccented syllables (from function words) and the syllable location of the initial peak. Vaissière (1997) claims that the initial rise is localized on the first or second syllable of the first lexical word, but when the initial syllable begins with a vowel and if there is no liaison consonant, the high target (i.e. initial accent) is delayed. In this respect, French is different from Korean, in which the underlying tonal pattern of an AP is LHLH, but the initial H is realized on the second syllable of the phrase regardless of the lexical information of the word (Jun, 1993, 1998). In our model, we temporarily assume that Hi in French is associated with the first syllable of the AP-initial lexical word, but it is realized on the preceding (if there is a function word) or the following syllables depending on the segmental context or many other factors mentioned above. Further study is needed to model the precise location of Hi and its phonetic realization rules.

In theory, Hi and H* should be considered to be pitch accents since they are associated with stressed syllables. However, we assume that Hi is not a pitch accent for two reasons. First, it is not always realized on a stressed syllable with a full vowel. Second, the Hi-toned syllable is not always significantly longer than non AP-final syllables (see Figure 10.5). In neutrally produced utterances, Hi is generally weaker than H* in its duration and pitch. The different status of these two tones is also implied in Beckman (1983). She notes that French is closer to English or Swedish than Japanese in that "the (normally final) accented syllable of each 'rhythm group' is longer in duration than the syllables preceding it." In producing a focused word, however, we found that Hi is often promoted to a pitch accent (See section 3 for more detail).

Finally, the AP-initial L tone in /LHiLH*/ is realized on the syllable before Hi, or as a low plateau if there are two or more syllables. When Hi is realized on the first syllable of an AP, however, the L is not always realized. The second L tone in the AP tends to be realized on the syllable preceding the H*-toned syllable, thus on the penultimate syllable if the phrase final syllable is full. However, as the number of syllables in an AP decreases, the L tends to occur in the final syllable, i. e., the same syllable as the H*-toned syllable. Figure 10.2 shows the position of the second L tone in a one-word AP, which is located in the subject position of a sentence. This data is from the same experiment investigating the location of initial H, as shown in Figure 10.1 above.

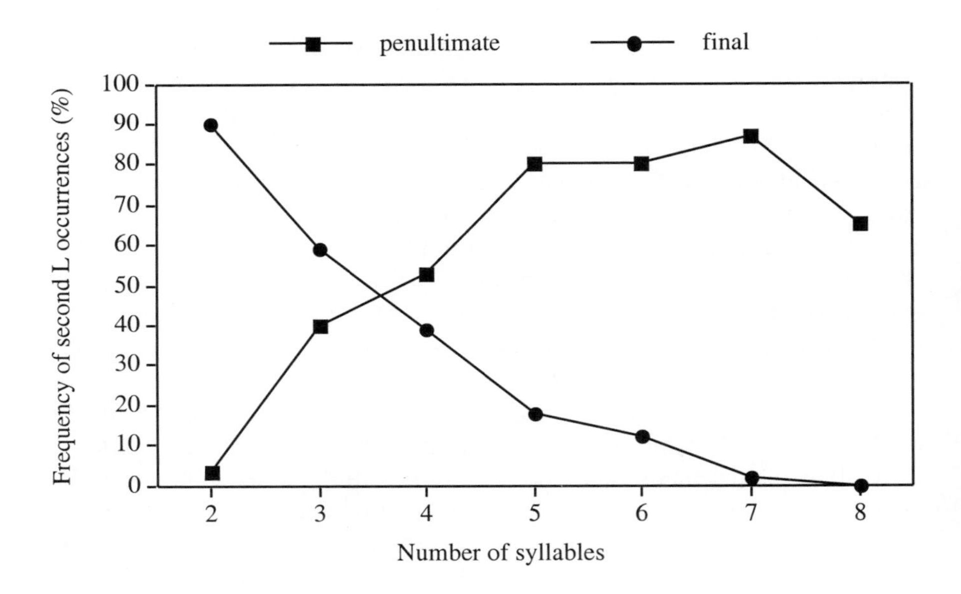

Figure 10.2. Position of second L in one-word AP located in sentence initial position. The location of L in each word is shown relative to the number of syllables in the word.

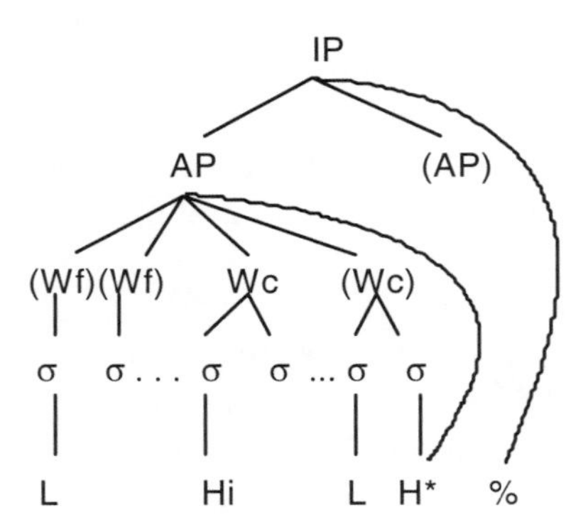

Figure 10.3. Schematics of underlying tones in AP (LHiLH*) and their affiliation.

In sum, when there are four syllables in an AP, each tone can be realized on its own syllable. But when there are more than four syllables in an AP, the first two tones are aligned with the first two syllables of the AP and the second two tones with the last two syllables of the AP (assuming that the AP begins with at most one monosyllabic function word, and ends with a full vowel). The syllable(s) between Hi and the following L are not specified with underlying tones. Their surface tonal values are determined by interpolation between Hi and the following L target. A schematic representation of the underlying tones in AP (LHiLH*) and their affiliation to a syllable or to a prosodic unit is shown in Figure 10.3. Next section describes various realizations of the underlying tones in AP.

10.2.1.2 Realization of AP

By assuming the underlying four tones associated with certain syllables within an AP and phonetic implementation rules (Beckman & Pierrehumbert, 1986; Pierrehumbert & Beckman, 1988) we can explain various realizations of French phrasal intonation.

The four underlying tones of an AP tend to be realized on the surface if the AP has more than three syllables. Example sentences parsed with AP boundaries, marked with { }, are shown in (1). Examples in (1a, 1b, 1c) show that the four underlying tones are all realized when there are at least four syllables in an AP with the LHi being realized at the initial two syllables and the LH* being realized at the final two syllables. When an AP has fewer than four syllables, all four underlying tones can still be realized if some of the syllables are lengthened. But without extra lengthening, either Hi, or the following L, or both may not surface. Examples in (1d) and (1e) illustrate the case when both Hi and L are not realized.

(1) Example sentences marked with AP boundaries (only the first AP is tonally specified)

(a) Européen est un mot utilisé par les français 'European is a word used
{L Hi L H*} { }{ } { } by French people'

(b) Le mauvais garçon ment à sa mère 'The bad boy lies to his mother'
{L Hi L H*} { }

(c) Le désagréable garçon ment à sa mère 'The unpleasant boy lies to his mother'
{L Hi L H*} { }

(d) Marie mangera des bananes 'Marie will eat some bananas'
{LH*} { } { }

(e) Le garçon coléreux ment à sa mère 'The irascible boy lies to his mother'
{L H*} { } { }

In our data, we have observed five surface tonal patterns of AP, as schematised in Table 10.1. The variability in surface realization of AP can be explained by assuming tonal undershoot linked to temporal constraints (cf. Lindblom, 1963, 1964). The tone in a parenthesis in Table 10.1 refers to the tones not realized due to undershoot. That is, when there are one or two syllables in an AP, the AP surfaces as [LH*] with Hi and the following L being undershot. When there are three syllables in an AP, the AP has four possible surface patterns: a. [LH*], b. [LLH*] when Hi is undershot, c. [LHiH*] when second L is undershot, and d. [HiLH*] when the initial L is undershot. Among these, the [LHiH*] pattern was the least frequent. The [HiLH*] pattern is not limited to a three-syllable AP, but is also observed in an AP of 4 or more syllables. This pattern is therefore not triggered by the number of syllables, but by the lexical components of an AP: the AP initial L may not be realized when the AP begins with a content word and Hi is realized on the first syllable of the word. Finally, an AP can have a [LHiL*] pattern (or [LHiLL*] when the AP has enough syllables to carry four tones) when the AP has a Hi tone and is followed by another AP which begins with Hi. In this case, the AP final H* is sometimes realized as an L* due to a constraint on tone sequences to avoid two or three consecutive H tones: *HH(H). I.e., H* -> L*/___Hi or /Hi___Hi. This phenomena of tone clash or tone repulsion is also observed in Korean (Jun, 1996) and other tone languages such as Shona (Stevick, 1965, cited in Kenstowicz, 1994) and Miya (Schuh, 1998).

Table 10.1. Five types of surface realizations of AP (/LHiLH*/) when not all four underlying tones are realized. The tone(s) in a parentheses refers to the tone(s) not realized due to undershoot.

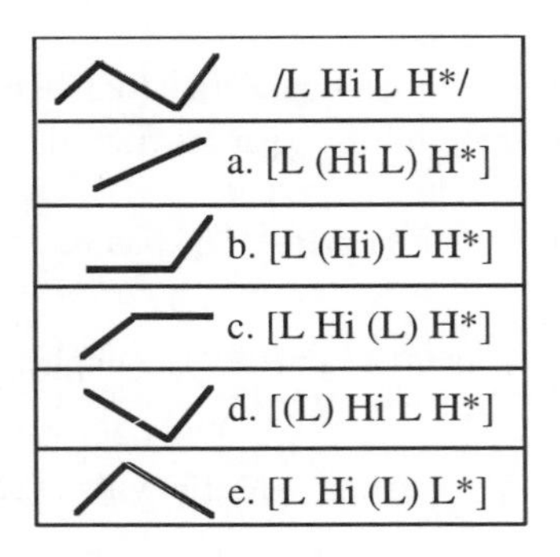

	/L Hi L H*/
	a. [L (Hi L) H*]
	b. [L (Hi) L H*]
	c. [L Hi (L) H*]
	d. [(L) Hi L H*]
	e. [L Hi (L) L*]

The last pattern [LHiL] is the most common pattern when an AP is final in an Intonation Phrase whose boundary tone is L%. Since the AP final syllable is also the Intonation Phrase final syllable, the AP final H* and Intonation Phrase final L% are supposed to be realized on the same syllable. In this case, the AP final H* is preempted by the higher level (Intonation Phrase) boundary tone.

In addition to the tonal shape, an AP is also marked by a lengthening of the H*-toned syllable. This syllable is significantly longer than non AP-final syllables (see Figure 10.5).

Finally, since an AP is delimited by a word final full vowel, in theory, every word-final full vowel could be the final syllable of an AP. But depending on many factors such as speech rate, length of the phrase, or syntactic and semantic constraints (Lucci, 1983; Pasdeloup, 1990; Delais-Roussarie, 1995; Fougeron & Jun, 1997), not every word final syllable becomes an AP-final syllable. For example, an AP tends to contain more words at fast speech than slow speech, and can contain more words when each word is short. In our data, we found that an AP contains on average 2.3~2.6 words (or 1.2 content words), and 3.5-3.9 syllables. This data is based on the story "La Bise et le Soleil" ('The North Wind and the Sun') read by three speakers at normal rate three times. The number of syllables in an AP is similar to that of Fónagy's (1980) "arc accentuel": 3.36 syllables in spontaneous conversation. The size of French AP is also very similar to the size of Korean AP which contains in average 1.2 content words and 3.2 syllables, based on the same story (Korean version) read by three speakers of Seoul Korean at normal rate four times.

10.2.1.3 Comparison of AP with Other Units

Due to undershoot of a certain underlying tones, our AP can contain one or two rising movements. In Di Cristo and Hirst's model (Hirst & Di Cristo, 1984, 1996; Di Cristo & Hirst, 1993a, 1996; Di Cristo, 1998) each rising movement has been analysed as one prosodic unit, a Tonal Unit. In their model, Tonal Unit (TU) and Intonation Unit (IU) are the only prosodic units defined by a tone. The TU is the lowest tonal unit and has a LH tonal pattern. The IU is the higher unit and has a phrase final boundary tone, L or H, and a phrase initial boundary tone, L.

When a lexical word is monosyllabic, an AP can be equivalent to one TU, as shown in (2a) (The examples 2a and 2c are modified examples from Hirst & Di Cristo, 1984, and the example 2b is from Di Cristo, 1998). This is because an AP is often realised as one rising tone when there are fewer than four syllables in the AP, or when the initial H is not realized. In other cases, our AP can be equivalent to two TUs as shown in (2b), since their TU delimits either the end of initial accent (= our Hi) or the final accent (= our H*). However, an AP is not a combination of any two TUs. As shown in (2c), the first AP matches the first TU, but the second AP matches the following two TUs. It is not possible to form an AP by combining the first two TUs because an AP boundary should match a Word boundary (observing the Strict Layer Hypothesis).

(2) a. Le CHIEN de mon FRÈRE 'my brother's dog'

```
(           )(                ): TU
{           }{                }: AP
```

b. Un FAbriQUANT de MAtéRIAUX de CONStrucTION 'a maker of building materials'

```
(      )(        ) (    )(      ) (       )(       ): TU
{                }{             }{                 }: AP
```

c. Le CHIEN du SEcréTAIRE de mon FRÈRE 'my brother's secretary's dog'

```
(          ) (    ) (        ) (            ): TU
{          }{                }{             }: AP
{                 }{         }{             }: not possible AP
```

It is argued in Hirst and Di Cristo (1996) that the Strict Layer Hypothesis (=SLH) is too demanding in French if one includes the Syllable as a phonological category, since a syllable boundary does not always match a Word boundary (e.g., in liaison or enchaînement). They further add that English Foot formation does not always observe Word boundaries (cf. Abercrombie, 1964), either. The fact that SLH is not strictly observed at levels below the Word has already been noted in Selkirk (1986:385). She limited her theory of the syntax-phonology relationship only to those constituents above the level of the Foot. Since an AP is higher than the Word level, we believe it observes SLH. If French TU is equated with the stress group, Foot, violating SLH would not be a problem for their model. But as mentioned earlier, a TU is not based on stress but tone (i.e. intonation), so its domain changes depending on how a phrase is uttered. A TU can include more than one word when each word is monosyllabic and semantically related to other words (Hirst & Di Cristo, 1984). Apart from the theoretical issue, however, a question would remain whether placing a prosodic boundary in the middle of a word agrees with native speakers' intuition.

A stronger reason for us to believe that AP is the lowest tonal unit comes from the fact that not every TU is the same in phonetic realization and in the degree of boundary level. A tonal transition between two TUs is not always the same, and some TU final syllables have more lengthening, thus providing a stronger sense of boundary, than others. Di Cristo and Hirst (1993a) capture the temporal difference by proposing a Rhythmic Unit (UR) (or the Prosodic Word in Di Cristo, 1998). But since their UR is not a tonal unit, but a rhythmic unit based on stress, it does not always correspond to our AP. Figure 10.4a illustrates a schematic F_0 contour of two falling movements, where the first fall occurs within an AP (indicated by '1') and the second fall occurs across an AP boundary (indicated by '2'). Figure 10.4b shows the duration of '1' (left panel, based on 374 data points) and '2' (right panel, based on 146 data points) against the number of syllables in the first and second AP, respectively. As shown in the figure, the duration of the fall is dependent on the number of syllables in the AP when it is AP internal, i.e. when the two surrounding rises belong to the same AP (fall '1'). In this case, the duration increases as the number of syllables increases. On the contrary, the duration of the fall is fairly constant, about 100-200 ms., when the two rises belong to different APs (fall '2'). Furthermore, the final syllable of the first rising (=Hi) is not consistently lengthened compared to non-final syllables, but that of the second rising (=H*) is significantly longer than other syllables (see Figure 10.5) for all cases. The tonal cohesion between the two LH rises and the different degrees of finality between these two would not be explained if we assume that each rising movement is a separate prosodic unit.

Our AP is similar to the "arc accentuel" of Fónagy (1980), or "mot phonologique" of Milner and Regnault (1987) in that it includes the beginning and the end of lexical or syntactic units (ex. ZEro HEUre, VINGT quatre HEUre, les ILes britannIQUE, la MAjeur partIE) (see Hirst & Di Cristo, 1996, for a review of our model[4]), and is similar to Rossi's (1985) "Intonème Mineur", or Vaissière's "prosodic word" (1974, 1992) or "Syntagme prosodique" (1997) in that these units are delimited by the final stress. Our AP is also similar to Mertens' Intonation Group (IG) (Mertens, 1987, 1993) since both units are the lowest prosodic unit based on tone, and include the final accent as well as the initial accent. The difference between our AP and Mertens' IG is in the number of underlying tones and the degree of abstractness: we assume four underlying tones with the initial and penultimate tones being L and the second and the final tones being H; Mertens assumes only one underlying tone (AF) and an optional tone after the final stressed syllable (IG=((NA)AI) (NA) AF (NA)), where NA is a non-accented syllable, AI is an initial accented syllable, and AF is a final accented syllable). Furthermore, each tone in Mertens' IG, if realized, can be L or H with four levels of height. His model is closer to the phonetic representation than ours, and has been implemented as the input to speech synthesis and recognition.

a.

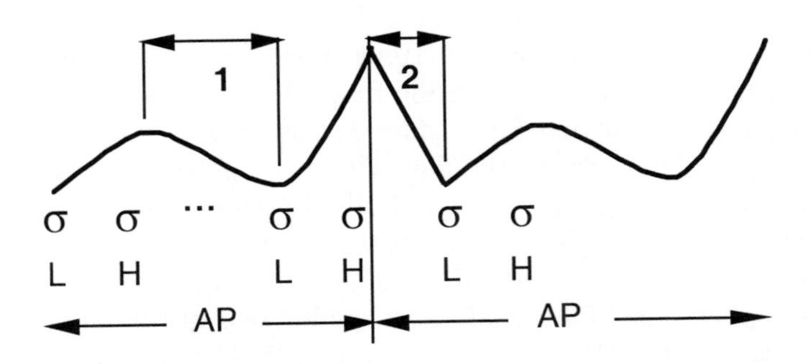

b.

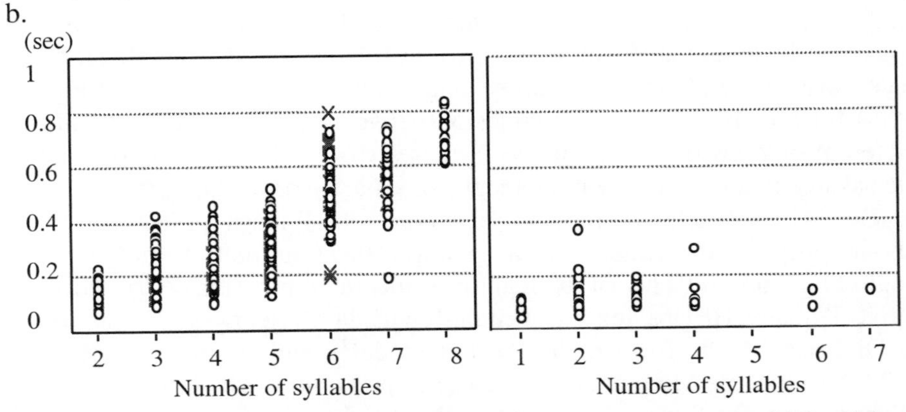

Figure 10.4. a) Schematics of F_0 contour of two APs. b) Left: duration of fall within an AP ('1' in (a)) against the number of syllables in the AP. Data from one-word APs are marked as 'o', and more-than-one-word APs are marked as 'x'; Right: duration of fall across two APs ('2' in (a)) against the number of syllables in the following AP.

10.2.2 Intonation Phrase

An Intonation Phrase (henceforth, IP) is a prosodic level higher than an AP, and the highest level in this model. This level in French roughly corresponds to the "Intonème Majeur" (Rossi, 1985, 1993), the "Unité Intonative" (Hirst & Di Cristo, 1984; Di Cristo, 1998), and the "Breath Group" (Vaissière, 1997).

The IP is marked by a major continuation rise or a major final fall. Following the notation developed for English by Pierrehumbert (1980), we will note the right tonal boundary of the IP by a H% or a L% tone, which is realized on the last syllable of the IP. This level is also marked by a large final lengthening and is optionally followed by a pause (e.g., Hirst & Di Cristo, 1984; Pasdeloup, 1990; Fougeron & Jun, 1998). Figure 10.5 shows the duration of IP final syllables, AP final syllables, Hi-toned syllables and unaccented syllables. The data is from "La Bise et le Soleil", produced by 3 speakers. The IP final syllable is significantly longer than the AP final syllable, and the AP final syllable is significantly longer than unaccented syllables. Here, unaccented syllables refer to all syllables other than the Hi-syllable and H* syllable. A lexical word final syllable is not lengthened if it is not AP-final. The duration of Hi is not consistently different from either H* or unaccented syllables. Some speakers show no difference between H* and Hi, while others show no difference between Hi and unaccented syllables. Thus, both prosodic levels, AP and IP, are cued by the duration of their final syllable. Finally, it was found that in the story, "La Bise et le Soleil", an IP contains an average of 2.0~2.7 APs and 7.3~10.1 syllables, averaged across speakers.

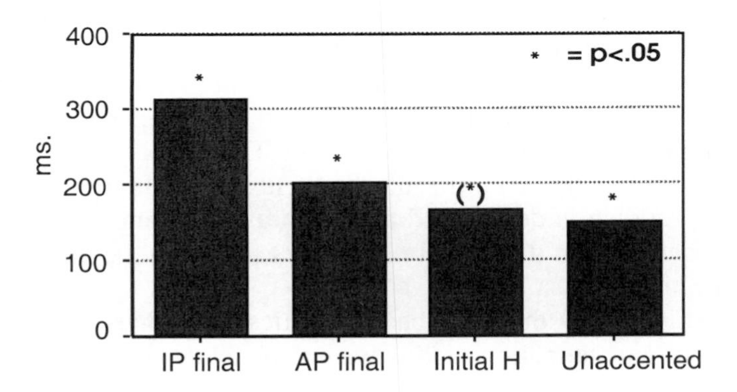

Figure 10. 5. Duration of syllable in different prosodic positions. See text above for explanations.

10.3 Experiment: Focus realisation

10.3.1 Background and Procedure

Narrow focus in French has generally been described as a large, sharp rise and fall of pitch contour (e.g. Rossi, 1985; Touati, 1987; Di Cristo & Hirst, 1993b; Clech-Darbon, Rebuschi & Rialland, 1997; Di Cristo, 1998). The location of the peak has been described to be realized on the initial accented syllable (Hi), or on the final accented syllable (H*), or on both, and more often on the initial syllable than the final syllable. Di Cristo (1998) notes that the timing of the peak on a focused item depends on the objective or expressive character of the focus. For an objective contrastive focus (e.g. *Le professeur a la clé* (pas l'étudiant) 'The teacher has the key (not the student)', the peak can be aligned with any syllable of the focused item except the last (i.e. *pro*, or *fe*, but not on *sseur*). In the case of an expressive contrastive focus, the peak occurs on the final stressed syllable of the focused item. He also notes that the fall from the peak is always associated with the rightmost full syllable of the item in contrast.

The sequence after focus has generally been described as "flat", "deaccented" or "dephrased", with a reduced register (Touati, 1987; Di Cristo, 1998; Clech-Darbon *et al.*, 1997). In declaratives, focus is followed by a low plateau until the end of the phrase, while in interrogatives focus is followed by a reduced copy of the pre-focus sequence (low or high plateau followed by a sharp rise at the end). Touati (1987) describes that in the sequence after focus, pitch accents are deleted, and in the sequence before focus, the amplitude of pitch accents is reduced. Since post-focus sequence has no tonal variation, the sequence is often perceived as being shorter than that of a neutral sentence. This kind of flat tonal pattern is also found in English after early focus in a phrase. Erickson and Lehiste (1995) showed that the duration of post-focus words in English is indeed reduced compared to that in the neutral sentence. This suggests that a post-focus sequence in English is deaccented and dephrased, meaning that the pitch accents are deleted and durational differences linked to the pitch accent or phrasal tone are also lost.

To investigate how the tonal characteristics of a focused sentence are represented in our intonation model, we examined the location of rise and fall for a focused item in a sentence, and the tonal pattern after and before focus, both in declaratives and interrogatives. Interrogatives are based on an incredulous echo-question, so that the order of words is exactly the same as that of declaratives. We also examined the duration of a post-focus sequence in a focus sentence compared to the same sequence in a neutral sentence to investigate whether a post-focus sequence is "deaccented and dephrased (i.e., loses both the tone and phrase boundary lengthening) or "deaccented" only (i.e. loses the phrase boundary tone only, but keeps phrase boundary lengthening). Di Cristo (1998) notes that in the post-focal part which is produced with a slightly declining parenthetical pitch pattern without pitch prominence, rhythmic units (UR) were signalled by temporal organizations. In our study, we test if post-focal APs are also maintained based on the temporal properties.

Five Parisian French speakers participated in this experiment, but most analysis reported here are based on two or three speakers. Speakers were given three sets of 15 sentences: Neutral Declarative, Focus Declarative, and Focus Interrogative. Sentences were varied in terms of the syntactic organization, the length of the focused word, and the location of contrastive focus. Each sentence was repeated five times. An example from each set is given in (3) below. The focused word is shown in bold capitals. In order to trigger narrow focus on the target word, a cue sentence was written in parenthesis just before each target sentence where the cue sentence has a form of 'Not A, but B' for declaratives and 'A or B' for interrogatives. This type of focus would be categorized as Di Cristo's (1998) objective contrastive focus. The nature of focus type (objective vs. expressive) was confirmed by the authors' judgments. For focus interrogative, a paragraph of background information was additionally given at the beginning of the whole list. Speakers were not given any other information regarding the type of focus or how to produce the sentences.

(3) **neutral declarative**
Marion mangera des bananes au petit déjeuner.
'Marion will eat some bananas at breakfast'

focus declarative
(Marion ne mangera pas des ananas au petit déjeuner, mais ...)
Marion mangera des **BANANES** au petit déjeuner.
'(Marion won't eat pineapples at breakfast, but...)
Marion will eat some BANANAS at breakfast'

focus interrogative
(je veux savoir si Marion mangera des bananes ou des ananas)
Marion mangera des **BANANES** au petit déjeuner?
'(I want to know if Marion will eat bananas or pineapples)
Marion will eat some BANANAS at breakfast?'

10.3.2 Results

10.3.2.1 The Tone-syllable Affiliation of Focus

As mentioned in previous studies, the focused item was always marked by a large peak on the first or the last, or on both syllables of the item, and was followed by a flat contour, low for the declaratives and mid or high for the interrogatives. The sequence before the focused item also differed from that of the neutral sentence in having a reduced pitch range and amplitude as well as a reduced number of phrase boundaries.

For focus declarative sentences, the location of the focus peak (focus H, henceforth Hf) differed depending on the speaker and the length of the word. Though the type of our focus sentences was close to Di Cristo's objective contrastive focus, we found many examples where the rise occurs on the final stressed syllable of the focused word. Figure 10.6 shows the location of Hf for five speakers: a white bar refers to the cases where Hf is on the initial syllable, Hi; a black bar refers to the cases where Hf is on the final syllable; and a mixed colour bar is for the cases where Hf is on both syllables. The upper panel shows when the words are short (two to three syllables: e.g., *Marion, bananes, mangera*), and the lower panel shows when the words are long (four to six syllables: e.g., *inadmissible, inimitable, irrémédiable*). For short words, Speaker 1F had Hf more on H*, while the other speakers had Hf more on Hi. However, this pattern was not the same for long words. Speaker 1F always focused Hi, and Speaker 2F and 3M focused H* more often than Hi. No speakers focused H* only. In general, speakers focus Hi more often than H*, and when they focus H*, they also put focus on Hi.

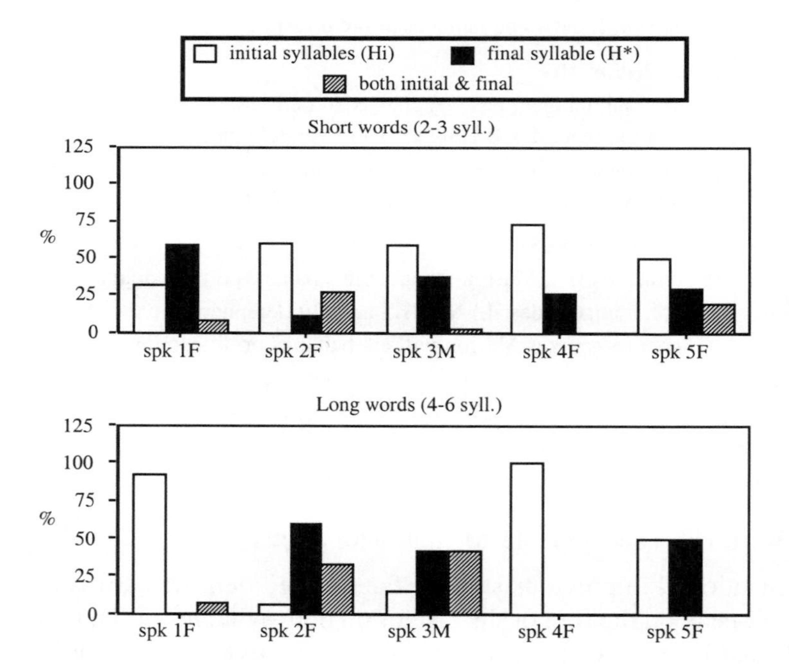

Figure 10.6. Frequency (in percent) of focus realization in declarative sentences, divided into two groups: short words, vs. long words. A white bar signifies focus H (Hf) on the initial accent (Hi), a black bar signifies Hf on the final accent (H*), and a mixed color bar signifies that Hf is on both.

The location of L tone after focus H (Hf) is mostly realized on the first syllable of the next word or the beginning of the next word regardless of the location of the peak (Hi or H*). Hence, the duration of the fall from Hf to the following L depends on the placement of Hf: it is short when Hf is on H*, but long when Hf is on Hi. In a few cases, L falls on the focused word final syllable or earlier (mostly for speaker 1F's long words). Table 10.2 shows the frequency of each pattern, depending on whether the focus is realized on H* or Hi.

Table 10.2. The location of L after focus H.

	short words (2-3 syll.)			long words (4-6 syll.)		
	1F	2F	3M	1F	2F	3M
Focus on H*						
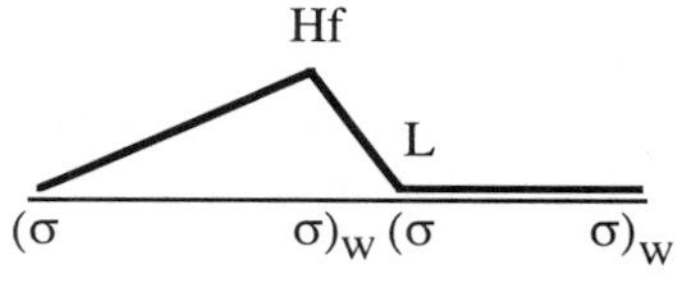 Hf L (σ σ)$_W$ (σ σ)$_W$	23 (100%)	11 (100%)	0 (0%)	1 (100%)	12 (85%)	3 (100%)
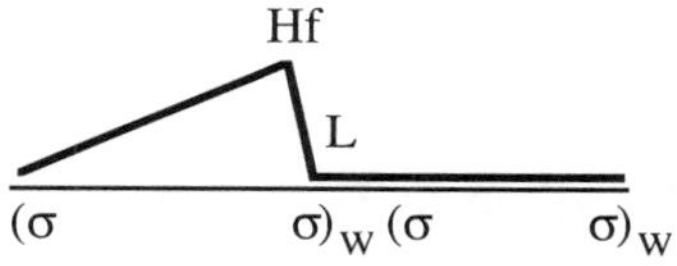 Hf L (σ σ)$_W$ (σ σ)$_W$					2 (15%)	
Focus on Hi						
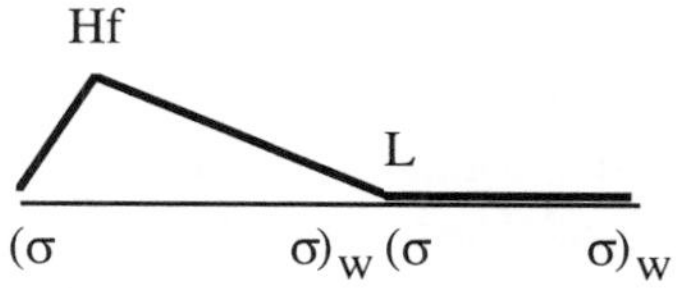 Hf L (σ σ)$_W$ (σ σ)$_W$	9 (100%)	20 (95%)	6 (100%)	4 (33%)	1 (100%)	1 (50%)
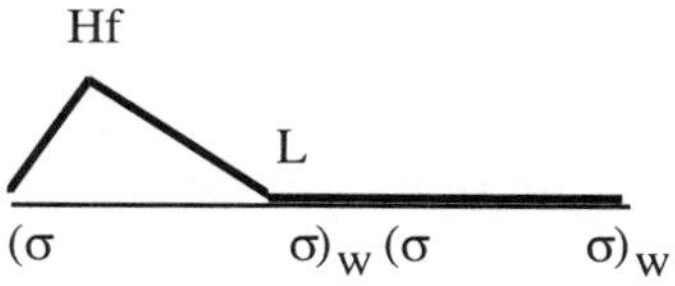 Hf L (σ σ)$_W$ (σ σ)$_W$		1 (5%)		7 (58%)		1 (50%)
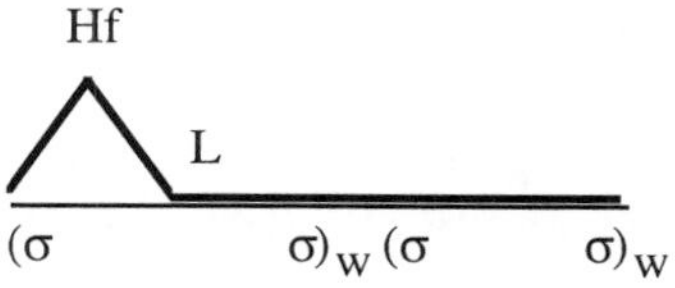 Hf L (σ σ)$_W$ (σ σ)$_W$				1 (8%)		

For focus interrogative sentences, the focused item was realized as a sharp rise on the final syllable of the item, reaching the peak at the end of the word, and sometimes even after the word. The location of Hf was always on the H* of all focused words except for *mangera* 'will eat' as produced by Speaker 3M. In this case, the speaker put a High Intonational phrase boundary (H%) after the focused word, and realized the initial syllable, *man-*, as Hf. That is, he produced a H on '*man-*', followed by a low tone and then rise at the end of the word with lengthening of the word final syllable. For other cases where speakers put an Intonational phrase boundary after focus, they produced a low tone on the initial syllable of the focused word and a high tone (in this case, Hf=H%) on the focused word final syllable.

In focus interrogatives, the peak of Hf was realized later than that in focus declaratives, i.e., mostly on the word immediately following the focused word. An example is shown in Figure 10.7, (a) declarative (b) interrogative. This is probably due to the type of tones following the focused word: a low plateau in declaratives, but a high or mid-high plateau in interrogatives.

10.3.2.2 Post-focus Sequence

Post-focus sequence has previously been described as a deaccented, flat, parenthetical, or subordinate contour (Rossi, 1985; Touati, 1987; Di Cristo, 1998). This description is confirmed in our study, but not always. Speakers sometimes paused after focus and made the post-focus sequence a new IP, with the same tonal variation (i. e., same phrasing) as in a neutral sentence, or they rephrased the sequence with fewer phrase boundaries than that of a neutral sentence. When there was no pause after focus, however, the post-focus sequence did not show any of the tonal variation observed in a neutral sentence, but was realized either as a low plateau in declaratives or as a high or mid-high plateau in interrogatives. In interrogatives, the plateau was either the same height as Hf or somewhat lower than Hf, then raised to a higher F_0 (i.e. upstepped H%) at the end of the sequence. Figure 10.7 shows examples of a pitch track for (a) a low plateau in declaratives, and (b) high and mid-high plateaus in interrogatives[5]. The height of the plateau in interrogatives varied depending on the length of the phrase and the speakers. The plateau was high when the post-focus sequence was short, one or two words, as in the first figure in Fig. 10.7(b). When the sequence became longer, the contour was either high as in the second figure in Fig. 10.7(b) or mid-high, as in the third figure in Fig. 10.7(b). In many cases, the contour showed a slight sagging in the center, or slowly declined toward the end of the phrase before rising to a H%.

(a) Low plateau after focus in a declarative.

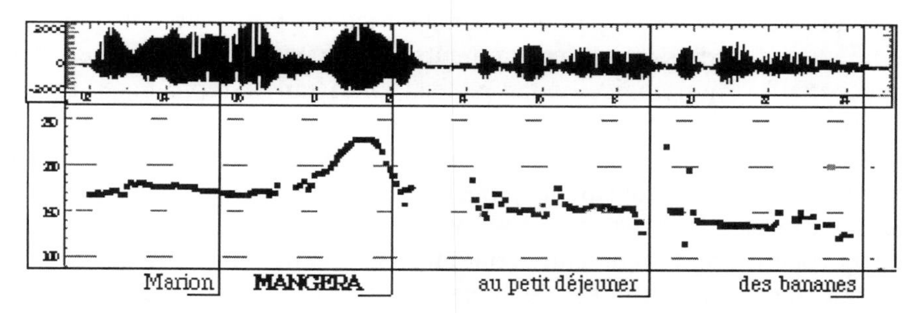

(b) High and mid-high plateau after focus in an interrogative.

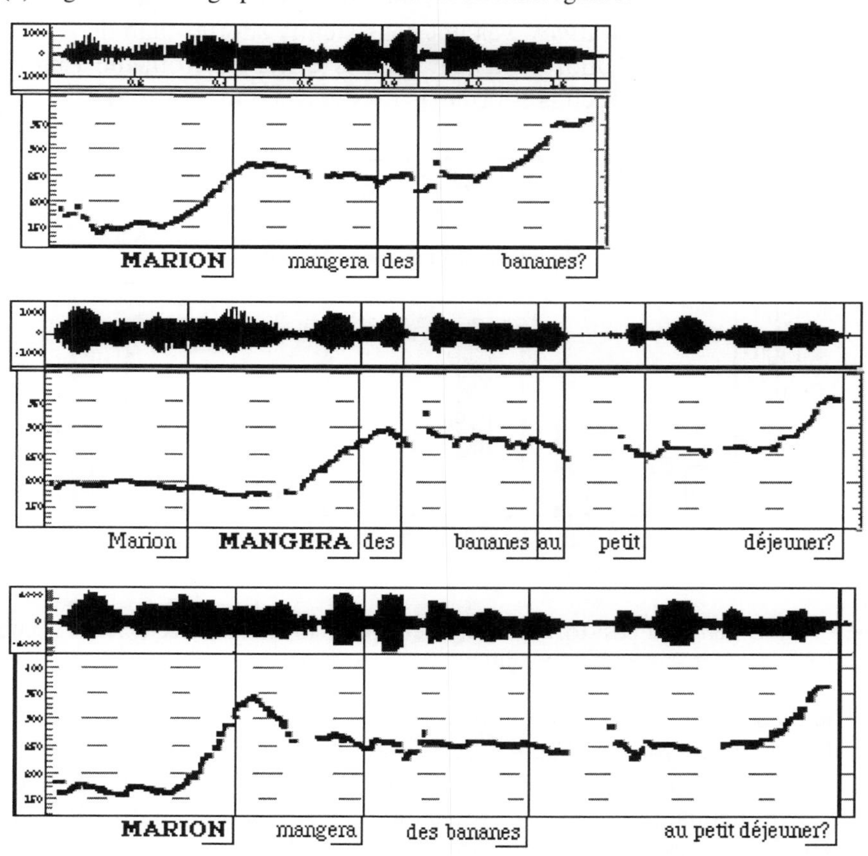

Figure 10.7. Example pitch tracks of (a) Low plateau in a declarative, and (b) High and mid-high plateau in interrogatives.

For both declaratives and interrogatives, the post-focus sequence (when it did not form a new IP separated from the focused word) did not show any of the pitch accents observed in neutral sentences; the sequence was "deaccented". Phonologically, the plateau tones can be represented by a L tone for a declarative or a H (phonetically high or mid) for an interrogative. There seems to be no pragmatic meaning contrast between high and mid-high plateau after focus in interrogatives, but there may be a meaning associated with a mid tone boundary which appears in intonational clichés such as vocative and implicature contours (See section 10.3.3.1).

Next, the possibility of dephrasing after focus is investigated by comparing the duration of the post-focus sequence in focused vs. neutral sentences. The results are shown in Figure 10.8. The duration of the post-focus sequence in declaratives was not consistently shorter than that of the same sequence in a neutral sentence. This is different from the English post-focus sequence, which is shorter (Erickson & Lehiste, 1995).

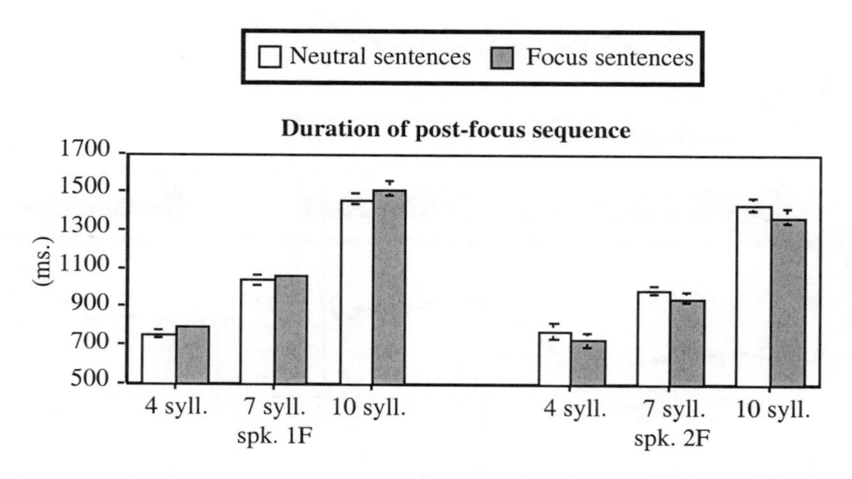

Figure 10.8. Duration of post-focus sequence compared to the same sequence in neutral sentences. Data for sequences containing 4, 7, & 10 syllables.

The data in Fig. 8 suggest that the post-focus sequence is not dephrased. But this does not prove that the AP final lengthening is still preserved within this sequence. It would be still possible that the duration of the sequence remains the same by reducing the AP final lengthening and increasing the durations of other non-final syllables. To confirm whether the post-focus sequence still preserves the AP boundaries in terms of durational cues, we compared the duration of AP-final syllables (H*-toned) in neutral sentences with the duration of the same syllables in focused sentences (i.e. potential AP-final syllables). This comparison is shown in the upper panel of Figure 10.9. The lower panel of the figure shows the duration difference between the AP-final syllable (or the potential AP-final syllable in focus sentences) and the preceding syllable in focused vs. neutral sentences. In both graphs, the duration data come from four words given in (4): H*-syllables are in bold, and the preceding syllables are underlined. The letters 'A' to 'D' in the graphs refer to the letters in (4).

(4) A. Neutral: Marion <u>man</u>**gera** des bananes. 'Marion will eat bananas'
Focus: MARION <u>man</u>**gera** des bananes.
B. Neutral: Marion <u>man</u>**gera** des bananes au petit déjeuner. '... bananas at breakfast'
Focus: MARION <u>man</u>**gera** des bananes au petit déjeuner.
C. Neutral: Marion mangera des <u>ba</u>**nanes** au petit déjeuner.
Focus: Marion MANGERA des <u>ba</u>**nanes** au petit déjeuner.
D. Neutral: Marion mangera des <u>ba</u>**nanes** au petit déjeuner.
Focus: MARION mangera des <u>ba</u>**nanes** au petit déjeuner.

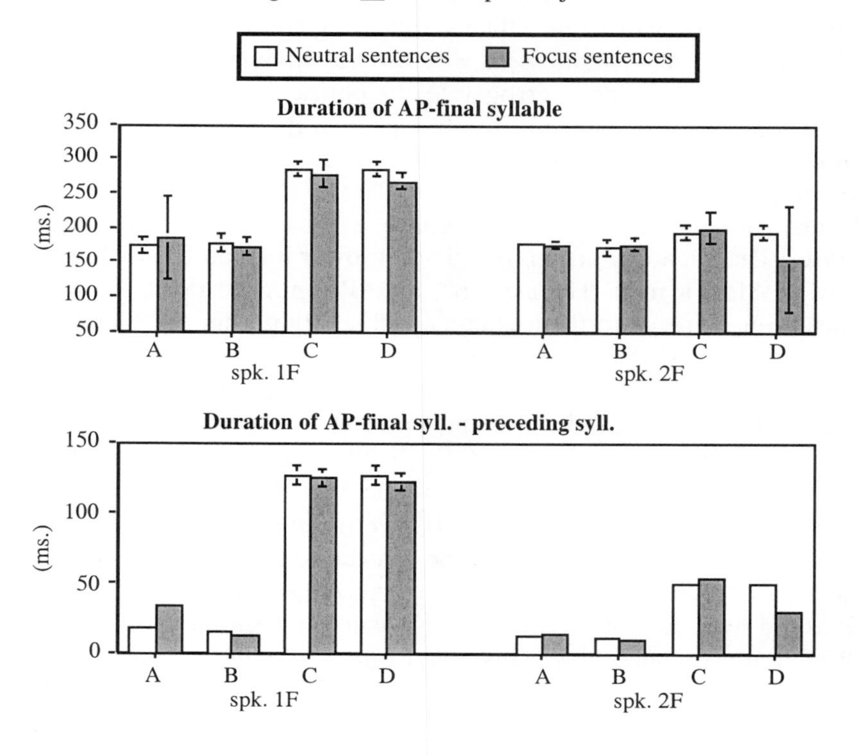

Figure 10.9. Durational comparison of AP-final syllable in neutral sentences with the same syllable in post-focus sequence, in terms of absolute syllable duration (upper panel) and relative final lengthening: i .e., final syllable minus preceding syllable (lower panel).

Results show that both absolute duration and relative lengthening of AP-final syllables in neutral sentences are maintained in focus sentences. The post-focus sequence seems not to be dephrased. Though it loses its intonational phrasal cues, it maintains its durational phrasal cues. The post-focus sequence in French is therefore *deaccented, but not dephrased.*

10.3.2.3 Representation of Focus Intonation

The intonation contour of a focused utterance is different from that of a neutral utterance. The peak of the focused word is not always on the final full syllable of the word, and the post-focus sequence does not show any of the tonal variation observed in the neutral rendition of the same sentence. Though the post-focus sequence preserves the AP boundaries of a neutral utterance in terms of duration, it is not parsable into a sequence of APs in terms of tones: the post-focus sequence is tonally flat, L for a declarative or H for an interrogative. This plateau tone covers syllables from the beginning of the word immediately following the focused word until the end of the phrase, regardless of whether the Hf is realized on Hi or H*. When the sentence is an interrogative, the phrase final boundary tone at the end of the phrase is realized even higher than the H-plateau. We assume that the plateau tone after focus in French is a phrasal tone associated with the phrasal node of an Intermediate Phrase (henceforth 'ip'), and the boundary tone at the end of the phrase is the property of an Intonation Phrase. That is, an 'ip' is a prosodic unit lower than an IP and higher than an AP. After a focus H tone (Hf), all AP tones are deleted, and the ip phrasal tone spreads to the toneless syllables right after the focused word. The ip tone is not associated with a particular syllable. Deaccenting after a focused word " is also found in English and Korean. In English, there is no pitch accent after focused word, and a plateau tone begins after the focused word as in French. The plateau in English is also analysed as an ip phrasal tone covering the syllables between the focused word and the phrase boundary (Beckman & Pierrehumbert, 1986). In Korean, all AP tones after focus are deaccented, but there is no plateau tone after the focused word. Rather, all syllables after focus are reorganized so as to belong to the same AP as the focused word. That is, in Korean, the focused word starts a new AP with an expanded pitch range, and follows the default tonal pattern of an AP: F_0 reaches the peak on the second syllable, and gradually falls over many syllables after focus, as in other long APs of a neutral utterance (Jun & Lee, 1998).

A schematic representation of deaccenting after focus, and a new prosodic structure including an ip node is shown in Figure 10.10. The upper half shows that either the initial or final syllable of the focused word is accented for focus, Hf, while all AP tones after Hf, if any, are deleted. When both the initial and the final syllable are focused, all AP tones after the final Hf are deleted. The lower half shows that the 'ip' phrasal tone, noted T- following the convention in Beckman and Pierrehumbert (1986), spreads to all toneless syllables after the focused word up to the final syllable which is realized with an IP boundary tone.

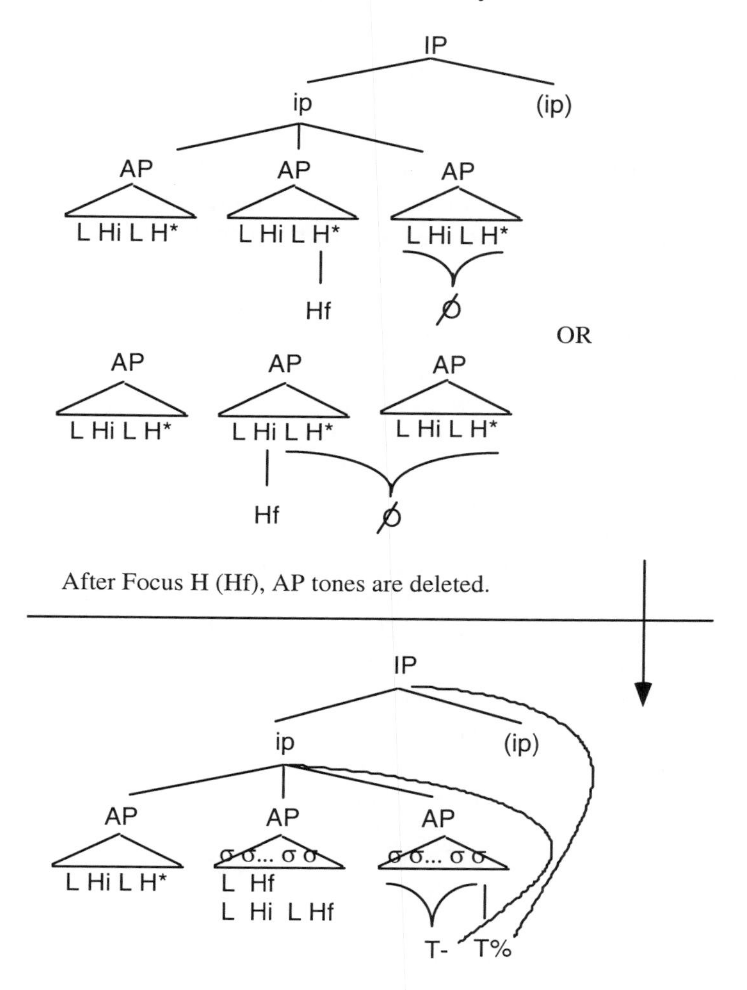

Figure 10.10. Schematics of AP tone deletion after focus (upper) and the realization of an intermediate phrasal tone, T- (lower).

10.3.3 Discussion

10.3.3.1 Further Evidence for an 'ip' Phrasal Tone

In addition to focus utterances, further evidence for an 'ip' phrasal tone can be found in intonation patterns of complex syntactic/semantic constituents such as tag-questions and dislocated theme/rheme structures, as well as in intonational clichés and wh-questions.

First, a sentence ending with a tag question separates the tag phrase from the main clause by a comma in orthography, but in speech, its boundary is cued prosodically by a low or high tone covering the tail part of the main clause and a final lengthening of the clause final syllable. The juncture between the main clause and the tag phrase is somewhat bigger than that of an AP boundary but smaller than that of an IP boundary. The intermediate nature of this juncture has also been captured by Di Cristo and Hirst (1996) as a unit called "segment d'UI". This means that the main clause of the tag question forms an IP by itself, but is embedded in a big IP, the whole sentence. For example, *une bonne bouteille de champagne, ça lui plairait?* 'a good bottle of Champagne, would he like it?' is parsed as [[*une bonne bouteille de champagne*]IP *ça lui plairait*]IP, just like a recursive IP (cf. Ladd, 1986). In our model, the embedded IP (the main clause) would form an 'ip' with a L- tone, and the tag-phrase itself would form another ip with a H- tone. That is, this sentence has two 'ip's within one IP: [[*une bonne bouteille de champagne*]ip [*ça lui plairait*]ip]IP[6].

A preposed or topicalised thematic constituent (or given information), whose trace is marked by a resumptive pronoun in the rhematic constituent (or old information), can also be analysed as two 'ip's within one IP. For example, an utterance such as *Mon voisin, il est toujours malade* 'My neighbour, he is always ill' (as an answer to a question about the health of a person's neighbour) can be analysed as one IP with two ips, when produced with no pause after the preposed constituent (example from Di Cristo, 1998): i. e., [[*Mon voisin,*]ip[*il est toujours malade*]ip]IP, with H- for the first ip and L- for the second ip. Here, the boundary between *voisin* and *il* is more likely to be stronger than the AP boundary between *voisin* and *est* in the simple sentence, {*Mon voisin*}AP{*est toujours malade*}AP 'My neighbour is always ill'. Di Cristo (1998) also implies this relative strength of boundaries by saying that "the preposed constituent can be pronounced either with a typical continuation rise or by a question pattern (i.e., *Mon voisin? Il est toujours malade*) which carries an additional character of emphasis. However, for the subject of a simple sentence, only the continuation pattern is possible".

Another construction whose intonation pattern is similar to focus intonation can be found in postposed thematic constituents and wh-questions (Di Cristo, 1998, 26-30). Di Cristo claims that a postposed thematic constituent (underlined in the following examples) is characterized by a low plateau in a declarative (e.g., [[*Il est toujours malade*] mon voisin] 'He is always ill, my neighbour'), and by a high plateau in an interrogative (e.g., [[*Tu l'as vendue*] ta belle maison]? 'Have you sold it, your beautiful house?' (in the context where 'your beautiful house' is already mentioned). A wh-question with a focused wh-word also shows a low plateau after the wh-word: e.g., [[Qui] va te rencontrer?] 'Who is going to meet you?'. The low or high plateau in these examples was analysed as a "segment d'UI" as in a post-focus sequence (e.g., [[Le professeur] a la clé] 'The teacher has the key (not the student)'). In our model, the plateau tone can be analysed as an 'ip' phrasal tone, L- or H-, starting after the last pitch accented word and spreading up to the phrase final syllable.

So far, we have shown that an ip boundary can occur as a plateau, starting from the syllable right after the last pitch accented word and up to the final syllable: e.g., in a focused phrase, wh-question, and postposed thematic constituent. We also showed that more than one ip boundary can occur within an IP: e.g. in a tag question and preposed thematic constituent. In Di Cristo (1998) and Di Cristo and Hirst (1996), all these example constructions are analysed as including a segment d'UI: 1. [[focused word] post-focus sequence]; 2. [[wh-word] rest of a question sentence]; 3. [[rhematic constituent] postposed thematic constituent]; 4. [[main clause] tag-Q]; 5. [[preposed thematic constituent] rest of a sentence with a resumptive pronoun]. Some of these constructions do have a plateau-like tonal contour, but we believe that the boundary (i.e. perceptual degree of juncture) between focus and post-focus sequence or that between a wh-word and the following word do not have the same degree of strength as that observed between a main clause and a tag-phrase or that between a preposed thematic constituent and the rest of the sentence with a resumptive pronoun. We believe that the boundary in the latter group (i.e., a tag-question and a preposed thematic sentence) is stronger than that in the former group[7]. This difference in boundary strength is reflected in our model by a grouping of ip. That is, the former group consists of one ip, while the latter group consists of two ips.

Next, there are intonational clichés where the final boundary tone after the last pitch accent shows either a mid plateau or a complex pitch movement, low-high (mid-plateau data below are from Fagyal, forthcoming).

First, the mid plateau at the end of a phrase can be found in several clichés: 'vocative or chanting' contour, 'list' contour, and 'implicature' contour. In a simple vocative (e.g. *Joanna!*), F_0 peak is on the *penultimate* syllable of the word, and is followed by a mid-plateau until the end of the word. In a complex vocative (e.g. *Bonjour, Madame Durand*!, *Venez-vous*!), F_0 peak can be on either Hi or H* of a vocative phrase initial word, *Bonjour*, or *Venez*, and is followed by a mid plateau until the end of the phrase. Figure 10.11 shows examples of F_0 tracks for (a) a simple vocative and (b) a complex vocative with the peak realized on Hi of the first word. This type of vocative contour has been modelled as a LHM sequence (Dell, 1984) or a sequence of LH followed by a lowered H boundary tone. Di Cristo describes the simple vocative as an 'upstep' followed by a fall, and the complex vocative, or cliché vocative, as rise-fall followed by a sustained plateau with the rise being aligned with the penultimate or the last syllable of the vocative word (Di Cristo, 1976, 1981). The same pattern occurs in English vocatives (Ladd, 1978, 1980, 1996; Pierrehumbert, 1980), and has been analysed as "H*+L H- L%" under Pierrehumbert's system. In this case, the 'ip' tone, H-, is realized as a downstepped H after the bitonal pitch accent, and the 'IP' tone, L%, is realized as upstepped L after H-, thus surfacing as a mid plateau. In our model, both simple and complex vocatives in French are represented as H* H- L%, capturing the similar shape of F_0 contours between these two vocative types. H* is for the F_0 peak before the mid plateau, H- is realized as a mid-high, and L% is realized as an upstepped L% after H-.

Similarly, the mid plateau boundary can be found in the 'list' contour or 'implicature' contour. As in a simple vocative, the peak (H*) occurs on the penultimate syllable of the phrase final word, and is followed by a mid plateau. We represent this sequence as H* H-L% (or as H* H- for an ip boundary). An example sentence of each type is shown in (5a) and (5b), respectively, and the tones relevant to H* and a mid plateau only are labelled. A pitch track of (5b) is shown in Figure 10.12 (data from Faygal's spontaneous speech corpus).

(5) a. Il y avait l'Allemagne, la Roumanie, le Venezuela, la Méditerranée.

H* H-L%	H* H-L%	H* H-L%	H* H-L% or
H* H-	H* H-	H* H-	H* H-L%

'There were Germany, Rumania, Venezuela, the Mediterranean.'

b. Alors, il y a tout de même les circonstances objectives.
H*H-L%
'So, there are still the objective circumstances.'

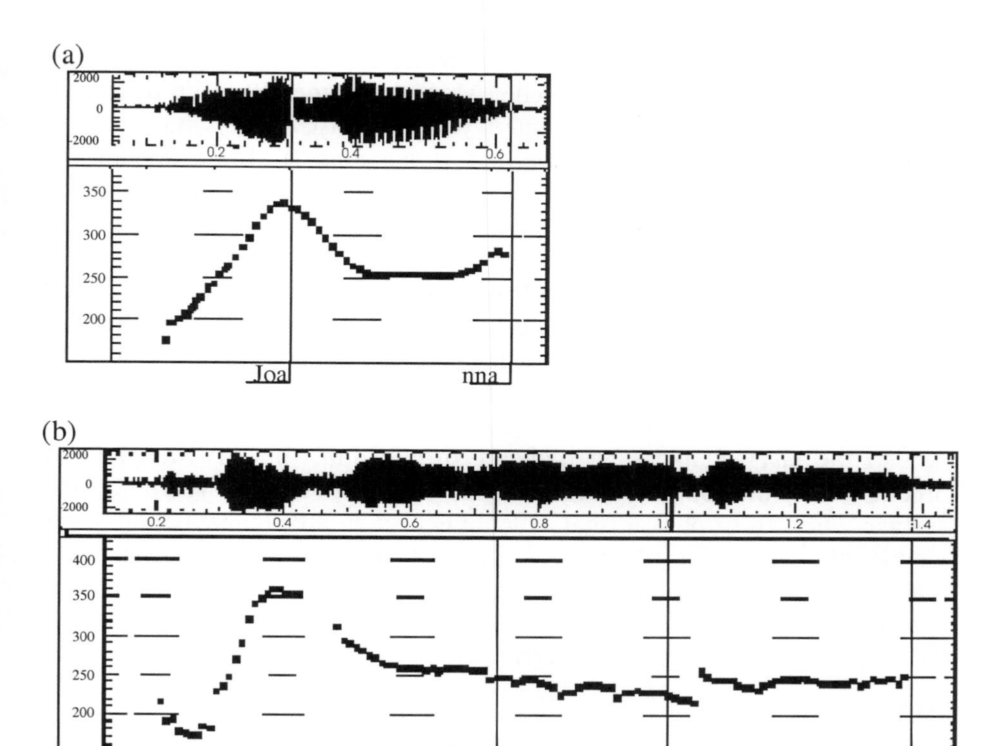

Figure 10.11. F_0 tracks of (a) a simple vocative and (b) a complex vocative.

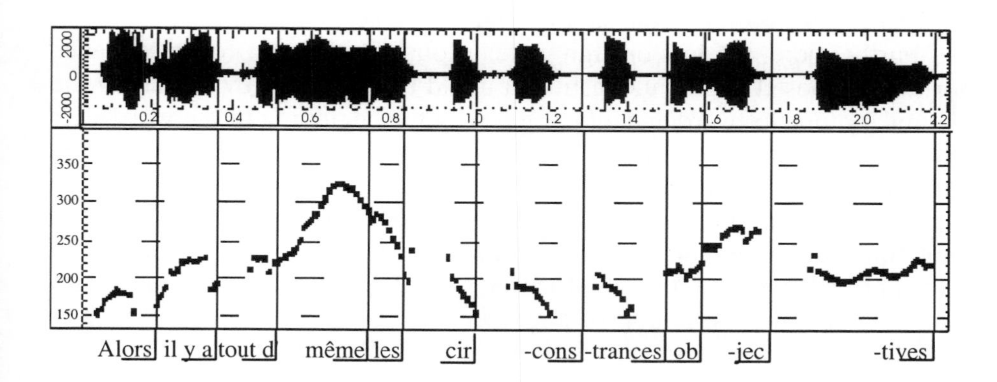

Figure 10.12. A F_0 track of the example in (5b) produced with 'implicature' meaning.

Second, the low-high complex pitch movement after the peak occurs in the well-known "incredulous" cliché, which has generally been described as a rise-fall-rise contour associated with the last three syllables of a sentence (ex. *Au garage*?, *comptabilité*?, *Jamais moi*?). In our model, the peak on the penult of the phrase is analysed as H*, and the following rise is analysed as a sequence of an ip tone, L-, and an IP tone, H%, on the same final syllable of the phrase.

In sum, by assuming three levels of prosodic units, AP, ip, and IP, and tones affiliated with each unit, we can represent the intonation pattern of a sentence with a high, mid or low plateau and a complex contour at the end of a phrase. In a neutral sentence produced without any stylised intonation, the final full syllable of an AP is realized with a H*, and when this syllable is also the last syllable of a higher prosodic unit, the AP final tone is preempted by the tone of a higher unit. An ip tone is not visible on the surface when an IP phrase final syllable is also an AP final syllable, but is visible otherwise. For intonational clichés, it seems that the AP final H* is realized on the penultimate syllable of the word, and the AP final syllable provides room for the realization of the phrasal tone (ip) and boundary tone (IP), the combination of which carries a different pragmatic meaning.

10.3.3.2 An alternative Analysis

As an alternative to the analysis of adding an 'ip' phrasal tone, the fall-rising tone or mid plateau tones occurring at the end of a phrase can be analysed as a complex boundary tone. The sequence of L- and H% could be represented as a LH% and that of H- and L% can be represented as a HL%. In this case, we would have to introduce two more boundary tones in French, in addition to the simple boundary tones, L% and H%.

Furthermore, the low or high plateau found in the post-focus sequence or the postposed thematic constituent could be explained by spreading the boundary tone leftward and aligning it with the beginning of the word after the peak. This non-peripheral realization of boundary tones has been proposed by Gussenhoven (1996). He claims that, in Roermond Dutch, the realization of an intonation phrase boundary tone after a focused syllable is not confined to the phrase final location, but spreads its domain up to the word or syllable right after the focused syllable. For the plateau boundary tone, he adopted the notion of alignment in Optimality Theory (Prince & Smolensky, 1993; McCarthy & Prince, 1993; Pierrehumbert, 1993), and proposed that the left-edge of the boundary tone is 'aligned' (or 'associated' if there are legitimate Tone Bearing Units) with the first stressed syllable after focused syllable, and the right-edge of the boundary tone is 'aligned' with the phrase boundary. Thus, the single boundary tone is subject to two alignment constraints, left edge and right edge, achieving the effect of leftward spreading of the boundary tone. He argues that an 'aligned' tone is less likely to have a precise location of the tone target, both in the time-dimension and the F_0 dimension, than is the 'associated' tone. In addition, the F_0 transition between the focus H and the aligned Intonational Phrase L boundary tone (his 'Li') is realized as a 'drooping' interpolation. In the case of a focus interrogative where a H plateau is followed by a L boundary tone, he proposed two boundary tones (his 'HiLi').

If we adopt Gussenhoven's model, the L plateau in French can be represented by a L% with two alignment constraints (left and right). The H plateau examples found in interrogative focus would be more challenging to explain since the end of the plateau is upstepped at the phrase final syllable. This can be represented either by an H% with three alignment constraints (left, penult, right) or by an HH% boundary tone with two alignment constraints for the first H, (left, right).

The advantage of this alternative analysis is that we do not need to add a new prosodic unit (ip). It seems that this alternative works for all examples whose IP boundary is preceded by a plateau (e.g. post-focus sequence) or whose IP boundary consists of a sequence of two tones (e.g. clichés). However, this alternative cannot explain the cases where an IP has more than one ip such as a tag question and a preposed or topicalised construction. Therefore, we assume that a better solution at this moment is to add an ip phrasal tone instead of modifying a boundary tone and adding alignment constraints. At the same time, however, we are aware of the fact that to further support an addition of an ip level to our model, we need to find more examples where an IP has more than one ip. We expect that spontaneous speech data will provide us with further evidence. In this paper, we cited examples from a tag question and a preposed or extraposed thematic constituent. Though the boundary within these examples seems to be intermediate between an AP and an IP, further evidence is needed to prove that there exists a categorically distinct intermediate prosodic unit in terms of perception data and the interaction with other tonal or segmental rules. It is also possible that there is a phonetic difference in the height or shape of F_0 rise between H* and H*H-, and/or durational cues for this intermediate phrase level.

10.4 Summary

In this paper, we first describe a phonological model of French intonation with two tonally defined prosodic units: Accentual Phrase and Intonation Phrase. These prosodic units are compared with the prosodic units of other models. This model is further developed to explain the intonation pattern of focus sentences and other intonational clichés. We propose that the plateau tones after focus can be represented as phrasal tones of an Intermediate Phrase, and show that the addition of a phrasal tone can also explain other intonation patterns. Under this newly developed model, we propose a phonological model of French intonation with three tonally defined prosodic units. The Accentual Phrase (AP) is the lowest tonal unit, and its underlying tonal pattern is /LHiLH*/. The Intermediate Phrase (ip) is the next higher level, and has a phrasal tone covering syllables or words after the last H* of the phrase and before the IP boundary tone. The presence of a phrasal tone is, therefore, apparent when H* is realized in non-phrase final position. Finally, the IP is the highest tonal unit and is demarcated by a boundary tone (H% or L%). Further research is needed to further support the level of an ip.

Notes

1. Fónagy (1980) discusses several factors conditioning the occurrence/appearance of the initial (also known as secondary or intellectual) accent: 1. modality - initial accent is more frequent in imperative or exclamation; 2. speaking styles - initial accent is more frequent in reading or conference style, or in TV or radio news, than in spontaneous speech (also in Vaissière, 1974); 3. segmental context - stops attract initial accent more than liquids; nasal vowels attract initial accent more than oral vowel; close syllables attract initial accent more than open syllables. Lucci (1980) includes several other factors: length of a word or rhythmic group (the longer, the higher tendency of having an initial accent), syllable structure (a syllable with an onset is more likely to have initial accent than a syllable with a vowel only), semantic conditioning (a word with high information load tends to have initial accent), and the location within a phrase. Pasdeloup (1990) also claims that various factors such as phonotactic, rhythmical, contextual constraints may account for this variability. In addition to all these factors, there are speaker dependent characteristics and variations (Vaissière, 1975; Duez, 1978 in political discourse).

2. The three 8-syllable words are: *panindoeuropéen*, *méditérranéisation*, and *incompréhensibilité*. Among these, *panindoeuropéen* very often had an initial accent on the prefix (*pan-*), the first syllable. Note that the frequency of Hi occurrence in Figure 10.10.1 is based on a sentence initial word, forming one AP. In general, Hi occurs more often when the word is in sentence initial than in sentence medial position. The high percentage of Hi in 2-syllable and 3-syllable words is due to the

fact that these words are lengthened; in 2-syllable words, the word initial syllable was often realized as LHi and the final syllable was realized as LH*.

3. We showed in Jun and Fougeron (1995) that the timing of the initial rising of French AP is not correlated with the number of syllables within an AP, but is fairly constant (i.e., taking about 200-300 ms) regardless of the number of syllables. This data was based on content words located at the initial position of a sentence, thus not preceded by any function word within an AP.
4. We are grateful to Hirst and Di Cristo for providing these two references.
5. In (a), the focused word 'mangera' is not followed by pause, but is immediately followed by 'au', and then a devoiced vowel in the first syllable of 'petit'. In the examples in (b), it is interesting to note that the plateau F_0 is around 250Hz whether it is phonetically high plateau or mid-high plateau. These are produced by the same speaker. It is possible that the target F_0 of the plateau is fixed for interrogatives, and the peak F_0 of a focused word varies, sometimes reaching higher than their neutral H target. Similar data are found in Chinese (Shih, this volume). We thank Chilin Shih for pointing this similarity to us.
6. Di Cristo and Hirst (1993b) and Sabio, Di Cristo, and Hirst (1995) treat this sentence as a left dislocated question. But in either analysis, this sentence is analysed as two ips in our model. We thank Di Cristo for providing this information.
7. Though we use this perceptual difference as evidence of an ip level, we are aware of the fact that a perceptually different boundary would not guarantee the existence of a prosodic level. As shown in Wightman, Shattuck-Hufnagel, Ostendorf and Price (1992), listeners can perceive more levels of juncture than the levels of a prosodic unit in English.

Acknowledgments

We would like to thank Mary Beckman for her valuable comments on the representation of focus intonation, Zsuzsanna Faygal for generously sharing her data with us and for stimulating many valuable discussions, Albert Di Cristo and Chilin Shih for their helpful comments, Victoria Anderson for her conscientious proofreading of this paper, and Russ Schuh for providing references for tone dissimilation. We also thank A. Di Cristo, E. Grabe, C. Gussenhoven, D. Hirst, and B. Post for their interest in our model and encouragement given at the ESCA Intonation Workshop, Athens, Greece, in Sept. 1997. Finally, we thank the speakers who participated in the experiment.

Références

Abercrombie, D. 1964. Syllable quantity and enclitics in English. In Abercrombie *et al.* (eds.), 216-222.

Abercrombie, D., D.B. Fry, P.A.D. McCarthy, N.C. Scott and G.L.M. Trim (eds.). 1964. *In Honour of Daniel Jones*. London: Longman.

Beckman, M.E. 1986. *Stress and Non-Stress Accent*. Dordrecht: Foris.

Beckman, M.E. and J.P. Pierrehumbert. 1986. Intonational structure in Japanese and English. *Phonology Yearbook* 3, 255-309.

Clech-Darbon A., G. Rebuschi and A. Rialland. Forthcoming. Are there cleft sentences in French? In Rebuschi and Tuller (eds.).

Delais-Roussarie, E. 1995. *Pour une Approche Parallèle de la Structure Prosodique*. PhD dissertation, Univiversité Toulouse le Mirail.

Dell, F. 1984. L'accentuation dans les phrases en français. In Dell *et al.* (eds.), 65-122.

Dell, F, D.J. Hirst and J.R. Vergnaud (eds.). 1984. *Forme Sonore du Langage*. Paris: Herman.

Delattre, P. 1939. Accent de mot et accent de groupe. *The French Review* XIII, 2, 141-46.

Duez, D. 1978. *Essai sur la Prosodie du Discours Politique*. Thèse de Doctorat, Université de Paris III.

Di Cristo, A. 1976. Des indices prosodiques aux traits perceptuels: Application d'un modèle d'analyse prosodiques à l'étude du vocatif en français. *Travaux de l'Institut de Phonétique d'Aix* 3, 213-58.

Di Cristo, A. 1981. Le vocatif en français. In Rossi *et al.* (eds.), 99-137..

Di Cristo, A. 1998. Intonation in French. In Hirst and Di Cristo (eds.), 195-218.

Di Cristo, A. and D.J. Hirst. 1993a. Rythme syllabique, rythme mélodique et représentation hierarchique de la prosodie du français. *Travaux de l'Institut de Phonétique d'Aix* 15, 9-24.

Di Cristo, A. and D.J. Hirst. 1993b. Prosodic regularities in the surface structure of French questions. *Proc. ESCA Workshop on Prosody* (Lund, Sweden), 268-71.

Di Cristo, A. and D.J. Hirst. 1996. Vers une typologie des unités intonatives du français. *XXIème JEP* (Avignon, France), 219-222.

Erickson, D. and I. Lehiste. 1995. Contrastive emphasis in elicited dialogue: durational compensation. *Proc. 13th ICPhS* (Stockholm, Sweden), vol. 4, 352-55.

Fagyal, S. Chanting Intonation in French. Forthcoming. *Proc. 21st Annual Penn Linguistics Colloquium.*

Fónagy, I. 1980. L'accent français: accent probabilitaire. L'Accent en français contemporain. *Studia Phonetica*, 15,123-233.

Fougeron, C. and S.-A. Jun. 1998. Rate Effects on French Intonation: Phonetic Realization and Prosodic Organization. *Journal of Phonetics* 26, 45-70

Grammont, M. 1914. *Traité de Phonétique*. Paris: Delagrave.

Gussenhoven, C. 1996. The boundary tones are coming: on the Nonperipheral realization of boundary tones. Paper presented in *Laboratory Phonology* V, Evanston, IL.

Hirst, D.J. and A. Di Cristo. 1984. French intonation: a parametric approach. *Di Neueren Sprache* 83 (5), 554-569.

Hirst, D.J. and A. Di Cristo. 1996. "Y a-t-il des unités tonales en français?" *XXIème JEP*, (Avignon, France), 223-226.

Hirst, D.J. and A. Di Cristo (eds.). 1998. *Intonation Systems*. Cambridge University Press.

Hirst, D.J., A. Di Cristo and R. Espesser. Forthcoming. Levels of description and levels of representation in the analysis of intonation. In Horne (ed.).

Horne, M. Forthcoming. *Prosody: Theory and Experiment*. Dordrecht: Kluwer Academic Publishers.

Jun, S.-A. 1993. *The Phonetics and Phonology of Korean Prosody*. PhD dissertation. The Ohio State University (published 1996, New York: Garland).

Jun, S.-A. 1996. The influence of the microprosody on the macroprosody: a case of phrase initial strengthening. *UCLA Working Papers in Phonetics* 92, 97-116.

Jun, S.-A. 1998. The Accentual Phrase in the Korean prosodic hierarchy. *Phonology* 15, 189-226.

Jun, S.-A. and C. Fougeron. 1995. The Accentual Phrase and the Prosodic structure of French. *Proc. 13th ICPhS* (Stockholm, Sweden), vol. 2, 722-725.

Jun, S.-A. and H.-J. Lee. 1998. Phonetic and phonological markers of Contrastive focus in Korean. *Proc. ICSLP* '98 (Sidney, Australia).

Kenstowicz, M. 1994. *Phonology in Generative Grammar*. Blackwell Publishers.

Ladd, D.R. 1978. Stylised intonation. *Language* 54, 517-540.

Ladd, D.R. 1980. *The Structure of Intonational Meaning*. Indiana University Press.

Ladd, D.R. 1986. Intonational phrasing: the case of recursive prosodic structure. *Phonology Yearbook* 3, 311-340.

Ladd, D.R. 1996. *Intonational Phonology*. Cambridge University Press.

Lindblom, B. 1963. Spectrographic study of vowel reduction, *J. Acoust. Soc. Am.* 35, 1773-1781.

Lindblom, B. 1964. Articulatory activity in vowels. *STL-QPSR,* 2, 1-5.

Lucci, V. 1983. *Etude Phonétique du Français Contemporain à Travers la Variation Situationnelle*. Publications de l'Université de Grenoble.

Marouzeau, J. 1956. Accent de mot et accent de phrase. *Le Français Moderne* 24, 241-248.

McCarthy, J.J. and A. Prince. 1993. *Prosodic Morphology 1: Constraint Interaction and Satisfaction*. (ms. University of Massachusetts at Amherst and Brandeis University).

Mertens, P. 1987. *L'Intonation du Français*. PhD dissertation, Katholieke Universiteit Leuven.

Mertens, P. 1993. Intonational grouping, boundaries and syntactic structure in French. *Proc. ESCA Workshop on Prosody* (Lund, Sweden), 155-159.

Milner, J.C. and F. Regnault. 1987. *Dire le Vers*. Paris: Seul.

Nespor, M. and I. Vogel. 1986. *Prosodic Phonology*. Dordrecht: Foris.

Pasdeloup, V. 1990. *Modèle de Règles Rythmiques du Français Appliqué à la Synthèse de la Parole*. Thèse de doctorat, Université de Provence.

Pierrehumbert, J.B. 1980. *The Phonology and Phonetics of English Intonation*. PhD dissertation, MIT (published 1988 by IULC).

Pierrehumbert, J.B. 1993. Alignment and Prosodic Heads. *Proc. Eastern States Conference on Linguistics* 10, 268-286. The Ohio State University.

Pierrehumbert, J.B. and M.E. Beckman. 1988. *Japanese Tone Structure*. Cambridge, Mass.: MIT Press.

Post, B. 1993. *A Phonological Analysis of French Intonation*. MA thesis, Univiversity of Nijmegen.

Prince, A. and P. Smolensky. 1993. *Optimality Theory: Constraint Interactions in Generative Grammar* (ms. Rutgers University, at New Brunswick and University of Colorado at Boulder).

Rebuschi, G. and L. Tuller (eds.). Forthcoming. *The Grammar of Focus*. Amsterdam: Benjamins.

Rossi, M. 1985. L'intonation et l'organisation de l'énoncé. *Phonetica* 42, 135-153.

Rossi, M. 1993. A model for predicting the prosody of spontaneous speech (PPSS model). *Speech Communication* 13, 87-107.

Rossi, M., A. Di Cristo, D.J. Hirst, Ph. Martin, and Y. Nishinuma. 1981. *L'Intonation, de l'Acoustique à la Sémantique*. Paris: Klincksieck.

Sabio, A. Di Cristo and D.J. Hirst. 1995. The prosodic structure of left-dislocated phrases in French interrogative utterances. *Proc. 13th ICPhS* (Stockholm, Sweden), vol. 2, 714-17.

Schuh, R.G. 1998. *A Grammar of Miya*. Linguistics, vol. 130. University of California Press

Selkirk, E.O. 1984. *Phonology and Syntax: The Relation between Sound and Structure*. Cambridge, Mass.: MIT Press.

Selkirk, E.O. 1986. On Derived Domains in Sentence Phonology. *Phonology Yearbook* 3, 371-405.

Stevick, E. 1965. *Shona Basic Course*. Washington, DC: Foreign Service Institute.

Sundberg, J., L. Nord and R. Carlson (eds.). 1992. *Music, Language, Speech and Brain*. Stockholm: Wenner-Gren Int. Symposium Series.

Touati, P. 1987. *Structures Prosodiques du Suédois et du Français*. Lund University Press.

Vaissière, J. 1974. On French prosody. *Res. Lab. Electr. Prog. Report* 115, 212-23. MIT.

Vaissière, J. 1992. Rhythm, accentuation and final lengthening in French. In Sundberg, Nord and Carlson (eds.), 108-120.

Vaissière, J. 1997. Langues, prosodies et syntaxe. *Revue Traitement Automatique des Langues*, 38, 53-82.

Wightman, C.W., S. Shattuck-Hufnagel, M.F. Ostendorf and P.J. Price. 1992. Segmental durations in the vicinity of prosodic phrase boundaries. *J. Acoust. Soc. Am.* 91, 1707-1717.

11

A Declination Model of Mandarin Chinese

CHILIN SHIH

11.1 Introduction

Declination has been studied intensively from production data and from perceptual experiments. There is a clear trend in many languages that F_0 values tend to drift down in an utterance, particularly in declarative sentences ('t Hart and Cohen, 1973; Maeda, 1976; Thorsen, 1980a; Cohen, Collier & 't Hart, 1982). Perceptually, listeners compensate for such a downtrend: given two peaks of equal F_0 value, the later one is interpreted as having higher prominence, or for two peaks to be perceived as equal, the second peak should have lower F_0 value than the first (Pierrehumbert, 1979; Gussenhoven and Rietveld, 1988; Terken, 1991; Ladd, 1993; Terken, 1993). There is also an extensive literature offering physiological explanations of declination (Lieberman, 1967; Titze, 1989; Strik and Boves, 1995). Nonetheless, there are some obvious difficulties in interpreting declination data. Most of the languages being investigated have constraints on accent combination so that the declination slope has to be estimated from sparsely located data points such as F_0 peaks or valleys, while the phonological status of the observed peaks and valleys may be unclear. In addition, an observable F_0 contour of a sentence is the result of the combination of many factors. There is no unique solution in decomposing an observed complex F_0 pattern into individual effects. To study one effect one often needs to make strong assumptions about others, which turns out to be a major cause of disagreements in the intonation literature.

This paper proposes an experimental design which uses Mandarin high level tones to study the declination effect. Mandarin offers an easy way out

A. Botinis (ed.), Intonation, 243-268.

of some of the problems cited above: it is possible to string sequences of high level tones together and thus allows for a syllable by syllable profile of the intonation topline. Some of the factors affecting intonation, to be discussed momentarily, are naturally absent in these high tone sentences, thus the experiment provides a unique opportunity to observe the declination effect more directly.

Previous studies of Mandarin intonation reported an overall downtrend in declarative sentences with mixed tones (Shen, 1985; Gårding, 1987) and in natural conversation (Tseng, 1981; Yang, 1995). There were also a few small-scale studies of Mandarin high tone sequences. Shih (1988) and Liao (1994) found declination effect in sequences of Mandarin high level tones, Shen (1985) emphasised the use of pitch raising to signal the beginning of syntactic boundaries, however, his data showed a downtrend within each syntactic phrase. Xu and Wang (1997) suggested that the observed downtrend comes from discourse effects. We will start with a brief summary of the major factors affecting surface intonation contours, and comment on how these effects may impact the current study and how at least some of the problems can be addressed in the experiment:

- Declination: A global downtrend referring to the tendency of F_0 to decline over the course of an utterance ('t Hart & Cohen, 1973; Maeda, 1976; Pierrehumbert, 1979; Pierrehumbert, 1980; Cohen *et al.,* 1982; Fujisaki, 1983; Ladd, 1984; Strik & Boves, 1995).
- Downstep: A lowering effect that is triggered by a low (L) accent or a L tone, resulting in a descending step-like function in F_0 contour (Liberman & Pierrehumbert, 1984; Pierrehumbert & Beckman, 1988; Shih, 1988; Prieto, Shih & Nibert, 1996).
- Final Lowering: An additional lowering effect near the end of a sentence (Maeda, 1976; Liberman & Pierrehumbert, 1984; Pierrehumbert & Beckman, 1988; Herman, 1996).
- Accents and tones: Accents and tones create local F_0 excursions. Each accent and tone could be emphasised or de-accented, reflecting the meaning and the structure of the sentence, and the speaker's rendition of the sentence (Liberman & Pierrehumbert, 1984; Eady & Cooper, 1986; Jin, 1996).
- Segmental effects: A lot of the observed F_0 movements are caused by segmental effects. Voiceless fricative and aspirated stops raise F_0, while sonorants typically lower F_0. Low vowels have intrinsically lower F_0 than high vowels (Peterson & Barney, 1952; Lea, 1973; Silverman, 1987).

- Intonation type: Sentence intonation such as declarative, exclamation or question intonation may interact with any of the aforementioned effects. Some intonation models treat declination as the property of declarative intonation (Thorsen, 1980a; Gårding, 1987).
- Discourse structure: Pitch is typically raised in the discourse initial position and lowered in the discourse final position, and topic initialisation is typically associated with high pitch (Hirschberg & Pierrehumbert, 1986; Sluijter & Terken, 1993; Nakajima & Allen, 1993; Yang, 1995).

There are many usages of the term "declination", associated with different schools of thought. Ladd (1993) classifies them as the *overt decline approach* and the *implicit decline approach*. The IPO model ('t Hart & Cohen, 1973) is the strongest proponent of the overt declination school, where the declination slope is directly observable from the surface intonation contour. A line fitted through the peaks (if prominence variation can be controlled) reflects the topline declination, while a line fitted through the valleys reflects the baseline declination. Other studies that can be subsumed under the overt approach include Maeda (1976) and Umeda (1982).

The implicit declination school includes Fujisaki's (1983) mathematical intonation models where the surface intonation contour is the sum of the phrase component and the accent component. Declination is not directly observable from the intonation contour, but is accounted for by the phrase command, which is a damped function responding to an impulse command, calculated logarithmically. Pierrehumbert's (1980) model offers another implicit decline approach, where declination refers to a global, gradual effect which is modelled as a declining straight line (the baseline) which creates a downward tilt of an intonation contour, and accounts for the perceptual equivalence of early and late peaks with different F_0. The baseline is not a line fitted through the valleys in the intonation contour, nor does it parallel the topline since the topline is subject to other lowering effects and prominence variation. In later models, declination is calculated as the residual downtrend when the downstep effect is factored out (Liberman & Pierrehumbert, 1984; Pierrehumbert & Beckman, 1988). Liberman and Pierrehumbert (1984) made the strong claim that after the downstep effect was factored out, there was no evidence of declination in English. This position was softened in Pierrehumbert and Beckman (1988) where both declination and downstep were incorporated in the modelling of Japanese intonation.

Mandarin offers a case where the implicit declination effect as defined in Liberman and Pierrehumbert can be observed overtly: First, in a sequence of high (H) level tones (tone 1), low (L) tonal targets are absent both phonologically and phonetically, so by definition there will be no trigger for any downstep effect. Secondly, the tone shape is level, which means that the contribution from the accent and the tone to the observed F_0 movement is minimal. There are still a few factors at work, of which final lowering and segmental effects are relatively easy to control, and the local prominence effect can be averaged out or smoothed out to some extent, when the sample size is large and the locations of uncontrolled local prominence distribute randomly in different sentences. Strictly speaking, isolated sentences read under experimental conditions are under the influence of the single-sentence discourse structure. However, the magnitude of discourse-raising or lowering in an isolate sentence, if they can be considered as such, is small comparing to what happens in the reading of long text with several paragraphs and in natural conversation. A natural extension of the current study will be to study paragraphs in high level tones to find out the difference between simple and complex discourse structure, and to model pitch range variations in complex discourse structure.

Under the current experimental conditions, the surface F_0 contour of the experiment sentences is a close approximation of the topline of the Liberman and Pierrehumbert model, and the observed downtrend on the surface is very close to the residue declination effect without downstep

Note that the proposed experimental materials do not offer a direct observation of the phrase command of the Fujisaki model. The observed F_0 values of a given syllable are the combination (in the log domain) of the phrase command and the accent command corresponding to the tone, plus any previous accent commands still in effect (Fujisaki, Hirose, Halle & Lei, 1990; Wang, Fujisaki & Hirose, 1990). If the tonal contribution (accent command) from each syllable is the same, the observed F_0 will be parallel to the phrase command after the initial effect subsides.

11.2 Experimental Design

This experiment investigates the pattern of declination in Mandarin and the possible interaction of declination with sentence length, final lowering and prominence. The database consists of 640 sentences: 10 test sentences with 10 length variations, 2 focus conditions, 2 final conditions, 4 repetitions, and 4 speakers (two females and two males, two from northern China and two from Taiwan). Each of the 10 test sentences starts with a two-syllable sentence frame in a low tone (tone 3) and a rising tone (tone 2) *Lao3 Wang2* "Old Wang". Tone 3 has a low target (L) and tone 2 rises from low to high (LH), so the frame provides a reference to the speaker's pitch range. Sequences of high-level tones (tone 1, H) ranging from 2 to 11 syllables long follow the frame. The sentences share the theme of a person *Lao3 Wang2* cooking a winter melon in various ways. Sonorant and unaspirated consonant are used as much as possible to minimise consonantal effects on F_0. Most of the syllables have mid or low vowels, and end in a nasal coda [n] or [ng].

The final condition refers to a set of the test sentences which have tone 1 in the utterance final position. A non-final condition is created by adding a sentence final perfective particle *le* to the test sentences so that the last tone 1 syllable is no longer utterance final. If there is a final lowering effect affecting the utterance-final unit, whether the unit being syllable, word, or a fixed time interval, the last tone 1 syllable in the sentences without *le* should be lower than the ones in the sentences with *le*.

These 20 sentences were presented in two ways to the speakers: unadorned, plain sentences intended to elicit unmarked reading style, and sentences with a leading question such as *Shei2 zheng1 gua1?* "Who steams the melon?", intended to elicit a narrow focus reading with the prominence landing on the frame *Lao3 Wang2*. In the focus sentences all test syllables (tone 1 syllables) are actually not in focus. The four test conditions are exemplified in Table 11.1 with the sentence with two tone 1 syllables *Lao3-Wang2 zheng1 gua1* "Old Wang steams the melon".

Table 11.1. Experiment sentences in four test conditions. In addition, there are ten length variations.

Plain, final		Lao3-Wang2 zheng1 gua1
Plain, non-final		Lao3-Wang2 zheng1 gua1 le0
Focus, final	(Shei2 zheng1 gua1?)	Lao3-Wang2 zheng1 gua1
Focus, non-final	(Shei2 zheng1 gua1?)	Lao3-Wang2 zheng1 gua1 le0

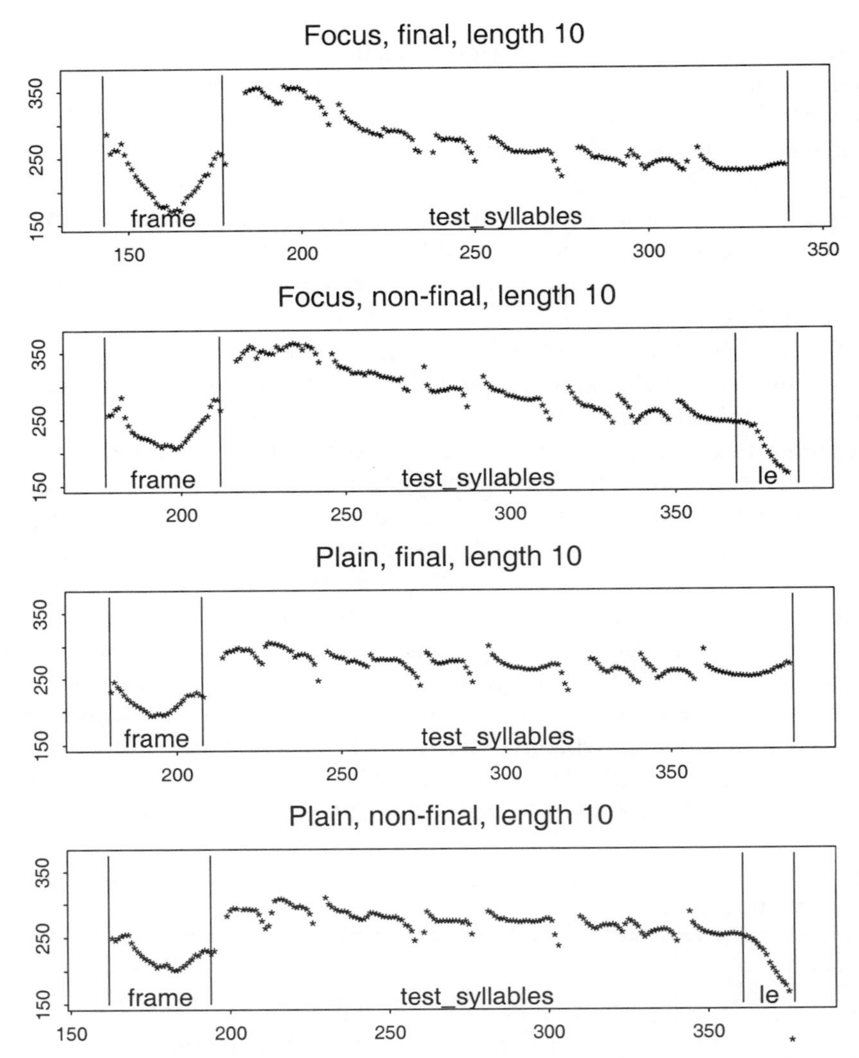

Figure 11.1. Pitch tracks of one set of experiment sentences in the focus/plain, final/non-final conditions. The sentences shown have ten test syllables.

Figure 11.1 shows the pitch tracks of a set of test sentences. Each sentence begins with the tone 3-tone 2 (L-LH) frame, labelled as "frame". There are ten test syllables in these sentences, which are labelled as "test syllables". All test syllables have tone 1 (H). Sentences in the *final condition* end in a tone 1, while sentences in the *non-final condition* end in a neutral tone syllable *le0* labelled as "le". A neutral tone syllable is unstressed and is realised with low pitch.

Although the end of the frame is tonally equivalent to the first test syllable, both have the tonal specification H, it is common in Mandarin that the end of a tone 2 doesn't reach the high level if the following tone is also high (Shih, 1988; Jin, 1996; Xu, 1997).

Three of Speaker C's sentences were read incorrectly and were discarded, resulting in a total of 637 sentences and 4139 tone 1 syllables in the database. One F_0/time measurement is taken from each syllable: the lowest point of tone 3, the highest point of tone 2, and near the center of the rhyme of tone 1, at least 50 msec after the onset of the rhyme.

11.3 Supporting Experiment: Segmental Effects

One disadvantage of using real, meaningful sentences is the complication of segmental effects, since different syllables must be used and measurements taken from various syllables may not be comparable. A supporting experiment was done to examine the effect of segmental perturbation to evaluate how comparable these syllables are to each other, and which regions are under the most heavy influence of segmental perturbation. If necessary, data from this study can be used to normalise segmental effects.

One of the speakers (Speaker B) read the sentences described in Table 11.2 in addition to the experimental sentences described earlier. The keywords were the test syllables used in the experiment *zheng1, dong1, gua1, gang1, zhong1, yi1, jin1, tian1, bang1* plus *ma1*, the favourite syllable used in reiterant speech. Each sentence was repeated six times.

Table 11. 2. Sentences for the supporting experiment on segmental effects.

	Ta1 He	shuo1 said	KEYWORD KEYWORD	san1 three	bian4 times	

The F_0 contour of each keyword were sampled in 30 points. Figure 11.2 plots average trajectories of the rhyme regions of some of the keywords with the F_0 value in Hz plotted as a function of sample number. Each line in the plot represent the average of six tokens.

The two top panels compare syllables with similar rhymes. As the initial consonant effect subsides, the remaining pitch contours are very similar. Syllables with a low vowel [a] (top left panel) have lower F_0 values in the vowel region than the syllables with mid vowel [e] or [o] (top right panel). The similarity of the three trajectories in the top right panel suggests that there is no difference between front and back mid vowel [e] and [o]. The lower left panel compares *gua* with *gang*, The [u] region of *gua* is high due to the high vowel intrinsic effect, while the [a] regions of these two syllables have comparable F_0 values even though one occurs early in the rhyme and one late in the rhyme. The lower right panel compares the syllable *jin* and the syllable *tian*. The first 30% of the rhyme region of *tian* has higher F_0 than *jin*, due to the raising effect of [t]. The main vowel of *tian*, a fronted low vowel, has lower intrinsic F_0 than the main vowel [i] in *jin*. The intrinsic F_0 differences of *jin* and *tian*, which form the word "today", become an issue that will be addressed in Section 11.4.2.

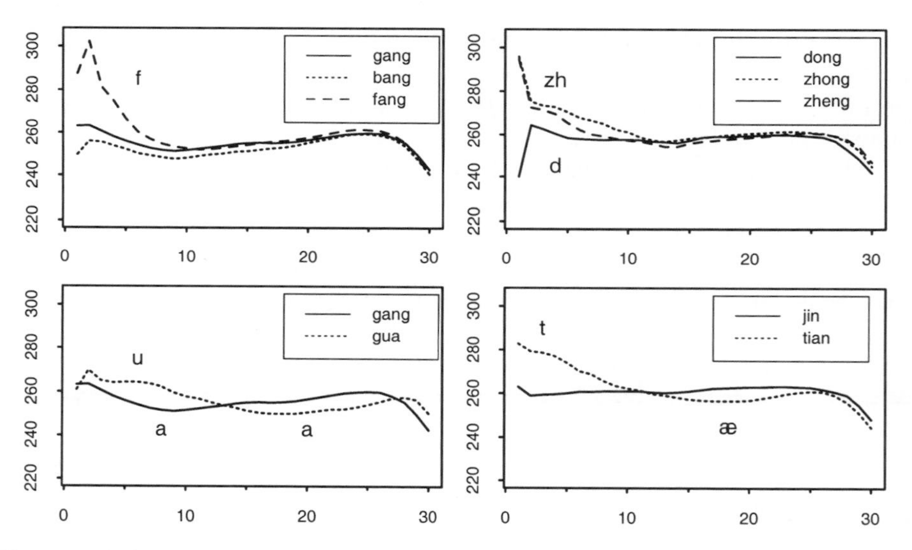

Figure 11.2. Segmental effects on high tone syllables.

Unaspirated voiceless stops, represented as [b], [d], [g], have rather short effects. Aspirated stop [t], retroflex affricate [zh], and fricative [f] raise the pitch considerably and the effects extend into the 10th sample, or the first 30% of the rhyme region. The intrinsic vowel effects were largely consistent with the previous finding, and were predictable from vowel height, with the low vowel [a] having the strongest lowering effect.

On the whole the result is encouraging in that at least the averaged consonantal effects appear to be manageable. The effects are predictable by phone classes and the strongest effects can be excluded from the measurement if the edges of the rhyme are avoided. The effects are surprisingly consistent, therefore they can be normalised if need be. For the purpose of this paper, we will avoid normalisation and use instead the raw unnormalised data.

11.4 Results and Discussions

ANOVA analysis was performed to test what factors have an effect on the scaling of F_0 values. Preliminary inspection of the data suggests that the location of word boundary and the location of the verb have some effects, so these two factors were coded in addition to the three factors in the experimental design. The five factors were used in the ANOVA analyses to model the difference in F_0 value between adjacent tone 1 syllables: sentence length (10 levels, from 2 to 11 tone 1 syllables), focus (2 levels, with or without narrow focus in the frame), final (2 levels, with or without final *le*), word (2 levels, whether the two syllables in question straddle a word boundary), verb (3 levels, the second syllable is a verb, the first syllable is a verb, or neither is a verb). The results are given in Table 11.3.

The ANOVA results show that sentence length, verb, and word have significant effect in at least three of the four speakers. Final condition has no effect, and focus has an effect on two of the speakers. These conditions are discussed in more detail below.

11.4.1 Final Lowering

Figure 11.3 plots the averaged F_0 trajectory of the four speakers, including the two frame syllables at position 1 (tone 3, L) and position 2 (tone 2, LH). The F_0 profiles in Figure 11.3 are obtained by averaging F_0 values of each speaker by position. The early positions represent more observations than later ones. A point at positions one to four represents 80 samples, while a point at position 13 represents only 8 samples. The zigzag pattern from position 11 to 13 is caused by both the smaller sample size and the similar syntactic construction used in those three positions.

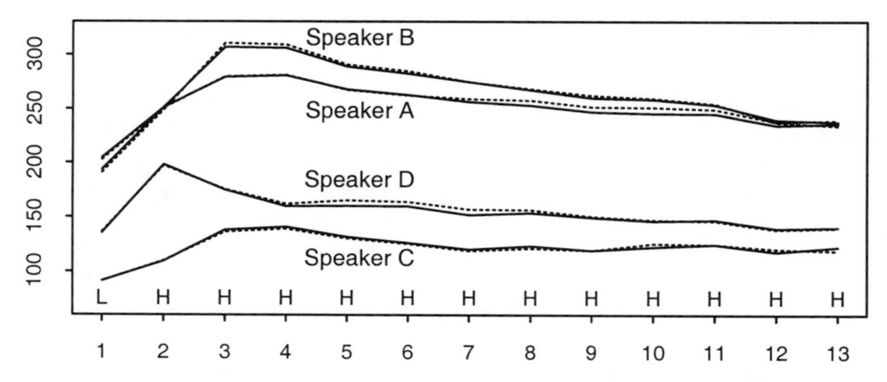

Figure 11.3. Averaged F_0 trajectories of sentences in the final (without *le*) and non-final (with *le*) conditions. The solid lines represent sentences in the final condition, the dotted line represent sentences in the non-final condition.

Tone 1 syllables (H) occur from position 3 and on. The final and non-final conditions are plotted separately: the final condition without *le* in solid lines and the non-final condition with *le* in dotted lines. The two populations match closely within each speaker. Not surprisingly, the difference between the final and non-final conditions are not significant in the ANOVA analysis. Since all data samples were included in the ANOVA analysis, and it is reasonable to assume that final lowering only affects the final section of an utterance, the lack of significance for final lowering in Table 11.3 could be a result of including sample points from other positions. However, when just the last tone 1 syllables from the final vs. non-final conditions are compared, the differences are still not significant for all four speakers. The results here show a lack of final lowering effect on Mandarin high tones under the assumption that final lowering has the strongest effect on the last syllable of an utterance, a reasonable assumption that is backed by plausible physiological causes (Maeda, 1976; Strik & Boves, 1992), and also in light of data from Kipare (Herman, 1996) where final lowering was found as a separate effect from declination which affects up to three syllables from the end, and the effect increases in magnitude as the sentence approaches the end.

Table 11.3. Results of ANOVA

Speaker A: Female from Beijing					
	Df	Sum of Sq	Mean Sq	F	Pr(F)
Length	9	750.89	83.43	1.79	0.07
Final	1	17.48	17.48	0.38	0.54
Focus	1	10.62	10.62	0.23	0.63
Word	1	113.96	113.96	2.45	0.12
Verb	2	3633.28	1816.64	38.98	0.00
Resid.	865	40315.07	46.61		
Speaker B: Female from Taiwan					
	Df	Sum of Sq	Mean Sq	F	Pr(F)
Length	9	4408.15	489.79	4.83	0.00
Final	1	489.67	489.67	4.83	0.03
Focus	1	32564.18	32564.18	321.09	0.00
Word	1	939.20	939.20	9.26	0.002
Verb	2	4559.81	2279.91	22.48	0.00
Resid.	865	87727.46	101.42		
Speaker C: Male from Taiwan					
	Df	Sum of Sq	Mean Sq	F	Pr(F)
Length	9	908.18	100.91	1.96	0.04
Final	1	55.25	55.25	1.07	0.30
Focus	1	75.26	75.26	1.46	0.23
Word	1	312.69	312.69	6.08	0.014
Verb	2	9672.35	4836.18	94.06	0.00
Resid.	847	43551.47	51.42		
Speaker D: Male from Tianjin					
	Df	Sum of Sq	Mean Sq	F	Pr(F)
Length	9	5654.57	628.29	9.35	0.00
Final	1	62.57	62.57	0.93	0.34
Focus	1	3299.44	3299.44	49.12	0.00
Word	1	6932.94	6932.94	103.22	0.00
Verb	2	4838.29	2419.15	36.02	0.00
Resid.	865	58098.01	67.17		

In the berry-list experiment of Liberman and Pierrehumbert (1984), the final lowering effect was estimated from the final stressed syllable of an utterance. Although they assumed that final lowering affects the final stretch of an utterance, the result was consistent with a hypothesis that final lowering affects the last *stressed* syllable in an utterance. This latter interpretation could in principle accounts for the lack of effect in the two conditions here: The added syllable *le* in the non-final condition is an unstressed particle, so the syllables being tested in the final and non-final conditions may be subject to the same amount of final lowering under this alternative hypothesis. The relevant test would then involve sentence pairs as in Table 11.4, where sentence (b) is formed by adding one stressed syllable to the end of sentence (a). Under this alternative hypothesis, the prediction is that the underlined syllable *gua* should have lower F_0 value in (a) than in (b). There are three sentence pairs like this in the experiments, but again no significant effect is found from any of the four speakers.

The conclusion here is that there is no final lowering effect on Mandarin high tones under the described experimental conditions. The final and non-final conditions are thus collapsed for subsequent analysis of the declination effect.

Table 11.4. Test sentences for the alternative hypothesis of final Lowering.

a.	Lao3-Wang2 zheng1 dong1 gua1
b.	Lao3-Wang2 zheng1 dong1 gua1 zhong1

11.5 Declination

We now turn to the plain sentences to investigate the declination effect. In this section, we will show that there is a declination effect in Mandarin sentences, and propose a model to account for the observed data. The model has some useful properties that makes it attractive to a text-to-speech system: First, it is pitch range independent. The model gives reasonable prediction under changes of pitch ranges. Second, it handles arbitrarily long text gracefully, which is an important property for a text-to-speech system that needs to process unrestricted text input. Finally, as we will discuss in Section 4.3 and Section 4.4, the model successfully captures the variations in declination slopes resulting from the use of focus and changes in sentence length. We believe that variations resulting from discourse structure can be accommodated as well.

Figure 11.4 show the average trajectories of the plain sentences for four speakers. All four speakers show a decline in tone 1 values starting from the first or the second tone 1 syllable, or the third or fourth position in a sentence. The last values in position 13, although phonologically equivalent to those in position 3, lie around the mid point between the initial L and the highest H. The declination rate is faster in the beginning and slows down as the sentence progresses. This pattern of declination is reminiscent of the downstep equation proposed in Liberman and Pierrehumbert (1984), where each successive peak is modelled as a fraction of the distance of the previous peak and the reference line plus a constant.

An exponentially decaying declination model can handle long sentences more gracefully than a time constant model which deducts a fixed amount of F_0 value over the same amount of time. Such a model allows pitch to quickly drop below a level that is possible for the human speech apparatus. In particular, the magnitude of pitch decline we observe here is much larger than the 10Hz/second drop previously proposed (Pierrehumbert & Beckman, 1988; Gooskens & van Heuven, 1995). The average declination rate calculated from the plain sentences are 34 Hz/sec, 25 Hz/sec, 18 Hz/sec, and 17 Hz/sec for speakers A, B, C, D respectively.

The equation below is used to estimate the F_0 value of a given syllable P_i from the F_0 value of the preceding syllable P_{i-1}. The coefficients α and μ fitted for each speaker are given in Table 11.5, together with the correlation of the observed values and the predicted values.

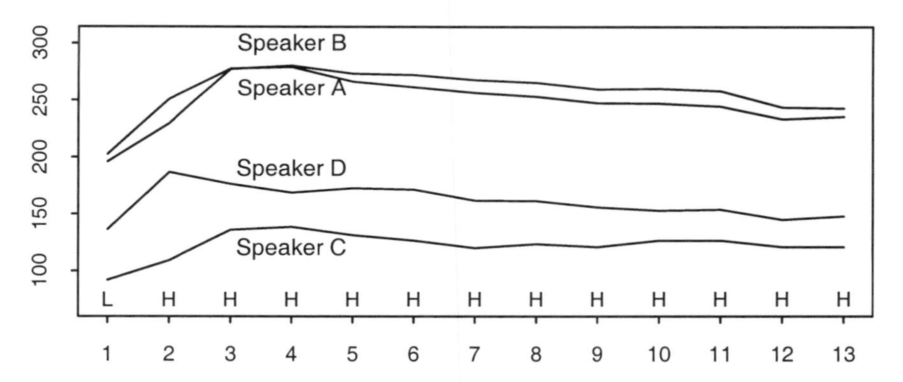

Figure 11.4. Averaged F_0 trajectories of sentences in the plain condition.

Table 11.5. Fitted coefficients obtained from the Plain sentences of the four speakers by Equation (1).

Speaker	α	μ	Correlation (R^2)
A	0.90	21.77	0.83
B	0.91	20.82	0.73
C	0.75	29.29	0.51
D	0.81	27.63	0.62

$$P_i = \alpha P_{i-1} + \mu \tag{1}$$

Figure 11.5 plots the predicted values by the four models for sentences 50 syllables long, using the averaged values of the first tone 1 of the plain sentences of each speaker as the initial values P_1. F_0 values decline faster in the early section of an utterance, then asymptote to a value which is appropriate for the speakers's pitch range.

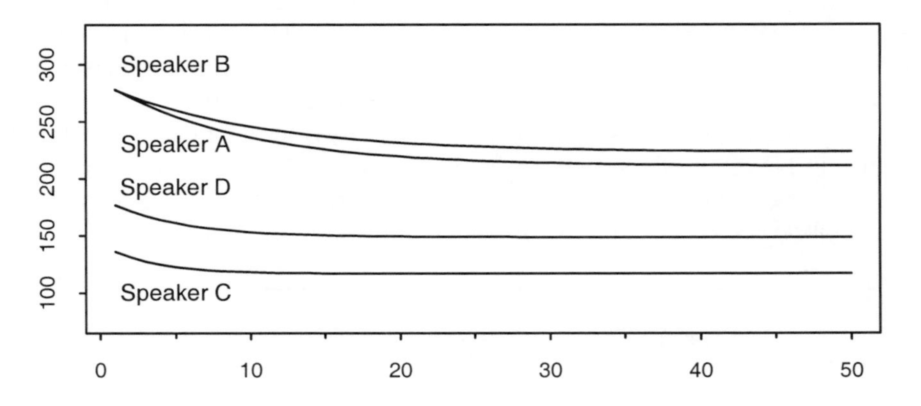

Figure 11.5. Declination slopes of the four speakers predicted by the declination models.

The coefficient α in Table 11.5 controls how fast F_0 value approaches the asymptote, the smaller the value of α, the steeper the decline. The coefficient μ controls the asymptote value: when μ equals 0 the F_0 values asymptote to 0; when μ equals the difference of the initial value P_1 and $\alpha(P_1)$, the asymptote value is P_1, with zero declination effect. The coefficients α and μ are fitted optimally with the observed value of P_1. The model is not very forgiving when the value of P_1 changes. For example, if P_1 is lowered so that μ is bigger then $P_1 - \alpha(P_1)$, F_0 values will rise instead of decline. This is not a desirable property for a text-to-speech system where the value of P_1 needs to be independently manipulated to model discourse effect (Hirschberg & Pierrehumbert, 1986), among other things.

A revised model in Equation (3) addresses this problem by explicitly incorporating P_1 in the equation, where μ is expressed as a fraction of $P_1 - \alpha(P_1)$. Given the P_1 value that Equation (1) is fitted for, the value μ in Table 11.5 can be straightforwardly transformed to the value β in Table 11.6 by Equation (2). The means of the initial tone 1 for each speaker are the P_1 values, which are included in Table 11.6. For the given P_1, the two models make the same prediction.

$$\beta = \mu / (P_1 - \alpha(P_1)) \tag{2}$$

$$P_i = \alpha P_{i-1} + \beta(P_1 - \alpha(P_1)) \tag{3}$$

Equation (3) gives more reasonable prediction than Equation (1) when the P_1 value changes. Minimally there will be no F_0 upstep when P_1 value decreases. Furthermore, P_1, α and β are intuitively linked to the initial value, the rate of decline, and the asymptote value, three parameters that can be fitted or hand-tuned to generate variations in the declination slope that correspond to naturally occurring intonation variations, including the lack of declination, where β equals 1.

Everything being equal, increasing P_1 gives higher asymptote value, which equals βP_1, consistent with the impression that when speakers raise their pitch register, the entire pitch contour floats on top of a comparable sentence spoken in a lower pitch register. A lower α value gives a steeper decline where the asymptote value is reached earlier. A lower β value gives a lower asymptote value. Being able to control these parameters is very convenient for a Text-to-Speech system. In later sections it will be shown that the variations in sentence length can by modelled simply by varying P_1. Focus, to be discussed in Section 4.3, is a complex phenomenon requiring the combination of raised/expanded pitch range for the word in focus, as well as steeper decline and lower asymptote value in the post-focus material, which can be modelled by increasing P_1 value while lowering α and β values.

Table 11.6. Fitted coefficient values for the four speakers by Equation (3), the revised declination model.

Speaker	α	β	P_1	**Correlation (R^2)**
A	0.90	0.76	278.46	0.83
B	0.91	0.80	277.86	0.73
C	0.75	0.86	135.88	0.51
D	0.81	0.84	176.78	0.62

Before we turn to the subject of focus and sentence length, we will address one unexpected peculiarity that can be seen in Figure 11.4. Most declination theories predict that the effect starts from the beginning of the sentence. It is not entirely clear why the declination effect seems to start one syllable late for speakers A, B, and C. On average the second tone 1 syllable appears to be at least maintaining the height of the first tone 1 syllable. The proposed model predicts that the biggest F_0 decline in terms of Hz value should be from the first to the second tone 1 syllable.

There are several possible explanations for this phenomenon. First, it could be an artefact: The syllable *tian1*was used as the second tone 1 syllable in sentences with five or more tone 1 syllables (sentence length seven syllables and up). It was the only aspirated consonant used in the experiment sentences, which was shown in the supporting experiment to have a strong consonantal effect raising F_0 in the first 30% of the rhyme, see Figure 11.2. It is possible that there are some remaining effect at the location of F_0 measurement. This consonantal effect could be at least partially responsible for the lack of observable declination effect on the second tone 1 syllable, since the problem was most obvious in sentences with five or more tone 1 syllables, exactly where the syllable *tian1* was used. See Figure 11.7 in Section 11.4.4 where the sentence trajectories are plotted by length.

Second, what we observe here may be a general property of like tone sequence (especially high tones) that the intended pitch height is reached later than what is expected from the lexical specification. This phenomenon can be observed in Mandarin in tone 2-tone 1 (LH-H) combinations (Shih, 1988; Xu, 1997). Phonologically the end of tone 2 is the same as tone 1, but the end of tone 2 rarely reaches the high level when the following tone starts high. In the experiment, the second syllable (the frame) is a tone 2 followed by tone 1, and in three of the four speakers, the end of tone 2 is considerably lower than the following tone 1, see Figure 11.3. A more dramatic version of this effect is clearly documented in Yoruba high tone sequences (Laniran, 1992) where the transition from a low tone to a string of high tones spans several syllables. In alternating high and low tone sequences, in Mandarin as well as Yoruba, no such delay is observed. It could be that the second tone 1 in this experiment represents the actual target of the high tone level while the first tone 1 is still in transition, and that the effect shows up only in long sentences because the slow transition is made possible only when the high tone sequence reaches a critical length.

Alternatively, there is a straightforward re-interpretation of this point from the Fujisaki model (Fujisaki *et al.*, 1990; Wang *et al.*, 1990). If the tone 1 syllables are sentence initial, the higher second syllable reflects the shape of the phrase command, which rises initially from the rest position after the impulse command becomes effective. If the tone 1 sequence follow a low or rising tone, as in our experiment sentences, the rising F_0 from the first to the second tone 1 is the result of the rising accent command still effective after the low or rising tone.

Third, two of the three speakers with non-declining second tone 1 also show less F_0 drop within a word. Since in most of the test sentences the first and second tone 1 syllables form a word *jin1-tian1* "today", it is possible that the effect is related to the within-word status.

A follow up experiment was done comparing 18 sentences: Two frames: one with two tone 1 syllables and one with the same frame as the main experiment (low and rising tones). Three lengths: 6, 8, and 10 syllables in the sentences. Three different syllables as the second tone 1, including one with *tian1*. Three different syntactic structures so that the first and second tone 1 form a word in two of the three sentences. One of the word conditions used *jin1 tian1* "today", the same as most of the long test sentences in the main experiment. The 18 sentences were randomised and repeated five times each by one speaker. The results show no effect from the frame and the length variations. The syllable *tian1* were higher than the other two test syllables in the same sentence position, and the F_0 decline is less in the second tone 1 if it is in the same word as the previous syllable. The follow up experiment suggests that the first and third explanations above were supported, while the slow transition hypothesis is not supported.

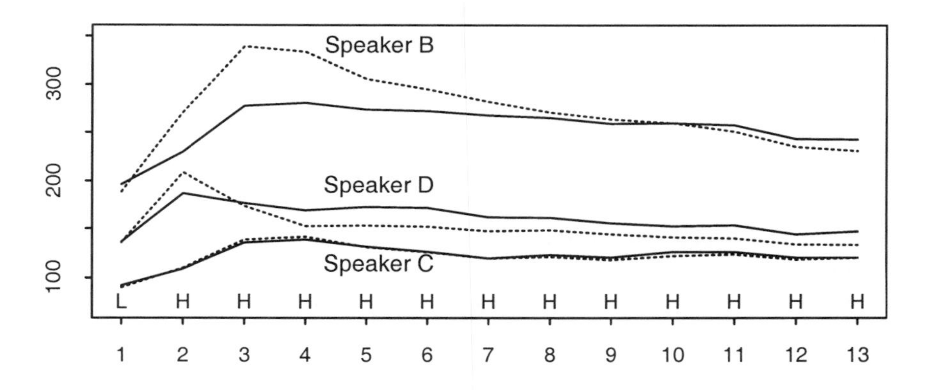

Figure 11.6. F_0 trajectories of sentences in plain (solid lines) and focus (dotted lines) conditions.

11.6 Focus

Figure 11.6 plots the averaged F_0 trajectory of the plain sentences (solid lines) and the sentences with focus (dotted lines) for three of the speakers. Speaker A's trajectories in this case are similar to the ones shown in Figure 11.3, so her data are omitted from the plot. Again, the two frame syllables are included in the plot.

Speakers A and C did not use F_0 to differentiate focus vs. plain reading, and the two focus conditions of these two speakers have nearly identical F_0

trajectories. Speakers B and D both raised F_0 to signal prominence. Speaker D raised F_0 on the word with narrow focus, and the highest F_0 value was reached by the end of the rising tone *Wang2*, after that F_0 started to decline and in two syllables it reached a level lower than the plain sentence and remained low throughout the sentence. Speaker B showed a similar pattern except that the highest point was on the first tone 1 syllable, or the third syllable position of the sentence, which is one syllable *after* the narrow focus. It was not until the 10th position, eight syllables after the narrow focus, that the F_0 trajectory crossed over the plain sentence trajectory. After the crossover the pitch value remained below the plain sentence until the end of the sentence. This observation is compatible with the claim that narrow focus is realised with expanded pitch range, and after the focus the pitch range is compressed (Gårding, 1987). However, one difference is in how soon the compression of pitch range occurs. Gårding's claim is that the compression happens immediately after the narrow focus, which is true if there is a low tone following the focus (Shih, 1988), or if there is a phrase break after the focus, but in the case of post-focus high tone sequences in the same phrase, the compression effect is gradual.

All four speakers lengthened the frame syllables *Lao3 Wang2*in the focus condition (Jin, 1996). The durational difference between the focus and the plain populations is statistically significant in three speakers except Speaker A, suggesting that speakers B, C, and D successfully followed the cues and put emphasis on *Lao3 Wang2*. Speaker A lengthened *Lao3 Wang2* in some of the focus sentences, but didn't do so consistently. However, only two speakers, B and D, raised the pitch range in addition to lengthening in their rendition of focus. This observation is consistent with the report that F_0 cues does not have to be present for the perception of emphasis in Mandarin (Shen, 1993).

The top panel in Table 11.7 compares the mean duration in msec for the syllable *Wang2* in the focus and the plain conditions. The duration of *Wang2* is longer in the focus condition than in the plain condition for all speakers. Also note that the duration of *Wang2* from speaker D is much longer than the other three speakers. This is because speaker D placed a short phrase break after the subject *Lao3 Wang2*, which led to considerable phrase-final lengthening.

Table 11.7. Top panel: Mean duration (in msec) of the frame syllable Wang. Lower panel: Mean duration of all test syllables.

Mean Duration of the Syllable Wang				
Speaker	**Focus**	**Plain**	**df**	**p value**
A	178	176	878	p=0.0985
B	170	145	878	p<0.05
C	167	146	860	p<0.05
D	262	238	878	p<0.05
Mean Duration of All Test Syllables				
Speaker	**Focus**	**Plain**	**df**	**p value**
A	174	173	878	p=0.7595
B	161	166	878	p=0.0084
C	166	168	860	p=0.618
D	166	170	878	p=0.0923

Table 11.8. Declination coefficients for the focus and plain conditions

Speaker	Focus Condition	α	β	P_1	Correlation (R^2)
B	Plain	0.91	0.80	277.86	0.73
B	Focus	0.89	0.42	339.16	0.87
D	Plain	0.81	0.84	186.97	0.62
D	Focus	0.59	0.78	208.00	0.47

The average duration data from all test syllables is included in the lower panel to show that Speaker D's speaking rate is otherwise comparable to Speaker B and C. The phrase break is the most plausible cause of the lengthening of the syllable *Wang*, and of the difference in the post-focus declination pattern of Speakers B and D.

There is also a trend to shorten the post-focus materials (Jun and Lee, 1998), shown by the shorter duration of the test syllables (lower panel) in the focus condition, in contrast to the plain condition. This effect is more robust in speakers B and D, the two speakers who happen to raise pitch range to signal focus. For Speaker B, who also keeps the post focus materials in the same phrase as the frame word in focus, the durational difference in focus and plain conditions is statistically significant.

For Speakers B and D, who consistently use expanded pitch range to convey focus, the post-focus declination slopes are different from the plain sentences. Models were fitted separately for the focus and plain conditions for these two speakers, and the results are summarised in Table 11.8 in the same format as Table 11.6.

For both speakers, the α and β values fitted for the focus condition are lower than those fitted for the plain condition, suggesting a steeper decline and a lower asymptote for the post-focus declination slope. The post focus F_0 patterns reported here is consistent with other studies of Mandarin (Shih, 1988; Liao, 1994; Jin, 1996; Xu & Wang, 1997). Speaker D reaches asymptote much faster in the focus condition, and this is captured by his very small α value.

11.7 Sentence Length

Sentence length has a clear effect on the scaling of Tone 1's. The question is what to change when sentence length changes. Figure 11.7 plots the sentence trajectories from all speakers. Sentences with different length are plotted separately in each plot. Plain sentences are plotted in the left column and focus sentences in the right column. Each line represents eight repetitions of the same sentence (final and non-final conditions combined).

In this plot, individual characteristics of some sentences show up clearly. We will discuss some effects such as the word boundary effect and part of speech effect in the next section.

The initial tone 1's are higher in long sentences than in short sentences, as speakers took a deeper breath to prepare for the delivery of longer materials. Although there are ten variations in sentence length in the experiment, the speakers seem to aim at roughly three pitch range settings. This is reasonable, since we can hardly expect speakers to count syllables before they start reading. The four syllable sentences with two tone 1 syllables, the shortest in the experimental sentence set, were a notch lower than the rest. The medium length sentences started with medium pitch range, while long sentences had the highest pitch range in the initial portion of the sentences. The same phenomena was reported in Thorsen (1980b). The low tones of the sentence initial tone 3's are slightly lower in short sentences, but the fluctuation is not as dramatic as in the high tone range.

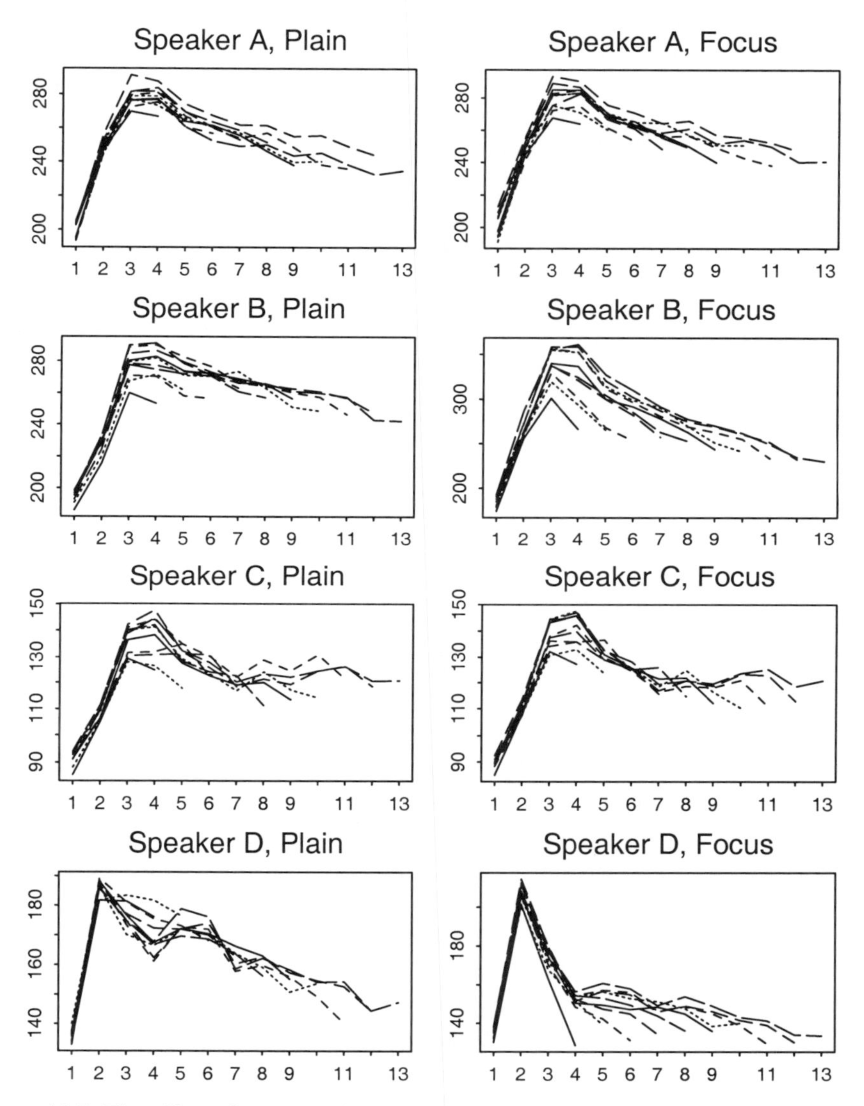

Figure 11.7. The effect of sentence length: F0 trajectories plotted by focus condition and by sentence length.

Comparing long and short plain sentences, the short sentences start low and end high. To model this phenomenon the starting point of each sentence (P_1 of Equation (3)) should be assigned with reference to sentence length. The higher ending of the short sentences and the lower ending of the long sentences fall out from the model, where longer sentences accumulate more declination effect.

Given the same sentence delivered many times in a similar style, the most common variations we see in the data is a difference in the overall pitch range. When speakers start higher, they also end higher. This aspect of the natural variation in speech can be modelled by Equation (3) with different P_1 values. One example is given in Figure 11.8, which shows eight repetitions from Speaker A for the sentence *Lao3 Wang2 zheng1 dong1 gua1 zhong1*, "Lao3 Wang steams the winter melon pot".

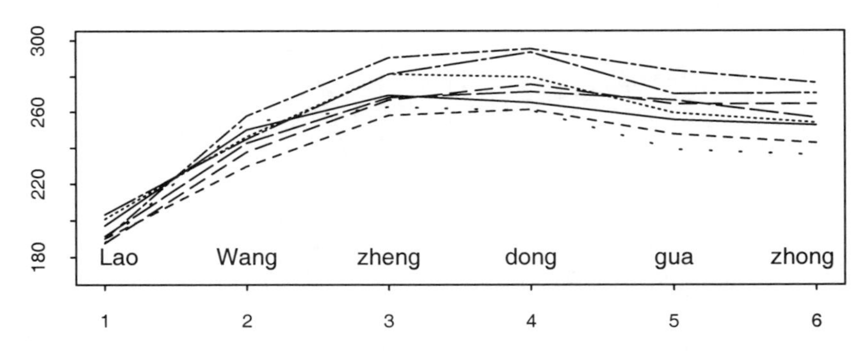

Figure 11.8. Pith range variations in eight repetitions of a sentence.

One interesting question regarding sentence length is whether longer sentences have more gradual declination slopes than short sentences as expressed by increasing values of α. We know that the slopes would be more gradual in longer sentence in our data if we follow Thorsen's model of Danish (1980b) by fitting a least squares regression line through all data points representing tone 1, since the data seems to suggest, and our model also predicts that most of the decline concentrates in the early section of an utterance, a well-known phenomenon that have already been accounted for in many different ways (Maeda, 1976; Cooper & Sorensen, 1981). There are two alternative ways to test this inquiry: fit the value of α by sentence length, and compare linear declination slopes by sentence length from the same sentence positions (say from the 4th to the 7th syllable). Both tests fail to yield a consistent picture that declination slope become more gradual as sentence length increases. It appears that P_1 value is the only parameter that varies with sentence length consistently, although as noted earlier, speakers vary initial pitch value according to a rough classification of sentence length.

Although there is a significant effect of sentence length on the F_0 value of the last syllable in plain sentences, as shown in the left column plots of Figure 11.7, the effect is weak in the focus condition for Speaker B and D, the second and the last plots in the right column of Figure 11.7. The ending

F_0 values in these plots are more similar to each other than the corresponding plain sentence plots to the left. A plausible explanation is that in the focus condition the utterance reaches asymptote sooner than in plain sentences, as a result of a lower α value in the model, therefore the ending F_0 values of long sentences are similar.

11.8 Lexical and Part of Speech Categories

Word boundary has a weak effect on F_0 scaling but the effects are not consistent across speakers: two speakers show more F_0 drop across word boundaries and two speakers show more F_0 drop within words. Also in the follow-up experiment (one speaker) where the effect of word boundary was specifically tested, it appears that the speaker attempted to maintain F_0 height within the word, and there is less F_0 drop within than across words.

The effect of verbs is consistent across all four speakers: verbs have lower F_0 values than surrounding polysyllabic nouns, creating shallow F_0 valleys in those cases. Many of the zigzag patterns in Figure 11.7 correspond to pitch lowering on verbs. The most plausible explanation of this phenomenon is that verbs are metrically weaker than object nouns in Mandarin. All verbs used in this experiment are monosyllables, while most of the selected nouns are disyllabic. This asymmetric distribution mirrors the natural distribution in the language: most of the colloquial verbs are indeed monosyllabic while nouns in Mandarin have been going through a disyllabification process since Middle Chinese. It is tempting to link both the monosyllabicity of verbs and their lower F_0 values to a weaker prosodic weight than nouns.

No F_0 valleys are observed in cases of monosyllabic verbs followed by monosyllabic nouns. In those cases the verb and noun construction may have been incorporated into one single prosodic word and is for all practical purposes being treated as a single word.

The verb effect can be captured in F_0 modelling by assigning a low default prominence setting to a monosyllabic verb which is followed by a polysyllabic object noun.

Furthermore, we note that the interesting humps in some sentences at positions 4 and 5 of Speaker D are caused by his occasionally emphasis placement on the adverb *gang1-gang1*, “just now”. Speaker C occasionally destressed the final syllable *gua* “melon”, changing its tonal category to neutral tone, which caused some of the final syllable to drop in pitch. This is a lexical issue, unrelated to final lowering.

11.9 Conclusion

In this paper we have shown that Mandarin has a clear declination effect. The F_0 decline is more pronounced near the beginning of the utterance, and the effect can be modelled as an exponential decay. The model has three parameters: P_1, the initial value; α, the parameter controlling the rate of declination; and β, which controls the asymptote value. The variations in long and short sentences are primarily controlled by varying the P_1 value, where longer sentences have higher initial pitch and shorter sentences have lower initial pitch. Declination also interacts with focus. In the data of one speaker, post-focus pitch range compression of high tone sequence appears to be gradual. Declination slope is steeper in post-focus materials than in plain sentences, which is reflected in a lower α value in the post-focus declination slope. The F_0 values eventually asymptote to a lower value than in plain sentences, which is expressed by the lower value of β.

Acknowledgements

I would like to thank Yi Xu and Sun-Ah Jun for their insightful comments on this paper.

References

Aronoff, M and R. Oehrle (eds.). 1984. *Language Sound Structure*. Cambridge, Mass.: MIT Press.

Cohen, A., R. Collier and J. 't Hart. 1982. Declination: construct or intrinsic feature of speech pitch? *Phonetica* 39, 254-273.

Cooper, W.E. and J.M. Sorensen. 1981. *Fundamental Frequency in Sentence Production.* New York: Springer-Verlag.

Eady, S.J. and W.E. Cooper. 1986. Speech intonation and focus location in matched statements and questions. *J. Acoust, Soc. Am.* 80, 402- 415.

Fujisaki, H. 1983. Dynamic characteristics of voice fundamental frequency in speech and singing. In MacNeilage (ed.), 39-55.

Fujisaki, H., K. Hirose, P. Halle, and H. Lei. 1990. Analysis and modelling of tonal features in polysyllabic words and sentences of the Standard Chinese. *Proc. ICSLP* '90 (Kobe, Japan), 841-844.

Gårding, E. 1987. Speech act and tonal pattern in Standard Chinese: constancy and variation. *Phonetica* 44, 13-29.

Gooskens, C. and V. J. van Heuven. 1995. Declination in Dutch and Danish: global versus local pitch movements in the perceptual characterisation of sentence types. *Proc. 13th ICPhS* (Stockholm, Sweden), vol. 2, 374-377.

Gussenhoven, C. and A.C.M. Rietveld. 1988. Fundamental frequency declination in Dutch: testing three hypotheses. *Journal of Phonetics* 16, 355-369.

Hart, J., 't and A. Cohen. 1973. Intonation by rule: a perceptual quest. *Journal of Phonetics* 1, 309-327.

Herman, R. 1996. Final lowering in Kipare. *Phonology* 13, 171-196.

Hirschberg, J. and J.P. Pierrehumbert. 1986. The intonational structuring of discourse. *Proc. 24th Association for Computational Linguistics* (Santa Cruz, CA., USA), vol. 24, 136-144.

Hyman, L. (ed.). 1973. *Consonant Types and Tones.* UCLA.

Jin, S. 1996. *An Acoustic Study of Sentence Stress in Mandarin Chinese.* PhD dissertation, The Ohio State University.

Jun, S.-A. and H.-J. Lee. 1998. Phonetic and phonological markers of contrastive focus in Korean. *Proc. ICSLP* '98 (Sidney, Australia).

Ladd, D. R. 1984. Declination: a review and some hypotheses. *Phonology Yearbook* 1, 53-74.

Ladd, D.R. 1993. On the theoretical status of "the baseline" in modelling intonation. *Language and Speech* 36, 435-451.

Laniran, Y.O. 1992. *Intonation in Tone Languages: The Phonetic Implementation of Tones in Yorùbá.* PhD dissertation, University of Cornell.

Lea, W.A. 1973. Segmental and suprasegmental influences on fundamental frequency contours. In Hyman (ed.), 15-70.

Liao, R. 1994. *Pitch Contour Formation in Mandarin Chinese.* PhD dissertation, The Ohio State University.

Liberman, M.Y. and J.B. Pierrehumbert. 1984. Intonational invariance under changes in pitch range and length. In Aronoff and Oehrle (eds.), 157-233.

Liebermann, Ph. 1967. *Intonation, Perception and Language.* Cambridge, Mass.: M.I.T. Press.

MacNeilage, P.F. (ed.). 1983. *The Production of Speech.* New York: Springer-Verlag.

Maeda, S. 1976. *A Characterisation of American English Intonation.* PhD dissertation, MIT.

Nakajima, S. and J.F. Allen. 1993. A study on prosody and discourse structure in cooperative dialogues. *Phonetica* 50, 197-220.

Peterson, G.E. and H.L. Barney. 1952. Control methods used in a study of the vowels. *J. Acoust. Soc. Am.* 24, 175-184.

Pierrehumbert, J.B. 1979. The Perception of fundamental frequency declination. *J. Acoust. Soc. Am.* 66, 363-369.

Pierrehumbert, J.B. 1980. *The Phonology and Phonetics of English Intonation.* PhD dissertation, MIT (published 1988 by IULC).

Pierrehumbert, J.B. and M.E. Beckman. 1988. *Japanese Tone Structure.* Cambridge, Mass.: MIT Press.

Prieto, P., C. Shih and H. Nibert. 1996. Pitch downtrend in Spanish. *Journal of Phonetics* 24, 445-473.

Shen, J. 1985. Beijinghua Shengdiao de Yinyu he Yudiao (Pitch Range and Intonation of the Tones of Beijing Mandarin). In *Beijing Yuyin Shiyan Lu (Acoustic Studies of Beijing Mandarin)*, 73-130. Beijing University Press.

Shen, X.S. 1993. Relative duration as a perceptual cue to stress in Mandarin. *Language and Speech* 36, 415-433.

Shih, C. 1988. Tone and intonation in Mandarin. *Working papers* 3, 83-109. Phonetics Laboratory, Cornell University.

Silverman, K. 1997. *The Structure and Processing of Fundamental Frequency Contours*. PhD dissertation, University of Cambridge.

Sluijter, A.M.C. and J.M.B. Terken. 1993. Beyond sentence prosody: paragraph intonation in Dutch. *Phonetica* 50, 180-188.

Strik, H. and L. Boves. 1992 Control of fundamental frequency, intensity and voice quality in speech. *Journal of Phonetics* 20, 15-25.

Strik, H. and L. Boves. 1995. Downtrend in F_0 and Psb. *Journal of Phonetics* 23, 203-220.

Terken, J.M.B. 1991 Fundamental frequency and perceived prominence of accented syllables. *J. Acoust. Soc. Am.* 89, 1768-76.

Terken, J.M.B. 1993. Baselines revisited: reply to Ladd. *Language and Speech* 36, 453-59.

Thorsen (Grnønum), N. 1980a. A Study of the perception of sentence intonation – evidence from Danish. *J. Acoust. Soc. Am.* 67, 1014-30.

Thorsen (Grnønum), N. 1980b. Intonation contours and stress group patterns in declarative sentences of varying length in ASC Danish. *Annual Report of the Institute of Phonetics* 14, 1-29. University of Copenhagen.

Titze, I.R. 1989. On the relation between subglottal pressure and fundamental frequency in phonation. *J. Acoust. Soc. Am.* 85, 901-906.

Tseng, C.-Y. 1981. *An Acoustic Phonetic Study on Tones in Mandarin Chinese*. PhD dissertation, University of Brown.

Umeda, N. 1982. F_0 declination is situation dependent. *Journal of Phonetics* 10, 198-210.

Wang, C., H. Fujisaki and K. Hirose. 1990. Analysis and modelling of tonal features in polysyllabic words and sentences of the Standard Chinese. *Proc. ICSLP* '90 (Kobe, Japan), 221-224.

Xu, Y. 1997. Contextual tonal variations in Mandarin. *Journal of Phonetics* 25, 61-83.

Xu, Y. and Q.E. Wang. 1997. What can tone studies tell us about intonation? *Proc. ESCA Workshop on Intonation* (Athens, Greece), 337-40.

Yang, L.-C. 1995. *Intonational Structures of Mandarin Discourse*. PhD dissertation, University of Georgetown.

12

A Quantitative Model of F_0 Generation and Alignment

JAN P. H. VAN SANTEN AND BERND MÖBIUS

12. 1 Introduction

Local pitch contours belonging to the same perceptual or phonological class vary significantly as a result of the structure (i.e., the segments and their durations) of the syllables they are associated with. For example, in nuclear rise-fall pitch accents in declaratives, peak location (measured from stressed syllable start) can vary systematically between 150 and 300 ms as a function of the durations of the associated segments (van Santen & Hirschberg, 1994). Yet, there are temporal changes in local pitch contours that are phonologically significant even though their magnitudes do not appear to be larger than changes due to segmental effects (e.g., Kohler, 1990; D'Imperio & House, 1997).

This paper starts out by addressing the following question: *What is invariant about the alignment of pitch accent contours belonging to the same class?* (We loosely define a *pitch accent contour* as a local pitch excursion that corresponds to an accented syllable.) We propose a model according to which pitch accent curves in the same class are generated from a common template using a common set of *alignment parameters.* These alignment parameters specify how the time course of these curves depends on the durations of the segment sequence with which a pitch accent is associated. The paper presents data leading up to this model, and defines precisely what alignment parameters are. We then proceed to embed this model in a more complete intonation model, that in key respects is similar to – but also different from – the superpositional model by Fujisaki (1983). Our model, besides serving the practical purpose of being used for most languages in the Bell Labs text-to-speech system, is also of

A. Botinis (ed.), Intonation, 269-288.

conceptual interest, because in a natural way it leads to a broader, yet reasonably precise, definition of the superposition concept. In our experience, discussions of tone sequence vs. superpositional approaches to intonation (e.g., Ladd, 1996) often suffer from too narrow definition of the superposition concept.

12.2 Accent Curve Alignment

To keep this section as empirical and theory-free as possible, the word "accent curve" is used very loosely in the sense of a local pitch excursion that corresponds to an accented syllable, not in the specific sense of the Fujisaki model (Fujisaki, 1983). In what follows, the term "accent group" (or "stress group") refers to a sequence of syllables of which only the first is accented. "Accent group structure" refers to the segments in an accent group ("segmental structure") with associated durations. Thus, renditions of the same accent group almost always have different structures because their timing is unlikely to be identical, but by definition they have the same segmental structure.

Our data base is an extension of the speech corpus described in a previous paper (van Santen & Hirschberg, 1994), and consists of speech recorded from a female speaker who produced carrier phrase utterances in which one or two words were systematically varied. The non-varying parts of the utterances contained no pitch accents.

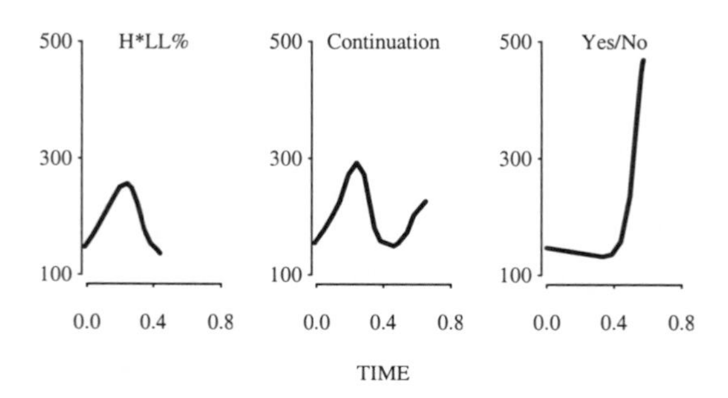

Figure 12.1. Averages of Declarative, Continuation, and Yes/No contours.

The earlier study focused on utterance-final monosyllabic accent groups, produced with a single "high" pitch accent, a low phrase accent, and a low boundary tone (Pierrehumbert label H*LL% (Pierrehumbert, 1980); Figure 12.1, left panel). The current data base also includes H*LL% contours for polysyllabic accent groups, continuation contours (H*LH%), and yes/no contours (L*H%). Continuation contours consist of a dual motion in which an early peak is followed by a valley and a final rise (Figure 12.1, center panel). Yes/No contours (Figure 12.1, right panel) consist of a declining curve for the pre-accent region (not shown), an accelerated decrease starting at the onset of the accented syllable, and then a steep increase in the nucleus. Unless stated otherwise, results are reported for H*LL% contours.

12.2.1 Effects of Accent Group Duration

As point of departure we take the most obvious analysis: Measure alignment of H*LL% accent curves in terms of peak location, and assume that accent group structure can be captured simply by total duration. There is indeed a statistically significant correlation between peak location and total duration, showing that peaks are not placed either a fixed or random millisecond amount into the stressed syllable. But the correlation is weak (0.57). We could stop here, and declare that accent curve timing is only loosely coupled to accent group structure. Or, as we do next, we can measure whether timing depends on aspects of accent group structure other than total duration.

12.2.2 Effects of Segmental Structure

In van Santen and Hirschberg (1994) it was shown that peak location strongly depends on segmental structure. For monosyllabic accent groups, peak location (measured from accented syllable start) is systematically later in sonorant-final accent groups than in obstruent-final accent groups (*pin* vs. *pit*), and later in voiced obstruent-initial accent groups than in sonorant-initial accent groups (*bet* vs. *yet*). Such effects persisted when we measured peak location from vowel start instead of syllable start, and when we normalized peak location by division by syllable or rhyme duration. Apparently, peaks are located at neither a fixed millisecond amount nor a fixed fraction of the accent group.

In our new data, we found that polysyllabic accent groups again behave differently. For example, peaks occur much later in the initial accented syllable (91% of syllable duration on average, and often located in the second syllable) compared to monosyllabic accent groups (35% of syllable duration). Relative to the entire accent group, peaks occur significantly earlier in polysyllabic accent groups (35% of accent group duration) than in monosyllabic accent groups (45% of accent group duration).

12.2.3 Effects of Accent Group "Sub-durations"

While these data undermine certain peak placement rules used in text-to-speech synthesis (e.g., the rule that peaks are placed a fixed percentage into the accented syllable), they do not unambiguously disqualify the overall accent group duration hypothesis: overall duration tends to be longer for *pin* than for *pit* (because of the lengthening effect of postvocalic consonant voicing), and longer for "bet" than for "yet" (because /b/ is longer than /y/). In addition, the hypothesis does not require that peaks are located at a fixed fraction into the accented syllable or its rhyme; it only requires that peak locations in accent groups of equal length are the same.

A better test concerns the prediction that changes in peak location do not depend on which *part* of an accent group is lengthened. To illustrate, consider two monosyllabic accent groups that have the same overall duration of 400 ms, but the first (*stick*) has a relatively long onset of 170 ms and a vowel of 180 ms, while the second (*woke*) has a short onset of 60 ms and a longer vowel of 290 ms. In both cases, the duration of the final /k/ is the same (50 ms). If total accent group duration is the sole variable that matters, than both accent curves should be the same. But if alignment depends on more detailed aspects of the temporal structure, than the curves could differ. Our analysis presented next will show that, in fact, the peak in *stick* lies 83 ms to the right of the peak in *woke* (198 vs. 115 ms from the syllable start).

We measure the effects on peak placement of different parts of the accent group by defining the parts, predicting peak location by a weighted combination (multiple regression analysis) of the durations of these parts ("sub-durations"), and inspecting the values of the weights:

$$T_{peak}(a) = \sum_j \alpha_{S,j} \times D_j(a) + \mu_S$$

Here, a is a rendition of an accent group with segmental structure **S**, $T_{peak}(a)$ is peak location, j refers to the j-th "part" of the accent group, $D_j(a)$ is the corresponding duration, and $\alpha_{S,j}$ its weight. We use three "parts": accented syllable onset, accented syllable rhyme, and remaining unstressed syllables (polysyllabic accent groups only). We include any non-syllable-initial sonorants in the accented syllable rhyme. For codas in monosyllabic accent groups, we include only the sonorants in the rhyme. Thus, the rhyme is *lan* in *blank*, /i/ in *seat*, /yu/ in *muse*, /in/ in *seen*, and /o/ in *off*; but in the word *offset*, the rhyme consists of /of/. We distinguish between four types of segmental structure: monosyllabic (coda *sonorant, voiceless, voiced obstruent*) vs. polysyllabic. This, *blank*, *seat*, and *off* have the same structure (monosyllabic, voiceless coda), while *muse* and *seen* are examples of the other two monosyllabic types; the final two syllables of *syllabic* have the polysyllabic type.

This unusual partition of the syllable is based on analyses where we applied Equation (1) for much narrower classes of phonemes, while varying which parts of the syllable were included in the onset or rhyme. We found that the proposed partition provided the most parsimonious fit.

Equation (1) is strong in that it assumes linearity [referring to the Σ sign]. Thus, it predicts that a change in onset duration from 50 to 75 ms has exactly the same effect on peak location as a change from 125 to 150 ms; in both cases, the size of the effect is ($\alpha_{S,j} \times 25$). Otherwise, it is quite general and subsumes many models and rules proposed in the literature. For example, the hypothesis that peak placement is solely determined by overall accent group duration corresponds to the statement that

$$\alpha_{S,j} = \alpha, \text{ for all } \mathbf{S} \text{ and } j.$$

Another rule often proposed is that peak are placed a fixed fraction (F) into the rhyme. For this, we let

$$\alpha_{S,1} = 1$$

$$\alpha_{S,2} = F$$

$$\mu_S = 0.0$$

The rule that the peak is placed a fixed ms amount (M) into the vowel (as IPO approach, 't Hart, Collier & Cohen, 1990) is given by

$$\alpha_{S,1} = 1$$

$$\alpha_{S,j} = 0$$

$$\mu_S = \mathrm{M}$$

The parameters of the model (α,μ) can be estimated using standard linear regression methods, because the quantities D and T_{peak} directly measurable. Consequently, the model in fact provides a convenient framework for testing these rules.

Results showed the following. First, the overall fit is quite good. The predicted-observed correlation of 0.91 (r^2 = 83%) for peak location explains more than 2.3 times the variance explained by overall accent group duration, where the correlation was 0.59 (r^2 = 35%).

Second, for all three contour classes, the weights $\alpha_{S,j}$ varied strongly as a function of part location (j = *onset, rhyme, remainder*), with the effects of the onset being the strongest and the effects of the remainder being the weakest. This violates the hypothesis that peak placement is solely determined by overall accent group duration (Eq.2), which requires that the weights should be the same. Third, setting the intercept μ_S to zero did not affect the fit (it reduced r^2 from 83% to 81%), suggesting that the accented syllable start plays a pivotal role in alignment. This analysis also contradicts the rule that the peak is placed a fixed ms amount into the vowel (Eq. 4).

Fourth, the values of the $\alpha_{S,j}$ parameters depended on segmental structure. Specifically, the values of the onset weights, $\alpha_{S,j}$, were smaller for sonorant *codas* than for non-sonorant codas; however, the onset weights were the same for all onset types, and ranged from 0.60 for sonorant codas to values in the 0.85-1.0 range for the other coda types (approximate values are given because the values dependent somewhat on details of the regression algorithm). The fact that $\alpha_{S,j}$ is less than 1.0 violates the rule that peak are placed a fixed fraction (F) into the rhyme (Eq. 3). A stronger violation of the same rule is, of course, that the peak is much later (measured as a fraction of the accented syllable) in polysyllabic than in monosyllabic accent groups, as was reported in Subsection 2.2.

To apply this model to the hypothetical "stick" and "woke" examples, using $\alpha_2 = 0.95$ and $\alpha_1 = 0.2$, we find that:

$$T_{peak}(stick) = 0.95 \times 0.17 + 0.2 \times 0.18 = 0.1975$$

$$T_{peak}(woke) = 0.95 \times 0.06 + 0.2 \times 0.29 = 0.115$$

In other words, in *stick* the peak is predicted to occur 197 ms after the start of the accented syllable and in *woke* 115 ms into the syllable. This difference is due to the effects of onset duration (0.95) to be much larger than the effects of the rhyme duration (0.2).

12.2.3.1 Importance of Accented Syllable Start

The key reason for measuring time starting at accented syllable onset is that in these data the pitch movement appears to start at this time point. When we re-did the analysis with a different starting point, the vowel start, results were far less clear cut. The lower prediction accuracy is in part due to the fact that we lose a free parameter ($\alpha_{S,j}$) when we remove onset duration from the equation. However, the more important reason for the poorer fit is that, even when we hold rhyme duration constant, peak location is not a fixed ms amount after vowel start; in fact, as we reported above for sonorant codas, each 1.0 ms lengthening of the onset causes only a 0.6 ms rightward shift in peak location. This effect cannot be captured by a model that ignores onset duration, such as a model assuming that we should measure time from vowel start.

Independent evidence for the assumption that the pitch movement starts at syllable onset was provided by our finding that the intercept μ was statistically zero.

The importance of the start of the accented syllable in rise-fall curves confirms earlier results by Caspers, who found over a wide range of conditions (in terms of speaking rates, intrinsic duration of the accented vowel, and presence versus absence of nearby pitch movements) that the start of the rise coincides with the start of the accented syllable (Caspers, 1994). The start of the rise was not strongly tied to other segment boundaries, and the end of the rise (which in most cases is the same as the peak) was not tied to any segment boundary. This is consistent with our model, because for polysyllabic accent groups peak location is given by:

$$\alpha_{S,1} \times D_{onset} + \alpha_{S,1} \times D_{rhyme} + \alpha_{S,3} \times D_{remainder} + \mu_S$$

the end of the rhyme by D_{onset}+D_{rhyme}, and the end of the onset by D_{onset}. Given our estimates of the $\alpha_{S,j}$ and μ_S parameters, peak location cannot coincide with, or be at a fixed distance from, either of the latter two boundaries.

In a study of Greek pre-nuclear accents, the onset of the rise of rise-fall accents (with peak location in the post-accentual vowel) was also found to coincide with accented syllable start (Arvaniti, Ladd & Mennen, 1998).

12.2.4 Anchor Points

12.2.4.1 Estimation of Anchor Points

The peak is only one point on an accent curve, and it is not clear whether it is the most important point – perhaps it is the start of the rise, or the point where the rise is steepest. In the tone sequence tradition following Pierrehumbert (1980), tone targets are elements of the phonological description, whereas the transitions between the targets are described by phonetic realisation rules. Taking the opposite view, the IPO approach ('t Hart *et al.*, 1990) assigns phonological status to the transitions, viz. the pitch accent movement types, themselves. In the model proposed here, both pitch movements and specific points characterising the shape of the movements matter. However, no particular status is reserved for peaks; our "target" are non-linear pitch curves, which are typically either bidirectional (H*LL% contour) or tridirectional (continuation contour). One way to capture the entire curve is by sampling many points on that curve ("anchor points"), and model timing of these points in the same way as peak location.

We have experimented with various methods for defining such points on an accent curve. For example, one can compute the first or second derivative of the accent curve, and define as anchor points locations where these derivatives cross zero or reach other special values. However, derivatives are not particularly well-behaved in the case of F_0 curves due to small local variations in periodicity. Among the methods that we used, the following proved to be the simplest and at the same time statistically most robust. We subtract a locally straight "phrase curve" from the observed F_0 curve around the area where the accent curve is located, and then consider the residual curve as an estimate of the accent curve (*estimated accent curve*). For the H*LH% curves, the locally straight "phrase curve" is computed simply by connecting the last sonorant frame preceding the accent group with the final sonorant frame of the accent group. We then sample the estimated accent curve at locations corresponding to a range of percentages between 0% and 100% (e.g., 5%, 10%, 25%, ..., 75%, 90%, 95%, 100%) of maximal height. Thus, the 100% point is the peak location, and the 50% pre-peak point is the time point where the estimated accent curve is half of maximal height. We call these time points *anchor points*. In Section 2.4.3 we will discuss how phrase curves are computed for the yes/no and continuation rise cases.

The model in Equation (1) can be applied to any anchor point by replacing the *peak* subscript by *i* (for the *i*-th anchor point) and adding i as a subscript to the parameters a and μ:

$$\mathrm{T}_i(a) = \sum_j \alpha_{i,\mathbf{S},j} \times D_j(a) + \mu_{i,\mathbf{S}}$$

We call the ensemble of regression weights ($\alpha_{j,\mathbf{S},i}$) or a fixed segmental structure **S**, the *alignment parameter matrix* (*APM*), and Equation (6) the *alignment model*. It could be said that an *APM characterises for a given pitch accent type how accent curves are aligned with accent groups.*

12.2.4.2 Alignment Parameter Results

Figure 12.2 shows the values of the alignment parameters for polysyllabic phrase-final accent groups (H*LL%). We note the following. First, the weights for the onset exceed the weights for the rhyme, and the latter exceed the weights for the remainder of the accent group. In other words, lengthening the onset duration of the stressed syllable by a fixed ms amount has a larger effect on any anchor point than lengthening the duration of the unstressed syllables by the same ms amount. Second, the curves are monotonically increasing. They initially diverge, and then converge. Early anchor points mostly depend on onset duration and hardly on the durations of the rhyme and the remainder, but late anchor points depend more evenly on all three subsequent durations.

A key point is that these alignment curves are well-behaved, and without a doubt can be captured by a few meta-parameters, e.g., two straight line segments per curve.

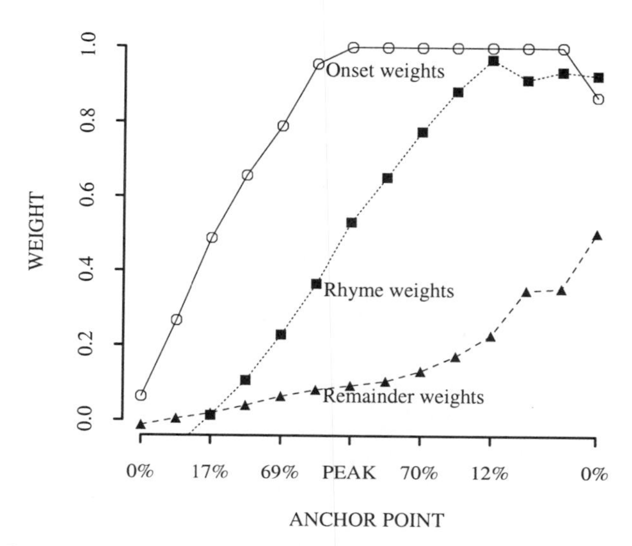

Figure 12.2. Regression weights as a function of anchor point, for each of the three sub-intervals of the accent group. H*LL% accent type. Solid curve: onset; dotted curve: rhyme; dashed curve: remainder.

12.2.4.3 Additional Results

Is the time course of yes/no curves at all related to the temporal structure of the phrase-final accent group? After all, traditionally the yes/no feature is attached to phrases, not to syllables. For example, in Möbius's application of the Fujisaki model, phrase commands are not tied to accent groups (Möbius, Pätzold & Hess, 1993). Our data for monosyllabic accent groups strongly suggest that there is a relationship (see Figure 12.1). We estimated yes/no accent curves by subtracting the descending line that passes through the approximately linear initial part of the F_0 curve in the accent group up to the point where a significant rise starts; the latter point was determined by a threshold on the first derivative. As with the standard H*LL% contours, we divided the result of the subtraction by the maximum to obtain a curve ranging from 0.0 to 1.0. Thus, at the time point where the estimated accent curve starts to rise (which we call the *rise start*), the F_0 curve reaches a local minimum due to the locally descending phrase curve.

We found that the rise start could be predicted quite accurately ($r = 0.96$, $rms = 17$ ms) from the onset duration and the sonorant rhyme duration, with respective alignment coefficients of 0.89 and 0.75. This means that the rise start is not located at some fixed or random amount of time before the phrase boundary, but varies systematically with the durations of the onset and the rhyme. In fact, the time interval between the rise start and the end of the phrase is given by:

$$t_{end\_of\_phrase} - T_{rise\_start} = (1 - 0.89) \times D_{onset} + (1 - 0.75) \times D_{rhyme}$$

Thus, as measured from the end of the phrase, the rise start occurs earlier as the rhyme and the onset become longer. Informal perceptual observations involving synthesis of polysyllabic accent groups suggest that it is important for the rise start to occur in the accented syllable, not in the phrase-final syllable. This further confirms our conclusion that yes/no rises are aligned with the phrase-final accent group in ways very similar to the alignment of declarative nuclear pitch accents. Of course, these data do not exclude an alternative account, according to which the phrase curve does not locally descend but stays flat instead, while the accent curve has an initial descending portion. In this case, it could be the local minimum and not the start of the rise whose timing is tied to the accented syllable.

Accent curves for continuation rises where estimated by subtracting a line that passes through the F_0 values at the start of the onset and at the minimum value attained before the final rise (see Figure 12.1). For continuation contours, an unexpected phenomenon was found – a zero (in fact, slightly negative) correlation between the frequencies at the H* peak and the H% phrase boundary. A closer look revealed that the anchor points can be dichotomised into a pre-minimum and a post-minimum group; all correlations between groups are close to zero, all correlations within groups are strongly positive. Correlations at similarly spaced anchor points for the H*LL% and the yes/no contours were all positive, typically strongly so. One interpretation of this pattern is that continuation contours involve two component gestures – one responsible for H* and the other for H%.

12.2.5 Theoretical Aspects of Alignment Model

12.2.5.1 Alignment Model and Time Warping

As is often the case, a given mathematical formulation can be interpreted in ways that differ conceptually. Above, the alignment model was described as a way to predict the locations of certain points on the F_0 curve from the durations of judiciously selected parts of an accent group, via multiple regression. Figure 12.3 shows how we can re-conceptualise the model in terms of *time warping of a common template.*

Panel (a) shows the percentages used to define the anchor points. Horizontal axis is the anchor point index, *i* in Equation (6). Panel (b) shows the alignment parameters, copied in simplified form from Figure 12.2. Panels (c) and (d) show the predicted anchor point locations using these alignment parameters, assuming onset and rhyme durations of 150 and 300 ms for the word *spot* and 100 and 350 ms for the word *noon*. These panels not only show the locations of the anchor points, but also display the *shape of the predicted normalised accent curves* by graphing corresponding percentage values from Panel (a) as heights of the vertical bars. These curves are called "normalised" because they range between 0.0 and 1.0.

Clearly, the two predicted normalised accent curves are similar in that they are both bell-shaped, yet one cannot be obtained from the other by uniform compression because, while the peak in "noon" is earlier, the overall horizontal extents of the accent curves are the same. It follows that there must be a *non-linear temporal relationship*, or *time warp*, between the two curves.

Finally, Panel (f) shows the time warps between the anchor point locations of the two words and anchor point indices (1-17), where we have taken the pattern in Panel (a) as the template. Note that these time warps are determined by the alignment parameters *and* the durations of the onset and rhyme. Formally, the predicted normalised accent curve is given by:

$$\hat{F}_0(t) = P[i(t)]$$

Here $i = i(t)$ is the index onto which location t is mapped, using an appropriate interpolation scheme. $P(i)$ is the percentage corresponding to the i-th anchor point. In Panel (*f*), t corresponds to the horizontal axis and $i(t)$ to the vertical axis. We further clarify the relationship between alignment parameters and time-warping by spelling out the steps involved in the computation of the predicted normalised accent curve for a rendition a of a given accent group:

Step 1: Measure the durations of all subintervals D_j of a.

Step 2: For each anchor point i, compute predicted anchor point location T_i using Equation (6).

Step 3: For each time point t, find i such that t is located between T_i and $T_i{+}1$.

Step 4: Retrieve values P_i and $P_i{+}1$, and obtain a value for t by interpolation.

In summary, the predicted normalised accent curve can be viewed as *time warped version of a common template*. The time warps for a given accent curve class vary from one utterance to the next, but they belong to the same *family* in that they are all are produced by Equation (6).

12.2.5.2 Phonological equivalence, alignment parameters, and templates

As Ladd (1996) states, a complete phonological description must specify how the categorical phonological elements map onto continuous acoustic parameters. In the IPO approach ('t Hart *et al.*, 1990), the pitch movement inventory includes a parameterisation of prototypical realisations, in particular the alignment of the movement with the segmental makeup of the syllable. This phonetic description is based on experiments that attempt to establish the limits of variation in the realm of perception; i.e., how different can two pitch movements be acoustically and still be perceived as the same? In analogy, we ask the question: how different can two pitch accent curves be and still count as representatives of the same pitch accent type, or be derived from the same template?

We propose that what pitch accent curves in the same class have in common is that they are generated from a common template using the same family of time warp functions. They "sound the same" not only because they have the same shape (e.g., bell-shaped), but also because *they are aligned in the same way with the syllables they are associated with.* In other words, we associate a pitch accent class with an *ordered pair*:

<template><APM>

We claim that two accent curves are phonologically distinct if, given the durations of the subintervals of their respective accent groups, they cannot be generated from the same template using the same APM.

Consequently, the relatively small temporal shifts causing phonological changes observed by Kohler (1990) and D'Imperio and House (1997) could be explained by our model as follows. In both studies the segmental materials and their durations were largely left unaltered. Because the predicted accent curve that results from applying a given APM to a given template is completely determined by the segmental materials and their durations, it is impossible that the alignments of both the original and its shifted variant fit the same <template><APM> pair. Hence they must be phonologically distinct[1].

What is somewhat difficult to grasp intuitively is that the time warp function is not constant for a given pitch accent class, but depends on the accent group sub-interval durations; what the time warp functions share is that they are generated from the same underlying APM. Our model thus provides a somewhat abstract and indirect definition for what it means for two curves to be aligned in the same way. This complexity is necessitated, of course, by the fact that all simpler alignment rules – which only applied to peak location anyway – were shown to fail.

12.3 Proposed Pitch Model

To use the predicted normalised accent curve for F_0 synthesis, three operations have to be performed.

1. Range of values has to be scaled to some appropriate range in Hz. Presumably, this scaling would reflect, among other things, the prominence of the associated pitch accent.
2. The range is scaled appropriately, the resulting curve has to be brought in line with other such curves, which requires rules for vertical placement. I.e., we must determine the location on the frequency axis of a representation of the F_0 curve which has time as horizontal axis. Vertical placement would be based on phrase-level phenomena such as declination and sentence mode.
3. Have to be applied for filling in gaps between successive accent curves and for resolving overlap between such curves.

Up to this point, we have refrained from using superpositional assumptions; the phrase curve was used merely as a device for computing anchor points, and we remarked that other methods that do not rely on phrase curves were rejected not on theoretical grounds but merely on the basis of our experience that they did not behave well statistically. However, we have found it difficult to do scaling, vertical placement, and successive-curve connections in any way other than in the superpositional framework. In the superpositional tradition, vertical placement is accomplished by adding accent curves to a phrase curve. Combination of successive accent curves follows as a side effect of this addition. Scaling would be accomplished by multiplying the predicted normalised accent curve with a scalar quantity.

We now discuss the details of our superpositional implementation.

12.3.1 Additive Decomposition

In the best-known superpositional model, the Fujisaki model (Fujisaki, 1993; Möbius *et al.*, 1993), the observed F_0 curve is obtained by adding in the logarithmic domain three curve types with different temporal scopes: phrase curves, accent curves, and a horizontal line representing the speaker's lowest pitch level. We likewise propose to add curves with different temporal scopes, but remove the base pitch line and include segmental perturbation curves instead.

12.3.2 Phrase Curves

For English, we found that phrase curves could be modelled as two-part curves obtained by non-linear interpolation between three points, viz. the start of the phrase, the start of the last accent group in the phrase, and the end of the phrase. The phrase curve model includes as special cases the standard linear declination line, and curves that are quite close to the phrase curve in Fujisaki's model. Moreover, some of the problems with the Fujisaki model, especially its apparent inability to model certain contour shapes observed in English (see discussion by Ladd, 1996, 30), can be attributed to too strong constraints on the shape of commands and contours. We prefer to be open to the possibility that phrase curves exhibit considerable and meaningful variability. For example, in our current work on Japanese (van Santen, Möbius, Venditti & Shih, 1998), phrase curves start with a rise culminating in a peak around the end of the second mora, a gentle decline until the start of the accented mora, followed by a steeper descent, and possibly terminated by a flatter region if several morae follow the accented mora. Phrase curve parameters are controlled by sentence mode and locational factors, such as sentence location in the paragraph.

12.3.3 Perturbation Curves

Perturbation curves are associated with initial parts of sonorants following a transition from an obstruent. We measured these effects, by contrasting vowels preceded by sonorants, voiced obstruents, and unvoiced obstruents in syllables that were not accented and were not preceded in the phrase by any accented syllables (van Santen & Hirschberg, 1994). The ratio curves (or, equivalently, difference curves in the logarithmic domain) resulting from these contrasts can be described by a rapid decay from values of about 1.30 to 1.0 in 100 ms. In our model, these curves are added in the logarithmic domain to the other curves.

12.3.4 Accent curve Height

In our model, accent curve height is determined via a multiplicative model by multiple factors, including position (in the minor phrase, the minor phrase in major phrase, etc.), factors predictive of prominence, and intrinsic pitch. Formally, the accent curve height parameter H(a) for accent group a is given by:

$$H(a)=A[location(a)] \times B[prominence(a)] \times C[nucleus(a)] \times \cdots,$$

where A, B, and C are mappings that assign numbers (multipliers) to discrete levels of the arguments (e.g.: *location = initial, medial, final*). The multiplicative model is often used in segmental duration modelling. It makes the important – and not necessarily accurate – assumption of directional invariance (van Santen, 1997): holding all factors but one constant, the effects of the varying factor always have the same direction. This may often be true in segmental duration; e.g., when two occurrences of the same vowel involve identical contexts, except for syllabic stress, the stressed occurrence is likely to be longer. However, one has to be very careful in which factors one selects and how one defines them. For example, if one were to use as factors the parts-of-speech of the word in question and its left and right neighbours, such directional invariance is extremely unlikely to occur.

12.4 General Assumptions of the Model

Both our model and the model proposed by Fujisaki can be seen as special cases of a much broader superpositional or "overlay" (Ladd, 1996) concept. Because discussions about the superpositional approach are often marred by focusing too narrowly on specific instances of this approach (e.g., Ladd, 1996, 26-30), we feel it is important to spell out the broader assumptions of our model. This is what we hope to do in this section.

12.4.1 Decomposition into Curves with Different Time Courses

The key difference between our model and the Fujisaki model is that accent curves are generated by time-warping of templates vs. by low-pass filtering rectangular accent commands, respectively. Nevertheless, the two models are both special cases of the *generalised additive decomposition* concept, which states that the F_0 curve is made up by "generalised addition" of various classes of component curves:

$$F_0(t) = \bigoplus_{c \in C} \bigoplus_{k \in c} f_{ck}(t)$$

C is the set of curve classes (e.g., {*perturbation, phrase, accent*}),c is a particular curve class (e.g., *accent*), and k is an individual curve (e.g., accent curve). The operator $\oplus$ satisfies some of the usual properties of addition, such as *monotonicity* (if $a \geq b$, then $a \oplus x \geq b \oplus x$) and commutativity ($a \oplus b = b \oplus a$). Obviously, both addition and multiplication have these properties.

A key assumption is that each class of curves, c, corresponds to a *phonological entity with a distinct time course*. For example, the *phrase* class has a longer scope than the *accent* class, which in turn has a longer scope than the obstruent-nonobstruent transitions with which perturbation curves are associated.

A central issue to be resolved for models in this class is which parameters of which curve classes depend on which factors. For example, in our model the alignment parameters do not depend on any phrase-level factors, and the perturbation curves are completely independent of accent status and location of the syllable containing the obstruent-nonobstruent transition.

As pointed out by Ladd (1996), computation of these curves from observed F_0 contours is not straightforward, and is often left unspecified (e.g., Thorsen, 1983; Gårding, 1983). Fujisaki and his colleagues have been successful in estimating these curves, because of the strong assumptions of this model. We were able to fit our model because of the extreme simplicity of the recorded F_0 curves, but significant statistical problems have to be solved to apply our model to arbitrary F_0 curves. However, we have little doubt that these obstacles can be overcome. But more relevant is the point that one should not confuse these estimation difficulties with the validity of the superposition concept.

Another point raised by Ladd is that, at times, in order to obtain a good fit of the Fujisaki model phrase or accent commands have to be put in implausible locations. Of course, this point is irrelevant for the broader superpositional concept, because this result might be due entirely to some of the specific assumptions of this model, such as the exact shape of the smoothing filters.

Many issues remain to be resolved. The least-researched issue of our model is the shape of the phrase curve. While the current shape produces decent synthetic F_0 contours, we are becoming increasingly more aware of challenges, such as the necessity of multiple levels of phrasing.

12.4.2 Sub-interval Duration Directional Invariance.

In the same way as addition of curves in the log domain is only a special case of a much more general decomposition principle (Equation 10), the linear alignment model is a special case of a more general principle: the *sub-interval duration directional invariance* principle. According to this principle, for any two accent groups a and b that have the same segmental

$$\textit{If } D_j(a) \geq D_j(b) \textit{ for all } j \textit{ then } T_i(a) \geq T_i(b)$$

Our alignment model is a special case, because when

$$D_j(a) \geq D_j(b) \textit{ for all } j$$

then, because all a parameters are non-negative:

$$\sum_j \alpha_{S,j} D_j(a) \geq \sum_j \alpha_{S,j} D_j(b)$$

and hence, by definition of our model (Equation 6):

$$T_j(a) \geq T_j(b)$$

The principle simply states that *stretching any "part" of an accent group has the effect of moving an anchor point to the right*, regardless of whether the stretching is caused by speaking rate changes, contextual effects on the constituent segments (e.g., emphasis), or intrinsic duration differences between otherwise equivalent segments (e.g., /s/ and /p/ are both voiceless and hence equivalent, but /s/ is significantly longer than /p/.)

An issue that needs to be addressed is the measurement of the sub-intervals. We found for the H*LL% curves that slightly different APM's were obtained depending on whether the coda was voiceless, voiced-obstruent, sonorant, or polysyllabic. It would be more elegant if the same

APM's were used. There are two ways of doing this. One is to alter the definitions of the sub-interval durations, in particular the definition of where an utterance-final sonorant ends; the latter is certainly reasonable because utterance-final sonorants have no well-defined endings; we used a somewhat arbitrary energy criterion. The other is to introduce a non-linearity that would reduce the effects of very long post-accentual regions in polysyllabic accent groups. Very crudely, one could set all durations in excess of 500 ms equal to 500 ms. Any of these changes would preserve sub-interval duration directional invariance.

12.5 Conclusions

This paper presented data on alignment that must be accounted for by any intonation model claiming to describe both the fine and coarse details of F_0 curves. We proposed a model that accurately predicts alignment of accent curves, defined as residual curves obtained from observed F_0 contours by subtraction of a locally linear phrase curve. The model provides a very good fit. We also showed that these data cannot be accounted for by some simple rules typical of current text-to-speech systems, which further justifies the more complicated rules embodied by the model. We described how this accent curve alignment model can be embedded in a complete superpositional model, that also incorporates phrase curves and segmental perturbation curves. Finally, we discussed generalisations of this superpositional model, and how it relates to the best-known superpositional model: the Fujisaki model.

Our model highlights the phonological importance of timing in intonation. This point has been made by many, in particular by Kohler (1990), but has been largely ignored under the assumption that the "*" notation – used by ToBI and its predecessors to indicate with which syllable a given tone as associated – is accurate enough.

Another aspect of our model relevant for phonology has to do with the problem current intonational phonology has with mapping from the phonological level to speech. Observed F_0 curves are complicated due to intrinsic pitch effects, perturbations of post-obstruent vowels, nasality effects, presence of voiceless regions, and temporal effects of segmental durations and other factors. Together, these effects can conspire to produce spurious local maxima and minima, perturb what otherwise might have been a straight line, or create an artificial straight line. This makes it difficult to determine the locations of true peaks and lows as is required for ToBI, or the locations of short linear rises or falls as is required for the IPO approach ('t Hart *et al.,* 1990). What either approach could use is a quantitative model that makes it possible to remove these effects from the observed F_0 curve; we believe that our model could play this role. What is not clear, of course, is what the relationship is between our <APM, template> pairs and the phonological entities in these approaches. So what needs further investigation is the phonological status of these <APM, template> pairs.

Finally, we discuss the relation of our work with an earlier paper by Silverman and Pierrehumbert (1990), in which they measured peak location in pre-nuclear high pitch accents; peak location was measured from stressed vowel start. As in our studies, Silverman and Pierrehumbert reported the effect of rhyme length on peak location; the effect was expressed either in milliseconds or as a proportion of the total rhyme length. They also found effects of pitch accent location, i.e., whether the pitch accent was nuclear as opposed to prenuclear. Silverman and Pierrehumbert's way of measuring or normalising peak location is problematic, because peak location is also affected by other parts of the accent group, for instance by onset duration and by the duration of the unstressed remainder of the accent group. We also have stated that we are not convinced that the peak should be the most important anchor point for alignment. Nevertheless, Silverman and Pierrehumbert's data show that different alignment parameters are likely to be needed as a function of pitch accent location. In fact, our text-to-speech system incorporates this effect. Overall, however, our model is more in line with Bruce (1990) who posits a more complex relationship between the F_0 contour and phonological categories than suggested by the work by Silverman and Pierrehumbert.

Notes

1. Of course, by the same token the alignments that were all perceived as being clearly declarative (or clearly interrogative) would also require different APMs. Thus, we must add the assumption that a given accent type is associated with a probability distribution over some APM space, and that different samples drawn from the same distribution are perceived as perceptually more similar than samples drawn from distinct distributions (e.g., the interrogative and the declarative distributions). This is not any different from other multivariate situations in which categories (e.g., *big* in size judgement, *blue* in colour vision, *autistic-spectrum* in clinical judgement, */ba/* in speech perception) correspond to somewhat hazy regions in multivariate spaces.

Acknowledgments

This paper has benefited from extensive discussions with Joseph Olive, Chilin Shih, Richard Sproat, and Jennifer Venditti. We thank Julia Hirschberg for help in the initial phases of this project. Finally, we wish to thank the reviewers – Gösta Bruce and Cinzia Avesani – for their challenging comments.

References

Arvaniti, A., D. Ladd, and I. Mennen. 1998. Stability of tonal alignment: the case of Greek prenuclear accents. *Journal of phonetics* 26, 3-25.

Bruce, G. 1990. Alignment and composition of tonal accents. In Kingston and Beckman (eds.), 107-114.

Caspers, J. 1994. *Pitch Movements under Time Pressure.* PhD dissertation, University of Leiden.

Cutler, A. and D.R. Ladd (eds.). 1983. *Prosody: Models and Measurements.* Berlin: Springer-Verlag.

D'Imperio, M. and D. House. 1997. Perception of questions and statements in Neapolitan Italian. *Proc. EUROSPEECH* '97 (Rhodes, Greece), vol. 1, 251-254.

Fujisaki, H. 1983. Dynamic characteristics of voice fundamental frequency in speech and singing. In MacNeilage (ed.), 35-55.

Gårding, E. 1983. A generative model of intonation. In Cutler and Ladd (eds.), 11-25.

Hart, J., t', R. Collier, and A. Cohen. 1990. *A Perceptual Study of Intonation.* Cambridge University Press.

Kingston, J. and M.E. Beckman (eds.). 1990. *Papers in Laboratory Phonology I.* Cambridge University Press.

Kohler, K. 1990. Macro and micro F0 in the synthesis of intonation. In Kingston and Beckman (eds.), 115-38.

Ladd, D.R. 1996. *Intonational Phonology.* Cambridge University Press.

MacNeilage, P.F. (ed.). 1983. *The Production of Speech.* New York: Springer-Verlag.

Möbius, B., M. Pätzold and W. Hess 1993. Analysis and synthesis of German F0 contours buy means of Fujisaki's model. *Speech Communication* 13, 53-61.

Pierrehumbert, J.B. 1980. *The Phonology and Phonetics of English Intonation.* PhD dissertation, MIT (published 1988 by IULC).

Silverman, K. and J.B. Pierrehumbert. 1990. The timing of prenuclear high accents in English. In Kingston Beckman (eds.), 72-106.

Thorsen (Grønnum), N. 1983. Two Issues in prosody of standard Danish. In Cutler and Ladd (eds.), 27-38.

van Santen, J.P.H. 1997. Prosaic modelling in text-to-speech synthesis. *Proc. EUROSPEECH* '97 (Rhodes, Greece), vol. 5, 2511-14.

van Santen, J.P.H. and J. Hirshberg. 1994. Segmental effects on timing and height of pitch and contours. *Proc. ICSLP* '94 (Yokohama, Japan), 719-722.

van Santen, J.P.H., B. Möbius, J. Venditti and C. Shih. 1998. Description of the Bell Labs intonation System. *Proc. ESCA Workshop on speech synthesis* (Jenolan, Australia), 93-98.

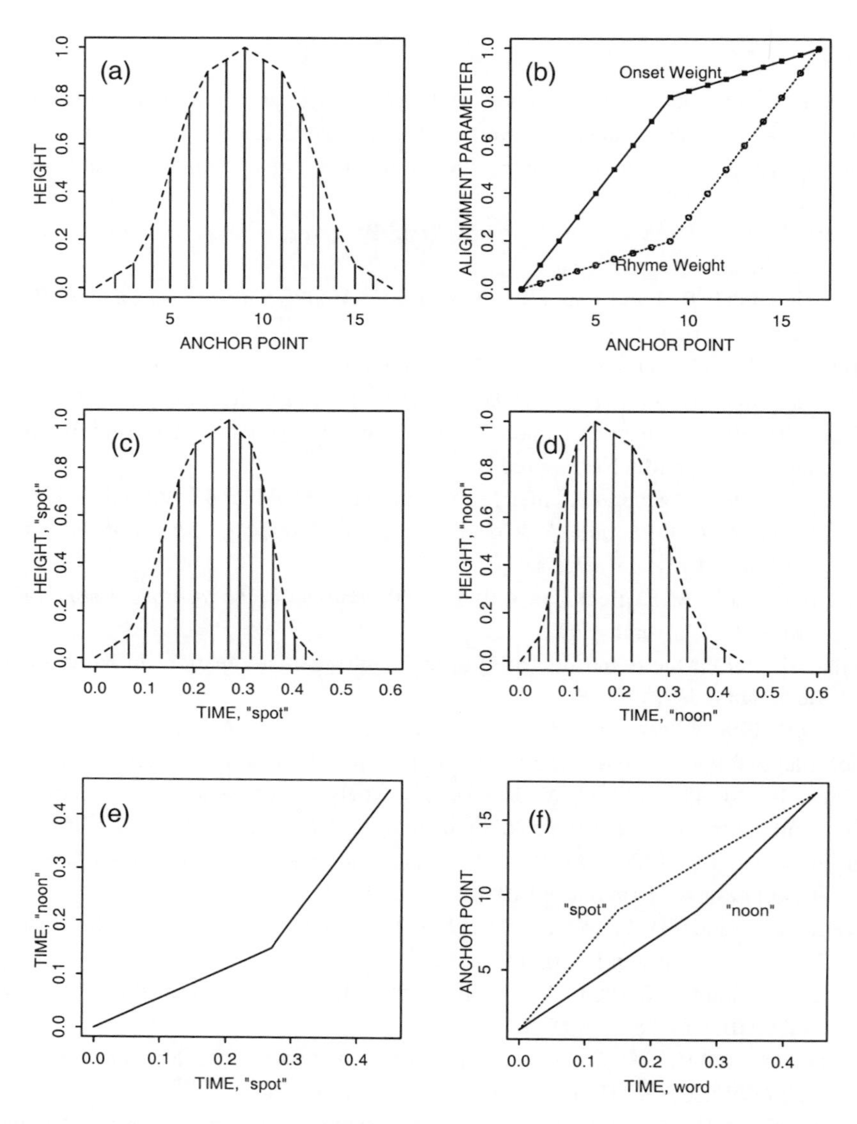

Figure12.3. Relationship between alignment parameters and time warping. (a): Template, defined as an array of fractions. (b): Hypothetical alignment parameters. (c): Predicted anchor points for the word "spot", using these alignment parameters and an onset duration of 300 ms and a rhyme duration of 150 ms. (d): Predicted anchor points for the word "noon", (onset duration: 100 ms, rhyme duration: 350 ms.) (e): Time warp of "noon" onto "spot", showing locations of corresponding anchor points. (f): Time warps of "noon" and "spot" onto the template.

Section V

Intonation Technology

13

Modelling of Swedish Text and Discourse Intonation in a Speech Synthesis Framework

GÖSTA BRUCE, MARCUS FILIPSSON, JOHAN FRID, BJÖRN GRANSTRÖM
KJELL GUSTAFSON, MERLE HORNE AND DAVID HOUSE

13.1 Introduction

In this chapter, we present a model for the analysis and synthesis of intonation in spontaneous conversations in Swedish. The model is an enhanced version of the model developed in Bruce (1977) and implemented in our text-to-speech synthesis. In our recent work we have developed the model from the perspective of discourse and multi-sentence texts. This takes us out of the restricted one-sentence/utterance analysis and synthesis into the living world of prosody in communication.

Our work has both a theoretical goal, that of a better understanding of Swedish prosody and of communicative prosody on the whole, and a practical one, that of incorporating a more effective prosodic module in working speech synthesisers and in practical man-machine dialogue systems.

Our work towards reaching this goal has consisted of analysis of speech data and a model-based analysis-by-synthesis methodology, as well as text-to-speech (TTS). The analyses are of various kinds, involving discourse and textual structure, as well as lexical semantics. The analysis-by-synthesis methodology has been an important part of the analysis work. It is implemented in a computational environment, making it possible to generate F_0 contours that can be imposed on a speech waveform taken from natural speech. The model-based resynthesis is based on the analysis of both the dialogue structure and the lexical-semantic and textual relations of spontaneous dialogues. Testing and exploration of this analysis-by-synthesis system on spontaneous conversational speech have in turn led to

A. Botinis (ed.), Intonation, 291-320.

extensions in the analytical framework, as well as to the extension of the prosody model itself.

Our TTS work exploits the results from our analyses of dialogues and studies of discourse and textual structure, as well as drawing on data from a database of man-machine dialogues.

13.2 The Basic Prosodic Model

In our prosodic analysis of Swedish, focus is on intonational aspects. The point of departure is a categorisation and symbolisation of the basic prosodic functions: prominence and grouping. Based on an auditory analysis, this results in a prosodic transcription involving two levels of prominence (accented, focussed), tonal junctures and two levels of grouping (minor, major phrase). The choice of word accent (accent I / accent II) in Swedish is lexically determined. The prosodic categories used are:

Tonal Structure		**Label**
accented	accent I	HL*
	accent II	H*L
focussed	accent I	(H)L*H
	accent II	H*LH
	compound	H*L...L*H
juncture	initial	%L; %H
	terminal	L%; LH%
Grouping		
boundary	minor	|
	major	||

The star notation (*) is used here in accordance with current usage in non-linear (autosegmental) phonology to denote the particular tone (H or L) to be associated with the stressed syllable. For focal accent I the word accent H may be missing (contextually determined), but the timing distinction between accent I and accent II is still maintained through the earlier timing of the focal LH gesture in accent I and the later timing in accent II.

13.3 Global Aspects of the Model

The intonation model was originally developed to cover single utterances and was based largely on data from 'lab speech'. As our attention turned toward spontaneous speech and dialogues, the model served as the starting point for a more elaborated version. When a speaker produces speech in a natural communicative setting, as in a dialogue, the speaker may structure his/her utterances in a manner that may be related to e.g. the topic structure and the interactive character of the speaker situation. This is reflected in the intonation used by the speaker, and this is what we term *global* aspects of intonation.

We have approached the study of global intonation along several different avenues, where we investigate the connections between discourse structure and the phonetics and phonology that are both a consequence of that structure and an instrument in the signalling of it. We have studied discourse structure from the angle of lexical-semantic relations and text analysis. During the course of this work we have developed an analysis-by-synthesis methodology which we exploit in our efforts to develop and refine our prosodic model.

The databases that we build on consist mainly of dialogues recorded from radio programmes, studio recordings of (quasi-)spontaneous dialogues and material collected as part of the Waxholm man-machine dialogue project (Bertenstam, Beskow, Blomberg, Carlson, Elenius, Granström, Gustafson, Hunnicutt, Högberg, Lindell, Neovius, Nord, de Serpa-Leitao & Ström, 1995).

13.4 Lexical-semantic Relations

A topic structure can be thought of as the result of the lexical, grammatical and semantic/pragmatic parameters that interact to create the text/conversation. A discourse is said to be 'coherent' if the topic structure is clear and easy to follow. The sentences or utterances are logically linked to each other and are relevant in the context where they are used. According to Halliday and Hasan (1976), coherence is created by the use of a number of 'cohesive devices'. One of these devices is coreference or cospecification, since, in order to know if one is still speaking or writing about the same topic, there must be some way of referring back in the discourse to a referent that has been mentioned earlier in the text/conversation. Content words are related to each other by morphological identity and lexical semantic relationships (synonymy, hyponymy, and meronymy) (Cruse, 1986). In Horne, Filipsson, Ljungqvist and Lindström (1993) it was shown how these relations can be tracked computationally in a linguistic preprocessor to a text-to-speech system. The information on cospecification can then be used in the F_0 generating component in order to appropriately assign focal and nonfocal word accents.

Although the tracking of these lexical relations was initially developed for predicting accent assignment within a restricted domain, it is possible to extend the modelling of lexical semantic relations to cover more domains and thus describe larger 'semantic frames' that define prototypical scenarios, institutions, etc. (cf. Metzing, 1981). One can imagine these frames as networks where there are connections between referents in different semantic fields.

In order to illustrate this type of frame analysis, we will show the lexical structure of the frame 'Recipe for a Hot Tunafish Sandwich' as it develops in a dialogue from a Radio Sweden programme where the guest is asked to present a favourite recipe for the listening audience. The following is a translation of an excerpt from the dialogue:

Guest: *(breathing)* uh I'm not really any kind of experienced cook uh but I do have a recipe for a hot sandwich which I've in fact develop developed a little *(breathing)* uh it's a tunafish sandwich and you make it like this you have white bread for example *(breathing)* and on that you put a mishmash of uh it makes rather a lot but you have to have a can of tunafish, some crème fraiche just tunafish in water uh is good otherwise there's so much fat *(breathing)* and then some crème fraiche and then about a third of a jar of mayonnaise preferably light mayonnaise there too since there's crème fraiche in [it] *(breathing)* and then just a tiny dab of mustard it can be strong mustard

Interviewer: French;

Guest: Yes;

Interviewer: Scanian ...

Guest: uh and then a lot of chopped leeks and a dash of Italian salladspice and then you stir all that up together nice and even you know *(breathing)* and then I think it's a good idea if you let it sit and rest a couple of hours so that the taste spreads through the whole thing and then you slap it on those slices of uh bread and then in the oven with them and if you want you can put ...

As can be seen in Figure 13.1, we are including inferences of the type *cook* ← *recipe* in the frame model as well as the traditional lexical semantic relations of hyponymy, meronymy (sometimes referred to as 'partonymy') and synonymy, so as to capture all the lexical relationships between referents (see also Litman & Passonneau, 1995 for the importance of modelling inference in algorithms for discourse segment boundary detection).

In order to include this discourse information in the database, we have developed a method of tracking and transcribing such relations The relations used are:

x	The discourse referent (DR) x is not related to any other DR
=x	The DR x is morphologically identical to or a pronominalised form of a preceding DR
x=y	The DR x is a synonym of the DR y
x<y	The DR x is a hyponym of the DR y
x>y	The DR x is a hyperonym of DR y
xEy	The DR x is a meronym of (part of) the DR y
xZy_1...y_n	The DR x is the sum of the DR's $y_1 \ldots y_n$
[x]	The DR x is a superordinate (non-basic) term
x ← y	The DR x is inferable from the DR y

Here, **x**, **y**, and **z** represent numbers. Each time a new referent is encountered in the text, it is given a unique number. Its relation with the earlier referent is expressed by stating the type of relation and the number of the earlier referent. In our analysis-resynthesis methodology, where we work in the ESPS/Waves+ environment, this information is encoded as labels in a separate discourse-structure tier along with the prosodic labels, as described below in section 13.9, *Model-based resynthesis*.

These lexical relations have subsequently been used to explain some unexpected patterns of accentual downtoning in the data. Downstepping of word accents, for example, has been observed to occur in a number of cases of lexically 'new' information in this dialogue (Ayers, Bruce, Granström, Gustafson, Horne, House & Touati, 1995). This is not what one would expect since new information is generally accentually highlighted. However, it has been seen that this downstepping correlates with certain aspects of lexical semantic structuring. Accentual downtoning is associated with lexically new information in two environments: the first is when the new information is realised by a superordinate 'non-basic' word such as *mishmash* ([x]), and the second when the new information is a specification which is in some sense non-central to the development of the topic, such as *tunafish in water* (x<yEz) - i.e. *tunafish in water* (x) is a specification (<) of *tunafish* (y), which is in turn a part of (E) the *mishmash* (z).

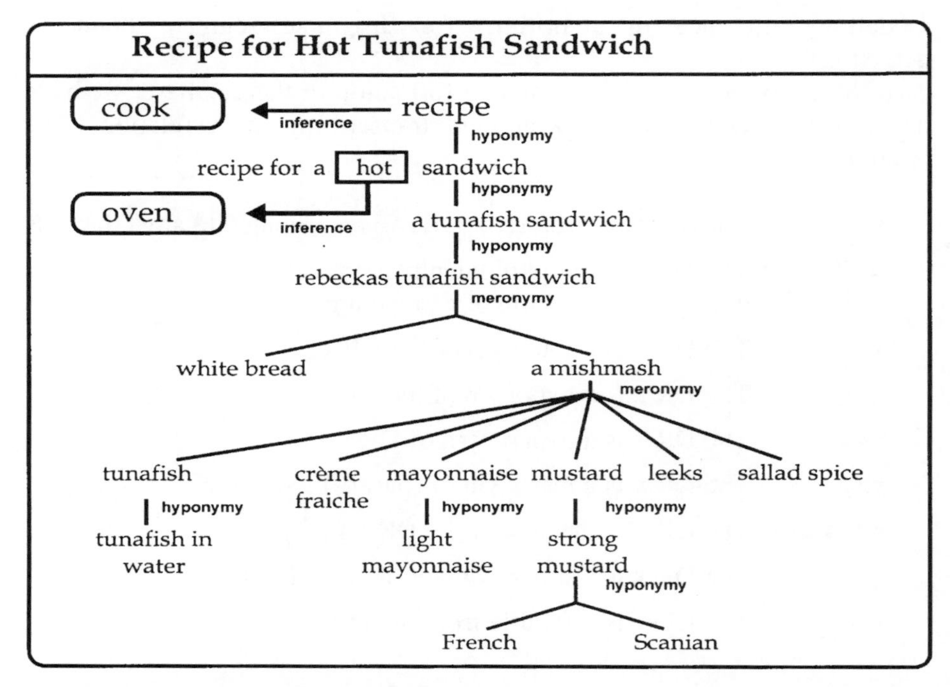

Figure 13.1. Frame semantic topic structure representation for dialogue fragment in section 13.4.

13.5 Text Analysis

There exists a variety of coding systems for text analysis, based on different principles, e.g., the Initiative/Response system proposed in Sinclair and Coulthard (1975). In an attempt to enhance the prosody model with a mechanism for including characteristics of the textual content of utterances, we have developed a simple textual topic analysis model that is based on a combination of the lexical-semantic analysis and functional grammar. A strictly textual approach is chosen since it avoids the circularity of including prosodic features in the definition of discourse structure.

The verbatim transcriptions of spontaneous dialogues are divided into segments. The lexical items are classified according to a simple model of functional grammar. The categories used are Subject, Predicate/Verb, Object and Attribute. Segmentation of the text is performed so that each segment maximally contains one subject and one (compound) verb. 'Connector' words (*and*, *or*, *if*, *so*, etc.) are used to locate the boundaries between segments. This means that a segment contains one subject, one (compound) verb, and the objects and attributes related to them. Classification of the segments is based on the word items that segments contain and their properties according to the lexical relations described in the previous section. The following categories are used:

- **Initial** (I), segment containing a DR not related to any other previous DR.
- **Expansion** (E), segment containing a DR related to a previous DR by hyponymy, meronymy, implication or summation.
- **Continuation** (C), segment containing coreference or iteration of previous referent; or a DR related to a previous DR by morphology or synonymy; or by pronominalisation, coordination (elliptic subject), or by a correlate; or an assignment of an attribute to a DR.
- **Follower** (F), segment not containing referential material, objects or attributes.
- **Summary** (S), evaluational expressions, judgements and opinions.
- **Mistake/Repair** (M/R), speech errors and the announcement of them ('... I almost said').

Each segment is assigned one of these labels, and then each label generates the values of global parameters used in the model-based resynthesis. This is further described below, in section 13.9.2, *Integrating discourse analysis and model-based resynthesis*.

13.5.1 Topic Structure

The idea that forms the main hypothesis in our work on textual influence on dialogue and discourse prosody is that the structuring of conversational topics, by means of signalling topic coherence (continuation) and topic boundaries (shifts), has important reflections in the intonation of utterances. Earlier attempts at this approach have been reported on in Bruce, Filipsson, Frid, Granström, Gustafson, Horne, House, Lastow and Touati (1996) and Bruce, Frid, Granström, Gustafson, Horne and House (1997), where we introduced a lexical-semantic analysis of the referent structure in a discourse. This accounts for the given/new structure of referents, and thereby the accentuation/deaccentuation in noun phrases. There are, however, relationships between other constituents in a discourse that can contribute to the overall structuring of the coherence of conversational topics. This can be illustrated by studying the intonation of the utterance in Figure 13.2. It consists of two phrases, which exhibit similar syntactic structures and discourse functions. In this speech segment, we would like to model the second phrase with a lower register than the first phrase, i.e., (in our model) with *Decreased* register. But there is nothing in the noun phrases that we can utilise to relate these phrases. There is, however, the similar syntactic structure, the conjunction *och* ('and') and the repeated attribute *jättebra* ('very good').

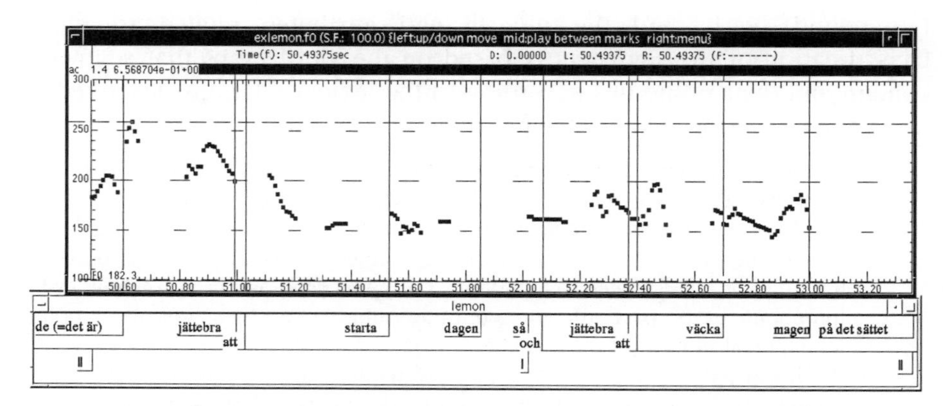

Figure 13.2. An utterance with two phrases, where the second one uses a lower register. See Figure 13.3 for a translation. The vertical lines in the lower tier symbolise major (||) and minor (|) phrase boundaries.

In order to capture features of discourse coherence that are not dependent on the lexical-semantic structure of discourse referents, an extended analysis scheme has been developed. It can be described as a hybrid of grammatical and semantic roles, as well as lexical-semantic structure. The initial step is an analysis of the lexical elements in a text, according to which they are classified in different categories. A description of the categories follows below.

- **Adverbials of Time, Manner and Space:** they give the setting of the events being talked about.
- **Subject:** the noun phrase with the referent that often performs an action (i.e., agent).
- **Verb:** the notional component of a verb complex that denotes the action, state or event.
- **Object:** the noun phrase with the referent that often undergoes an action (i.e., patient).
- **Phrasal attributes:** text units altering or evaluating the relationships between subjects, verbs and objects.
- **Phrase starters, Phrase enders:** transitional elements, conjunctions, relative pronouns that occur at the beginning and/or end of phrases.

The categories form a partitioning of the dialogue text into analysis units, corresponding to grammatical functions. Whenever a new main verb is found a new analysis unit is formed. A new analysis unit is also formed if any piece of text is analysed as a category that already exists in the current analysis unit. Relative pronouns and conjunctions also signal the start of a new unit, since they often coincide with an intonational phrase boundary. The infinitive marker *att* ('to') and the subordinating conjunction *att* ('that') are associated with the end of a new analysis unit (cf. Horne & Filipson, 1996). However, the following unit is not assigned an intonation label of its own if it lacks a subject, i.e. after the infinitive marker. In summary, an analysis unit consists minimally of a verb. Subjects, objects and attributes that are grammatically linked with a verb are put in the same unit as that verb. The boundaries between units are determined by the phrase starters and enders, as well as certain function words, and upon encountering a new verb or another category that is grammatically linked with another verb than the one in the current analysis unit.

Connections between categories in different analysis units are then identified by means of a set of rules, e.g. the lexical-semantic relations. The connections are then used to generate intonational labels that are used in the model-based resynthesis, see sections 13.6.3 and 13.9. The basic principle of label assignment is: as long as there is an element in an analysis unit that has a connection with an element in the immediately preceding analysis unit, it gets the label *Decreased*, but if there is no connection it gets *Mid*. If there is a textual link between two consecutive units, their coherence is signalled by using a non-resetting phrase intonation. If there is no obvious semantic relation, a boundary is signalled by resetting the register in the second phrase. In the example in Figure 13.2, the second phrase will be assigned the label *Decreased*, because of the repeated attribute *jättebra*, which is lexically identical to the attribute in the first phrase. Figure 13.3 shows an example of the analysis of a dialogue fragment. Units containing speech errors, mistakes, and hesitations are not counted as previous units but are disregarded. In these cases the previous unit is thus defined as the one prior to the unit containing the error or hesitation.

Performing this kind of analysis on spontaneous speech is by no means as trivial as it may appear here, but these categories are believed to represent a reasonable blend of grammatical detail and analytical ease. We do not claim to have found the ideal set of constituents, but these categories have emerged as a working system for our purposes.

Dialogue

EH hur funkar de då
EM det gör att *em* det hjälper till med matsmältn när man *eh* alltså smälta maten det är jättebra att starta dagen så och jättebra att väcka magen på det sättet

(Translation:)

EH how does it work then
EM it makes *um* it helps with the food-digest when you *uh* I mean digesting food it's really good to start the day like that and very good to wake up your stomach that way

Analysis

PHRASE STARTER	SUBJECT	NOTIONAL COMP. OF VERB	ATTRIB.	OBJECT	PHRASE ENDER	LABEL
hur	det	funka			då	Mid
	det	göra			att	Mid
	det	hjälpa till med		matsmältn		Mid
när	man					Decr.
alltså		smälta		maten		Decr.
	det	är (=vara)	jättebra		att	Mid
		starta		dagen	så	-
och			jättebra		att	Decr.
		väcka		magen	på det sättet	-

Figure 13.3. Example of analysis of a dialogue fragment.

13.6 Global Features of Intonation: Phonetic and Phonological Correlates of Discourse Structure

13.6.1 Tonal Downtrends and Downstepping

One of the aspects of our modelling of dialogue intonation involves analysis of intonational downtrends. See for example Pierrehumbert and Beckman (1988) and Grønnum (1992). According to earlier studies of Standard Swedish in controlled laboratory experiments, the occurrence of a focal accent is a pivot for tonal downstepping (Bruce, 1982). Before a focal accent, non-focal accents do not appear downstepped, while after focus downstepping of successive accents within the same phrase or utterance is a characteristic feature. Thus an early focus in an utterance will typically trigger downstepping, while a late focus will tend to arrest this tonal downtrend. That is to say, downstepped accents occur on information that is given in the context as in (1b), which constitutes an answer to the question in (1a). The underlined words in (1b), which are contextually given in the preceding question, are characterised by downstepped accents. In this example and in the ones that follow, the underlined words are characterised by downstepped accents, and words in bold print are focussed. Phrase boundaries, major and minor, are also indicated, where present.

(1) a. Vem lämnar ungen nallar? 'Who gives the kid teddy bears?'
b. **Mamman** lämnar ungen nallar 'The **mamma** gives the kid teddy bears'

In our earlier study of spontaneous speech within the KIPROS project, this regularity was found to appear in spontaneous dialogue as well (Bruce, Touati, Botinis & Willstedt, 1988; Bruce & Touati, 1992). Several typical examples of downstepped pitch patterns were observed which seemed to be triggered by the placement of focal accent in the same way as described above.

In our current analysis of spontaneous speech, downstep has also been seen to correlate with contextually given information. However, it has also been observed to correlate with information that cannot be classified as given. The following examples are taken from our analysis of a dialogue recorded from a Swedish radio program about jazz music called 'What's cooking?'.

In (2), the clause after the word *macka* 'sandwich' does not contain given information in the same sense as in example (1). Yet, it is characterised by downstep in the same way as the underlined words in (1).

(2) Jag har ett litet recept på en varm **macka** | som jag faktiskt har har utvecklats utvecklat en aning

'I've got a recipe for a hot **sandwich** | which I in fact have have improved improved a little'

It should also be noted that in the examples there is no focal accent present in the underlined phrase itself as a direct trigger of the downstepping. Instead, each successive non-focal accent within the phrase is downstepped.

Thus, the generalisation that it is given information that gets downstepped is perhaps too narrow to cover all cases of the phenomenon in spontaneous speech. It would, however, be insightful if one could relate examples like those in (1) and (2) to some more general discourse/semantic parameter(s).

One idea which we would like to pursue in this respect is to relate downstepped information to the development of discourse topics. In this regard, one could say that the downstepped information in (2) is similar to the nonfocal material in (1), in that it can be considered as information which is not central to the development of the topic. By 'central to the development of the topic', we then mean related to the specification of the lexically important/'generic' referents in the semantic field under discussion as well as the specification of the relationships among these referents. In the specific dialogue under consideration, the central topic involves the description of the ingredients in a recipe for a hot tunafish sandwich. Just as the downstepped information in (1), being contextually given, does not provide anything new as regards the relationships between the referents *ungen* 'the kid' and *nallar* 'teddy bears', the downstepped information in (2) likewise does not lead to the development of the central topic in the discourse from which it is extracted, i.e. to the description of the discourse referent *macka* 'sandwich'. The downstepped material constitutes parenthetical information that is unrelated to the specification of the ingredients in the sandwich, i.e. is non-central to the development of the topic.

Another example of the use of downstep is given in (3):

(3) ... man har vitt **bröd** förslagsvis | och på detta lägger man en röra av ...
'... you take white **bread** for example | and on that you put a mishmash of ...'

In this case, noncentral can be related to the level of specificity of the discourse referents. The word *röra* 'mishmash' is semantically nonspecific or non-generic as regards its status in terms of lexical hierarchies. That is to say, it is not "at the level of ordinary everyday names for things and creatures" (Cruse, 1986) as regards its relation to the other referents which are mentioned in the dialogue, e.g. *bröd* and the ingredients that make up the 'mishmash': *tonfisk* 'tunafish', *majonnäs* 'mayonnaise', etc. Thus, one can hypothesise that the downstepping in the utterance in (3) is related to the nonspecificity or nongenericness of the referent mentioned. In other words, one can hypothesise that referents that are central to the topic are ones that are relatively more 'generic' or more 'basic'.

A similar situation can also lead to prosodic downtoning, i.e. when the speaker 'comments on' or specifies an already introduced 'generic' discourse referent as in (4-5):

(4) ... man måste ha en burk **tonfisk**, en burk **crème fraiche** | tonfisk med vatten bara, eh, är bra ...

'... you have to have a can of **tunafish**, a package of **crème fraiche** | just tunafish in water, uh, is good ...'

(5) ... en tredjedels burk **majonnäs** || gärna lätt majonnäs där också för att crème fraiche i ...

'... a third of a jar of **mayonnaise** || preferably light mayonnaise there too since (there's) crème fraiche in (it) ...'

Here the downstepping characterising the second occurrences of *tunafish* and *mayonnaise* and their respective specifications can be interpreted as reflecting the noncentrality of that information for the development of the topic, i.e. the central referents (generic terms), *tunafish, mayonnaise* have already been mentioned; the comments concerning the fact that it is good if it is tunafish in water, and that the mayonnaise should preferably be light, although new information, are relatively unimportant as regards the development of the central theme.

Since our material is restricted, we cannot be sure of the generality of the observations made here concerning downstepping. Nevertheless, by pinpointing the environments in terms of the topic structure and related lexical semantic correlates, we can test these hypotheses against more extensive data in future studies.

Section 13.8, *Incorporating global features in the model*, describes how downstepping is introduced in our intonation model and F_0 contour generation system.

13.6.2 Register and Range

Many utterances in spontaneous speech can be analysed as consisting of several phrases. Our model uses two units: major and minor phrases. One way of analysing global aspects of intonation is to characterise them by means of two concepts that inherently belong to each phrase: Register and Range. Register denotes the starting level of a phrase's intonation, roughly corresponding to the F_0 level of the valley preceding the first accented syllable. Thereafter it is often gradually lowered at each accent as a result of downstepping. Another parameter represents the speaker's floor, and limits the lowering of Register. In model terms, this will be the levels of the base and floor parameters as described below, section 13.9.3, i.e., the level of the L tonal turning points.

By Range we mean the height of a pitch gesture of an unfocussed accent, starting from the Register level. This corresponds to Ladd's concept *tonal space*, which is defined as: "... a subset of the overall speaking range which is available for realising tonal distinctions at any given point in the utterance." (Ladd 1996, 73). In our model this is the level of the H tonal turning points and the focal multiplier (see section 13.9.3). As the range of the pitch gesture usually is higher for a focal accent, we use two separate parameters for the two levels of prominence.

13.6.3 Analysis

These global features are analysed for each intonational phrase, alongside with the analysis of prominence and grouping. The analysis is performed according to a model recognising three Register levels (*High*, *Mid* and *Low*) and four Range categories (*High*, *Mid*, *Low* and *Flat*). The register can also be *Decreased*, which indicates a more gradual (less than a full shift to a lower level) decrease of register. We thereby combine absolute and relative approaches to phrase intonation. The absolute levels (*High*, *Mid* and *Low*) are absolute compared to other phrases, but relative to the speaker. The *Decreased* level is relative to the preceding phrase.

The choice between an absolute or relative label is related to the grammatical and semantic content, as well as the topical coherence of the phrases. The stronger the relation between the phrases, the more likely it is that the relative level is used. The question of how to measure the topical coherence is naturally an intricate matter, involving many aspects of semantic and grammatical relationships, as was discussed in more detail in sections 13.4, *Lexical-Semantic relations* and 13.5, *Text analysis*.

The result is a detailed notational device which covers the intonation over several phrases. Together with a transcription of accentuation and grouping, and rules for mapping the different categories to pitch levels, we can extend the intonation model so that it can create phrase contours with different characteristics over a stretch of several phrases instead of stereotypically repeating the same parameters.

13.7 Global Features of Intonation: Example Measurements

In order to base the mapping rules between global categories and pitch levels on data from spontaneous dialogues, another aspect of our work has been to investigate some acoustic features of the global F_0 characteristics of real conversations. For phrases in isolation, it is a well known and often occurring phenomenon that F_0 declines. The perceptual impression over the course of a number of consecutive phrases is that a similar lowering occurs, especially if they are semantically coherent (van den Berg, Gussenhoven & Rietveld, 1992; Bruce, 1982; Grønnum, 1992). This is also visible when inspecting F_0 contours. Our expectations were thus to find a reflection of this lowering in the F_0 measurement points within a speaker turn.

13.7.1 Method

In order to investigate this, we first analysed the dialogues according to our model in order to determine where the phrase boundaries were. Resynthesis was performed to validate the analysis. Then, a computer program calculated the F_0 Min, Mean and Max for each phrase. The Means were calculated using all voiced portions of the utterance. Technical errors of the F_0 analysis, such as octave jumps, for example due to creaky voices, could be removed by setting a local F_0 range for each phrase, thus excluding the data points outside this range.

13.7.2 Results

Table 13.1 shows the differences in Min, Mean and Max F_0 of initial and final phrase within the same turn in three different dialogues, averaged over the total number of turns in each dialogue. This shows the average decrease per dialogue turn of each measurement parameter. Figures are in Hz.

Table 13.1. Average F0 decrease in Hz per dialogue turn of three measurement parameters.

Dialogue 1: Conversation among four participants			
	Min	*Mean*	*Max*
Number of turns=21	18	29	36
Dialogue 2: Conversation between two participants			
	Min	*Mean*	*Max*
Number of turns=15	8	2	12
Dialogue 3: Two different speakers in the same conversation			
	Min	*Mean*	*Max*
Number of turns=20	8	6	1
Number of turns=15	18	22	21

As can be seen in these tables, there are differences both between dialogues and between speakers, both regarding the extent of lowering and in the manner in which Min, Mean and Max interact with each other. However, a common trend is that the summed-up phrase differences are negative in all dialogues and for all measurement parameters.

13.7.3 Discussion

These results are interpreted as a trend that speakers use a lower F_0 turn-finally than turn-initially. A possible explanation of the 'negative' trend within each turn is that there are fewer accents later in the turn. This would result in fewer potential F_0 peaks. However, even though this has not been explicitly examined, our impression of the distribution of accents over the phrases in a turn does not support that explanation.

In view of these statistical indications, it seems necessary that F_0 contours should be modelled so that corresponding measurements of the modelled F_0 contours should approach these results, i. e. a lower Max, Mean and Min F_0 in turn-final phrases than in turn-initial ones. Listeners presumably expect this, and it is plausible that a similar lowering in our modelling of F_0 contours contributes to the naturalness of the utterance. This is an important consideration when implementing the model in a man-machine dialogue system, for instance.

13.8 Incorporating Global Features in the Model

The modelling of F_0 generation has been developed to take the findings and suggestions of the previous sections into account. The downstepping feature within each phrase has been elaborated and the model can now be used to specify an F_0 contour for multiple phrases, corresponding to one major phrase that consists of one or more minor phrases. The initial minor phrase always gets a specific value - dependent on whether it has *Low*, *Mid* or *High* register - whereas the following phrases get *Decreased* register as default. More variation may then be generated as described in sections 13.4, *Lexical-semantic relations*, and 13.5, *Text analysis*. A major phrase may correspond to a speaker turn, but need not necessarily do so.

Each accent is modelled according to the principle of downstepping, where the valley succeeding the accent is lowered compared to the valley preceding the accent. Downstepping occurs after a focal accent in a phrase, and continues after a minor phrase boundary, unless the next phrase also contains a focal accent. A major boundary, however, breaks the downstepping.

Downstepping is realised by setting the level of the post-accent valley to a new value. This value is determined by multiplying the level of the pre-accent valley by a fraction. Thereby the fall of the accent gesture becomes greater than the rise. The fraction is added as a phrase-dependent parameter alongside e.g. the base, floor and range parameters. The level for the following L tonal turning point thus becomes lower after each accent (see Figure 13.4 and also Ladd 1996, 75).

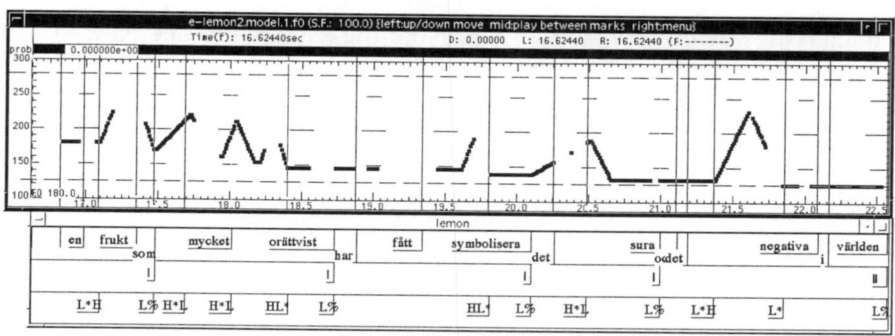

Figure 13.4. Modelled F_0 contour of the utterance fragment ... *en frukt som mycket orättvist har fått symbolisera det sura och det negativa i världen* ('... a fruit that very unfairly has come to symbolise what is sour and negative in the world'). The three tiers show the orthographic, the phrasal, and the tonal transcriptions.

As stated previously, phrase boundaries are classified as either minor or major, and different F_0 characteristics of each phrase boundary marker have been implemented. A minor boundary is marked by a fall to a level somewhere between the starting register and the speaker's floor. This level is determined by a 'Minor Floor' parameter. No register reset is made at the beginning of the next phrase, instead it continues on the same level. This results in a lower initial F_0 than the preceding phrase. A major phrase boundary is marked by a fall to the speaker's floor. The following phrase, which is initial in the following major phrase, starts at a register level specified for that phrase.

Two phrases with a minor boundary between them are thus modelled so that the parameter values for the register of the second phrase are lower compared to the first. When more than two phrases follow each other, the result is a gradual lowering for all the phrases, since each phrase will start at a register level equal to the final level of the preceding one.

These extensions of the model capture the global characteristics of spontaneous speech described in previous sections. The modified model can capture cases where both register and range are decreased over a number of minor phrases, with resetting occurring at a major phrase boundary. This pattern can be related to the information structure of a phrase; there may, for example, be a difference in intonational pattern between phrases with a content that refers backwards to the preceding phrase and phrases where this is not the case.

An example of an F_0 contour generated by the model is shown in Figure 13.4. The utterance is taken from the four-speaker conversation, and is spoken by a male speaker. The parameters are set so as to copy his acoustic characteristics as closely as possible. Note the asymptotic decrease in the downstepping of the pitch accents.

13.9 Model-based Resynthesis

Throughout our work we have used model-based resynthesis as a tool to verify our hypotheses and to refine the model itself. The building blocks in our system for performing model-based resynthesis are a prosodic analysis as described above, a computational method for generating F_0 contours from a prosodic transcription, and the facility of performing resynthesis using the generated contours. These are all realised in the ESPS/Waves+ environment.

13.9.1 Analysis-by-synthesis

The prosodic transcription of the dialogues is made by an expert, based on a purely auditory analysis. The transcription results in a sequence of boundary and tonal labels (see Figures 13.5 and 13.6). The alignment of the tonal labels is with the CV boundary of the stressed syllable, and the alignment of the boundary labels is with the start and end points of the speech or group boundaries. The labels for the orthographic words are right-aligned in the space that they occupy.

The prosodic transcription thus represents a phonological analysis of prominence and grouping. It also constitutes the input to the resynthesis module. In the implementation of the intonation model, the prosodic information contained in the transcription is supplemented with phonetic rules that take care of the specific timing and realisation of F_0 events, as well as concatenation between them. These include pitch level and range (including focus realisation), F_0 drift (downdrift, downstep, upstep), as well as the interpolation between tonal turning points. For the resynthesis we are using an implementation of the PSOLA synthesis algorithm in the ESPS/Waves+ environment (Moulines & Charpentier, 1990, Möhler & Dogil, 1995).

The use of the analysis-by-synthesis method has a double purpose in the present framework. The first, more direct goal is to verify/falsify the prosodic transcription. This will give us feedback on the correctness of the transcription and will reveal any incorrect auditory judgements about focus placement, prominence levels, phrasing and the like. The second, more long-term goal is to use the analysis-by-synthesis tool for developing our intonation model in a dialogue prosody framework, as a first step towards improving the generalised model used in the text-to-speech system. As our prosodic transcription covers only prominence and grouping, other aspects of intonation are thus not explicitly modelled in our current resynthesis.

The analysis-by-synthesis system presupposes that the speech waveform is divided into discrete units of different lengths. This partitioning may for instance be based on the locations of the boundary labels (see also the section 'Analysis of global intonation' below). For each unit, certain parameters can be set in order to represent other intonational features and to explicitly state the physical identity of F_0 events. In a speech synthesis perspective, the values of these parameters can be determined automatically for each unit using different methods, e.g., by applying models of topic structure and/or the Given/New distinction, as discussed in section 13.5.1.

13.9.2 Integrating Discourse Analysis and Model-based Resynthesis

A method of performing the parameter setting automatically from phrase to phrase is also implemented in the model-based resynthesis system. This is done by introducing a separate tier of labels, whose categories are associated with different sets of parameter values. This is a useful configuration, since it enables us to use different types of dialogue analysis to affect the modelling and synthesis of intonation. Currently, this is exploited to implement the textual model described in section 13.5. By means of the textual analysis, the verbatim transcription of a spontaneous dialogue is divided into segments, and each segment is consequently assigned one of the six categories on the basis of their lexical-semantic properties. Each segment then receives its own set of parameter values from rules associated with the textual category the segment belongs to.

In writing the rules for the influence of the textual categories on intonation, we perform an acoustic analysis, where we exploit the model-based resynthesis. For each segment, pitch contours are modelled using two different techniques:

a) the 'default' approach, in which the same frequency parameters are used for all phrases. The exact parameters vary with each speaker, and they are dependent on a model of the speaker's average register and range. All phrases thus have the same register and range. This results in an overview of how the prosodic model (just accentual and phrasing information) performs.

b) the 'fine-tuning' approach, where the frequency parameters are set so that the resulting pitch contour is as similar to the original of a phrase as possible. This gives us an indication of what the optimal, or desired, parameter values for each particular phrase are.

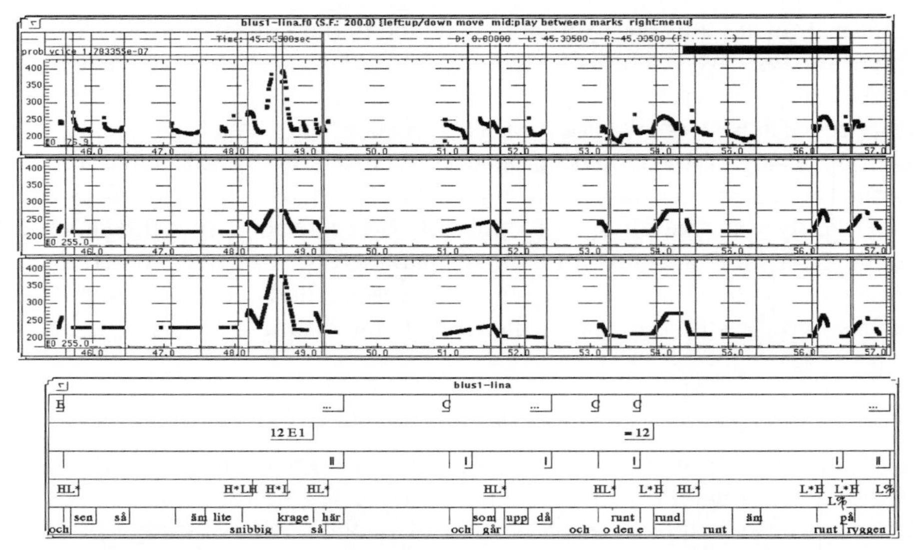

Figure 13.5. From top to bottom: Original, default-modelled and discourse-modelled F_0 contours of the utterance *och sen så äm lite snibbig krage så här och som går upp då och runt och den e rund äm runt på ryggen* ('and then eh a little pointed collar like this that goes up then and goes round and it is round eh round on the back'). Below is shown the corresponding labelling for text segmentation relations, lexical relations, phrasing, prominence and boundary tones, and orthography. Note the different ranges between segments of accent peaks and floor in the discourse-modelled contour.

The methodology is illustrated in Figure 13.5. By comparing these two kinds of contours, we are able to see deviations from the current prosodic model, which does not contain regulation of discourse prosody, e.g., topic structure, dialogue features, etc. The places of major deviation need additional modelling that can govern the setting of parameter values appropriately. The idea is to relate the textual analysis of discourse to intonational patterns in dialogues and extract possible correlations. On the basis of the results, a model of the textual influence on intonation is created, i.e. rules for parameter value generation are constructed. Our extended prosody model is based on these results. In this model we can control both local (accents and boundary tones) and global aspects of F_0 (F_0 trends within phrases, relations between phrases and influence from discourse structure). We are, of course, well aware that this modelling represents a simplification of the dialogue situation. At present, we are not, for instance, taking into account prosodic variation due to attitudinal factors or the emotional state of the participants, or to other factors extraneous to the dialogue situation.

13.9.3 F_0 Contour Generation

The process of generating an F_0 contour involves transforming the phonological prosodic transcription of the analysis into a set of labels marking the location of the turning points of the F_0 curve. The labelling conventions that we use during this transformation process are the same as we use in the general description of the prosody model. This means that the basic labels are H and L, which can be followed by * or by % to mark the presence of an accent or a boundary. The transformation is achieved by means of simple rules. It is important to note that the only input to the system is comprised of the transcription labels and their time alignment. This gives us a segmentation of the speech signal into stretches corresponding approximately to stress groups or feet (aligned with the CV boundary of the stressed syllable). We have no other segmental information available, nor any information about voiced/voiceless distinctions. This, of course, limits the amount of information which can be included in the rules for placing the turning points, because we cannot refer to e.g. vowels or syllables as points of reference in the speech signal. In general, the starred (*) turning point is placed at the location, or very near (30 ms after) the location of the transcription label. Preceding H's are placed a fixed number of milliseconds (30) before the location of the label. Succeeding turning points are spaced equally between the locations of the current transcription label and the next label. This solution may seem to be *ad hoc* but is not without motivation in production data. In a study by Bruce (1986) variability in the timing of the pitch gesture for focal accent relative to segmental references was demonstrated. Instead, disregarding segmental references and using the beginning of the stress group as the line-up point, there appeared to be a high degree of constancy in the timing of the whole focal accent gesture. It should be noted that the actual numbers in milliseconds of the implementation are at this point chosen as test values partly based on earlier work on tonal stylisation (House, 1990; House & Bruce, 1990). However, there is a possibility to change the timing parameters as a function of the phrase type (see below). A summary of the prosodic labels used, the rules for splitting them into turning points and their temporal realisations is given below. HF is used to indicate the focal accent location in the intonation contour.

1. Accent I, **HL***
   H 30 ms before the label.
   L* at the label.
2. Accent II, **H*L**
   H* 30 ms after the label.
   L one half of the distance in time to the next label.
3. Focal accent I, **(H)L*H**
   H 30 ms before the label.
   L* at the label.
   HF one half of the distance in time to the next label.
4. Focal accent II, **H*LH**
   H* 30 ms after the label.
   L one third of the distance in time to the next label.
   HF two thirds of the distance in time to the next label.
5. Last part of focal accent II compound, **L*H**
   L* at the label.
   HF one half of the distance in time to the next label.
6. Phrase final, **LH%**
   L 100 ms before the label.
   H% at the label.
7. All others, **%L**, **%H**, **L%**, **H%**
   At the label.

Besides these rules for the distribution and localisation of transcription labels, an additional set of rules has been developed to capture some of the variations of the general patterns that we have found in our production data. One such rule concerns the realisation of an L* of accent I in focal positions as H*, resulting in an upstep, another generates a plateau by having a post focal H in a phrase realised as an HF.

Finally, there is a large set of optional rules or settings in the system, which may be set independently for each phrase. These options are specified parametrically and control such things as range, floor, downdrift and height of focal gestures. Relevant parameters are: base level (Hz), floor (Hz), range (Hz), focal multiplier, maximum time for a rise within a label, maximum time for a fall within a label, maximum time for a rise between two labels and the temporal alignment of labels. The base level is the starting value of the F_0 baseline, and the floor is the end value of the base line. All low turning points (L) are placed along this line. By varying these values, a downdrift or updrift can be created. The drift will then be introduced at the first focal high, and will persist to the end of the phrase (see also section 13.6.1, *Tonal downtrends and downstepping*). If there are no focal highs, the slope will start at the first tonal accent. The range sets the location of the normal highs (H). The focal multiplier will be applied to all focal highs (HF) to achieve an extra wide range at those points. The three maximum time options make it possible to constrain the timing of turning points. The rules given above space the turning points equally between the transcription labels. This may sometimes result in an unnaturally shallow rise or fall if duration is considerable between two labels, and is corrected by the time options.

Next, when all turning points are placed properly, a contour is generated by interpolation. At the moment we are using simple linear interpolation, but other options may be exploited, such as cosine functions and splines.

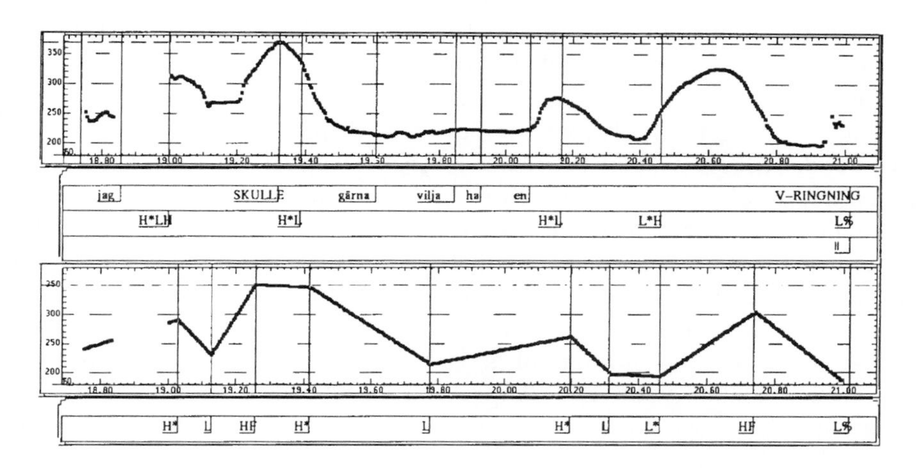

Figure 13.6. An example using the resynthesis approach on an utterance from a spontaneous dialogue: an original F_0 contour (upper part) and the model-generated F_0 contour (lower part). The utterance is: *Jag skulle gärna vilja ha en V-ringning* ('I would quite like to have a V-neck').

In the top part of Figure 13.6 the original F_0 contour of a test sentence can be seen. Below are an orthographic transcription and a prosodic transcription with the tonal and the boundary tiers. Further below is a model-generated F_0 contour of the same sentence, with the turning points as they have been placed by the algorithm.

The analysis-by-synthesis tool enables us to evaluate the auditory analyses and, indirectly, the model itself, thereby supplying data for further enhancements of the model, along with the results of other types of analysis, as discussed in section 13.8, *Incorporating global features in the model*.

13.10 Text-to-speech

In parallel with the analysis-resynthesis approach, we have been developing a text-to-speech (TTS) implementation of our enhanced prosodic model. The TTS version has a wider scope than the analysis-resynthesis approach, as it is intended to be able to model inter-speaker characteristics of various kinds, such as socio-geographic variation and sex and age characteristics, as well as intra-speaker variation, reflecting, for instance, emotional and attitudinal aspects of the speech. Therefore, the TTS implementation has been designed with greater freedom of expression in mind than has the analysis-resynthesis method, which is meant primarily as a research tool; appropriate constraints can be placed on this freedom, in order to make the output realistic. The TTS implementation is meant to provide a working model both in traditional TTS systems and in more advanced man-machine dialogue systems.

The enhanced model is an extension of a simpler model that is used in our standard TTS system (Carlson, Granström & Hunnicutt, 1990) and aimed to provide adequate prosody in 'neutral' utterances, in one variety of Swedish, the standard speech of Stockholm; (most of) its features could be implemented by using elements that were basically phonological and discrete in nature. The current approach, on the other hand, is a step towards a system that can model the variability encountered in the speech of Swedes from a wide variety of geographical and social backgrounds, of both sexes and a wide age range, and produced in widely different situations. In order to achieve this, it was necessary to introduce a mechanism that can manipulate freely the acoustic parameters that correlate with the speaker variability parameters. The parameters that are necessary in order to model the various kinds of speaker variability fall into three categories:

- F_0 (level, range, contour details)
- timing and duration of segments, and duration of pauses (i.e. rhythm-related parameters)
- spectral characteristics

Variations in spectral characteristics have, however, not been included in the current implementation, but were an important part of the development of the Swedish synthesis under the VAESS project (Bertenstam, Granström, Gustafson, Hunnicutt, Karlsson, Meurlinger, Nord & Rosengren, 1997), which ran in parallel with our work and whose results we will be able to draw on in future extensions that seek to integrate the results of the two projects.

13.10.1 Implementation

In order to achieve the variability goal, the enhanced model is parametric and employs, both in the analysis-resynthesis approach and the TTS implementation, certain gradational elements that can be manipulated experimentally or by rule. The gradational elements enable us to model non-categorical variation, for instance related to individual speaking styles and to dialogue situations, thus paving the way to realistic synthesis of speech as produced by a wide variety of speakers, and in a wide variety of conditions. In addition, they supply a tool for the experimental manipulation of the phonetic realisation of the synthesised utterances.

The TTS implementation and the analysis-resynthesis approach build on the same theoretical foundations, using the same prosodic model, but the practical implementations are different. This is a reflection of the different historical backgrounds and traditions of the two laboratories involved in this work.

The organisational basis of the TTS implementation, like that of the analysis-resynthesis approach, is the major phrase unit. Major phrases are related prosodically to other major phrases, and they can be decomposed into two or more minor phrases. Each phrase can be divided into a number of domains: initial juncture domain, (optionally) prefocal domain, focal domain, (optionally) postfocal domain, terminal juncture domain. As a variant, a non-focal main domain may take the place of a prefocal + focal domain. These domains are an expedient device to localise various prosodic events of the synthesis.

Within each domain, the salient prosodic features are encoded parametrically. This applies to both F_0 and duration events. The F_0 events can be specified both in terms of frequency and timing. The coordination of these two dimensions is important in order to achieve the nuances and naturalness characteristic of spontaneous and read human speech in its various manifestations. The timing parameters are an important part of the modelling of speaker variation, especially in terms of geographical background. The implementation allows of both absolute and relative specification of the parameters. Absolute parameter values are useful in an experimental setup, for instance, in that they enable us to specify the exact manifestation that is desired. A relative specification is, however, more realistic when it comes to a general TTS system.

The F_0 parameters define, among other things, F_0 high (H and H*), F_0 low (L and L*), F_0 slope, and the phonetic details of initial and terminal junctures. The timing parameters regulate the exact timing of specified types of turning points, relative to a default time value.

Having established a methodology for manipulating this variability, it was necessary to create a way to encode the phonologically and phonetically significant variation in the TTS system. The parametric TTS implementation basically specifies the phonetic properties of the prosody of utterances. This has to be linked to a mapping procedure whereby relevant phonological and discourse-related categories can be mapped to specific settings of the phonetic parameters. The mapping is encoded in a set of symbols that will generate the various prosodic contours when inserted in orthographic texts. The mapping and coding were developed on the basis of our studies of different kinds of speech material. This comes partly from the man-machine dialogues of the Waxholm database (Bertenstam *et al.*, 1995), partly from the speech material used in our work with the analysis-resynthesis approach. The categories used to describe dialogue prosody were derived as described above in the sections on lexical semantics, text analysis and global features of intonation, sections 13.4-13.8.

13.10.2 TTS in Dialogue Systems

One of the goals of the TTS implementation is to provide the prosody of working dialogue systems. Admittedly, we have yet far to go, not least because we still have a lack of basic knowledge of many aspects of spoken dialogue prosody. But as these gaps are filled in, we will be able to fine-tune our model. For the system to work properly and automatically, it would have to be coupled to a mechanism supplying it with the parameters necessary to produce prosody that is adequate from a textual and dialogue point of view. When coupled to an automatic dialogue system, the dialogue manager should be able to provide these parameters and their values, and this is in principle possible today. If for some reason these values cannot by provided by a dialogue manager, we must resort to manually inserting adequate codes in the texts to be synthesised. This is also useful in an experimental setup. The symbol codes are automatically expanded to give the appropriate parameter values. These parameters can be both of the relative and the absolute kind, and, if carefully chosen, can be used to synthesise prosody that is judged relevant for various dialogue situations, both in terms of F_0 and durational patterns.

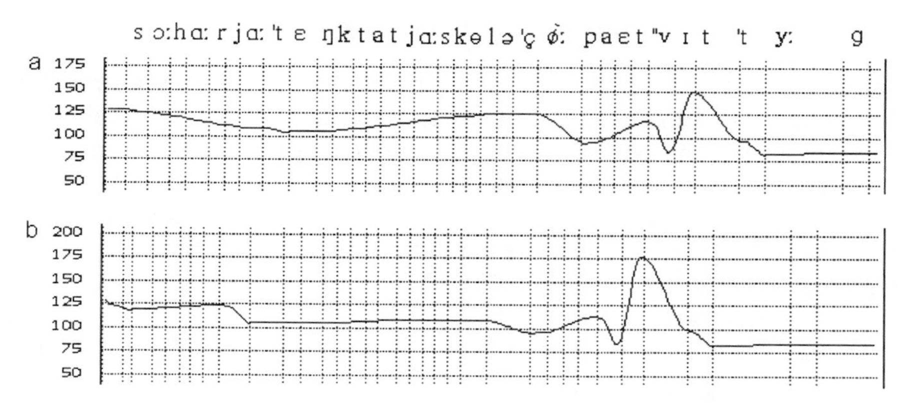

Figure 13.7. TTS without and with dialogue-specific markers. Figure 13.7a shows the default F_0 contour for the utterance *Så har jag tänkt att jag skulle köpa ett* ***vitt*** *tyg* ('Then I had thought I might buy some *white* material'), Figure 13.7b shows the same utterance with markers inserted into the orthographic text. The markers are intended to convey both the feedback-seeking and the contrastive stress that the original utterance in the database is judged to convey.

Examples of the kind of situationally conditioned prosodic realisations that we have encoded symbolically in the system are:

- reduced range and lowered F_0 in parenthetic phrases
- high F_0 level pre-focally as a device for forced turntaking in dialogues
- reduced F_0 range, high F_0 level and faster tempo pre-focally and a wider F_0 range focally, as indicative of a lively contrast to what the previous speaker has been saying
- a reduced F_0 contour in the post-focal domain typical for turn-final positions
- a gently rising final F_0 contour as indicating politeness or non-assertiveness in a dialogue

Figure 13.7 shows the F_0 contour of two synthetic versions of an utterance from one of the dialogues in our corpus - without dialogue-specific markers (Figure 13.7a) and with dialogue-specific markers inserted in the text (Figure 13.7b). See figure captions for further details.

13.11 Future Uses of the Prosodic Model

We envisage that our enhanced prosodic model will in the future become an integrated part of dialogue systems for Swedish. When the dialogue analysis tools become sufficiently powerful to supply reliable prosodic markers for the text, we will be able to generate high-quality synthesis in real time with a prosody relevant for the current dialogue situation. This will be essential in order to achieve the kind of naturalness that is required if man-machine dialogues are to win the hearts of ordinary people in everyday applications.

Man-machine dialogues require high-quality speech recognition to be viable. In real applications, as in natural man-man dialogues, covering a multitude of conversational topics, prosody is a vital cue to speaker intentions and serves to disambiguate linguistic messages. Any speech recogniser that does not make sufficient use of the global prosody of the utterances will eventually fail to make correct judgements in many dialogue situations. One of our aims, and hopes, is that our prosody model, with the associated text and discourse analysis tools, will find its way into powerful speech recognition systems for Swedish forming an integral part of man-machine dialogue systems.

As a research tool, we can see further important uses of the model in tests of perceptual aspects of Swedish prosody and of the quality of synthetic prosody generally.

A further future use is to investigate variability aspects of speech synthesis. One problem with traditional approaches to the prosody of speech synthesis is the lack of variation. Variation is part of natural human speech, not only in terms of inter-speaker differences, but also in terms of a more or less systematic variation within the speech of single speakers. Our model can be used to investigate the limits and functions of these kinds of variation.

Acknowledgements

This work has been supported by grants from The Swedish National Language Technology Programme (HSFR-NUTEK).

References

Ainsworth W.A. (ed.). 1990. Advances in Speech, Hearing and Language Processing. London: JAI Press.

Ayers, G., G. Bruce, B. Granström, K. Gustafson, M. Horne, D. House and P. Touati. 1995. Modelling intonation in dialogue. *Proc. 13th ICPhS* (Stockholm, Sweden), vol. 2, 271-281.

Bertenstam, J., J. Beskow, M. Blomberg, R. Carlson, K. Elenius, B. Granström, J. Gustafson, S. Hunnicutt, J. Högberg, R. Lindell, L. Neovius, L. Nord, A. de Serpa-Leitao and N. Ström. 1995. The Waxholm system - a progress report. *Proc. ESCA Workshop on Spoken Dialogue Systems* (Aalborg, Denmark), 81-84.

Bertenstam, J., B. Granström, K. Gustafson, S. Hunnicutt, I. Karlsson, C. Meurlinger, L. Nord and E. Rosengren. 1997. The VAESS communicator: a portable communication aid with new voice types and emotions. *Proc. Swedish Phonetics Conference Fonetik* '97 (Umeå, Sweden), PHONUM 4, 57-60.

Bruce, G. 1977. *Swedish Word Accents in Sentence Perspective*. Lund: Gleerups.

Bruce, G. 1982. Developing the Swedish intonation model. *Working Papers* 22, 51-116. Department of Linguistics, Lund University.

Bruce, G. 1986. How floating is focal accent? *Nordic Prosody* IV, 41-49.

Bruce, G., M. Filipsson, J. Frid, B. Granström, K. Gustafson, M. Horne, D. House, B. Lastow and P. Touati. 1996. Developing the modelling of Swedish prosody in spontaneous dialogue. *Proc. ICSLP* '96 (Philadelphia, USA), vol. 1, 370-73.

Bruce, G., J. Frid, B. Granström, K. Gustafson, M. Horne and D. House. 1997. Prosodic segmentation and structuring of dialogue. *Nordic Prosody* VII, 63-72.

Bruce, G. and P. Touati. 1992. On the analysis of prosody in spontaneous speech with exemplification from Swedish and French. *Speech Communication* 11, 453-58.

Bruce, G., P. Touati, A. Botinis and U. Willstedt. 1988. Preliminary report from the KIPROS project. *Working Papers* 33, 23-50. Department of Linguistics, Lund University.

Carlson, R., B. Granström and S. Hunnicutt. 1990. Multilingual text-to-speech development and applications. In Ainsworth (ed.), 269-296.

Cruse, D.A. 1986. *Lexical Semantics*. Cambridge University Press.

Docherty, G.J. and D.R. Ladd (eds.). 1992. *Papers in Laboratory Phonology II*. Cambridge University Press.

Grønnum (Thorsen), N. 1992. *The Groundworks of Danish Intonation*. Copenhagen: Museum Tusculanum Press.

Halliday, M.A.K. and R. Hasan. 1976. *Cohesion in English*. London: Longmans.

Horne, M. and M. Filipsson. 1996. Computational extraction of lexico-grammatical information for generation of Swedish intonation. In van Santen, Sproat, Olive and Hirschberg (eds.), 443-457.

Horne, M., M. Filipsson, M. Ljungqvist and A. Lindström. 1993. Referent tracking in restricted texts using a lemmatised lexicon: implications for generation of prosody. *Proc. EUROSPEECH* '93 (Berlin, Germany), vol. 3, 2011-14.

House, D. 1990. *Tonal Perception in Speech*. Lund University Press.

House, D. and G. Bruce. 1990. Word and focal accents in Swedish from a recognition perspective. *Nordic Prosody* V, 156-173.

Ladd, D.R. 1996. *Intonational Phonology*. Cambridge University Press.

Litman, D.J. and R.J. Passonneau. 1995. Combining multiple knowledge sources for discourse segmentation. *Proc. Association Computational Linguistics*, 33, 108-115.

Metzing, D. 1981. Frame representations and lexical semantics. In Eikmeyer and Rieser (eds.), 320-342.

Möhler, G. and G. Dogil. 1995. Test environment for the two level model of Germanic prominence. *Proc. EUROSPEECH* '95 (Madrid, Spain), vol. 2, 1019-22.

Moulines, E. and F. Charpentier. 1990. Pitch-synchronous waveform processing techniques for text-to-speech synthesis using diphones. *Speech Communication* 9, 453-67.

Pierrehumbert, J.P. and M.E. Beckman. 1988. *Japanese Tone Structure*. Linguistic Inquiry Monograph 15. Cambridge, Mass.: MIT Press.

Sinclair, J. and M. Coulthard. 1975. *Towards an Analysis of Discourse*. Oxford University Press.

van den Berg, R., C. Gussenhoven and A.C.M. Rietveld. 1992. Downstep in Dutch: Implications for a model. In Docherty and Ladd, (eds.), 335-359.

van Santen, J., R. Sproat, J. Olive and J. Hirschberg (eds.). 1996. *Progress in Speech Synthesis*. New York: Springer-Verlag.

14

A Prosodic Model for Text-to-speech Synthesis in French

ALBERT DI CRISTO, PHILIPPE DI CRISTO, ESTELLE CAMPIONE AND JEAN VÉRONIS

14.1 Introduction

Text-to-speech synthesis, so long confined to small industrial applications, is now being opened up to many new areas of general interest, especially since the tremendous upsurge of multimedia applications. Endowing speech synthesis with the kind of quality that makes it acceptable and attractive to the general public is thus one of the major challenges of today's speech technology research. A second challenge is the capability of processing unrestricted running texts, regardless of their length or content.

The quality of speech is often associated with its naturalness. However, the term "natural" has multiple meanings since it refers not only to various aspects of acoustic signal production, but also to linguistic and pragmatic factors. In fact, the concept of quality in synthetic speech is a complex one. It encompasses a variety of different characteristics including intelligibility, pronunciation accuracy, fluidness and pleasantness of voice, which can be assessed through diagnostic and evaluation tests of variable levels of sophistication (Pisoni, Greene & Logan, 1989; Pols, 1991; Cartier, Emerard, Pascal, Combescure & Soubigou, 1992; Monaghan, 1992; van Santen, 1993; Sorin & Emerard, 1996). This complexity calls for interdisciplinary collaboration between natural language processing (NLP) specialists, signal processing specialists, phoneticians, and psycholinguists (Benoît, 1995).

From the production standpoint, it seems to be generally agreed that among the three most common synthesis methods – articulatory, formant, and concatenation of digitised speech segments – the concatenation technique produces the most "natural" speech because it is capable at a

A. Botinis (ed.), Intonation, 321-355.

relatively low cost of incorporating a wealth of information (particularly regarding coarticulation) that is very difficult to model. One problem with this method, however, is that the fixed diphone dictionary is incompatible with the manipulation of prosodic parameters (especially F_0) required for improving naturalness. In this approach, prosodic parameters are manipulated using signal processing and encoding techniques such as PSOLA (Moulines & Charpentier, 1990), whose detrimental effects on naturalness need no further demonstration. In an attempt to overcome this obstacle and thus to obtain a degree of prosodic variability that is more representative of different speech styles, Campbell (1997) proposed an approach consisting of extracting the basic segments from recordings of speech produced under less constraining conditions than reading. A major difficulty with this approach, however, as the author points out, is determining labels that will capture linguistic phenomena as well as the speaker's communicational and discursive strategies.

In a perspective similar to Campbell's which attempts to "reconcile" the segmental and prosodic levels, it would thus be desirable to have a unified theory of the interactions between these two components, known to be interdependent (Kohler, 1997). Such a theory would most likely be instrumental in defining the set of labels to use in encoding spontaneous speech corpora. It is also likely that the management of the label set would require the development of a multilinear system far more powerful than current systems like ToBI (Silverman, Beckman, Pitrelli, Ostendorf, Wightman, Price, Pierrehumbert & Hirschberg, 1992) and INTSINT (Hirst & Di Cristo, 1998).

Solving the problem of interaction between the prosodic and segmental levels is not the only reason for using spontaneous speech samples. Another motive is the desire to give voice synthesis the kind of variability it still so sadly lacks despite the fact that this is one of the most salient aspects of a natural sounding voice. "Prosodic monotonousness" is one of the major flaws of synthesis systems. It is indeed possible today to generate an isolated utterance whose production quality is so good that it can barely be differentiated from natural speech, but as soon as the prosodic rules are applied iteratively to longer texts, the repetition of the same prosodic patterns makes the output sound artificial, and contributes to rejection by listeners (Collier & Terken, 1987).

The compilation of a "prosodic lexicon" (Vaissière, 1971; Emerard, 1977; Aubergé, 1992; Morlec, Aubergé & Bailly, 1995) would be another means of enlarging the available database of prosodic patterns, and thus of enhancing variability. This approach also has its limitations, however. Firstly, it usually operates on mean contours, which is an obstacle to variability. Secondly, the problem is not so much one of prosodic pattern diversity in itself as it is of how to use the patterns. This partially depends on syntactic, semantic, and pragmatic information which is out of the reach of current text processing techniques. Ideally, the results of a deep syntactic analysis of the text would be needed, along with enunciative, pragmatic, and interactional information, in short, with all forms of conceptualisation which can account for how speech functions in the real world. Admittedly, these goals are more those of the "synthesis of the future" (near?) than of the situation as it stands today (Bailly, 1996).

Several partial solutions have been proposed for counteracting prosodic monotonousness. These include generating realistic, context-specific pauses (Pardo, Martinez, Quilis & Muñoz, 1987), modifying the amplitude of intonation variations, varying the declination slope (Collier & Terken, 1987), and even inserting hesitations. But these are merely stop-gap measures. One practical solution is to rely on automatic learning systems, which are usually based on quantitative techniques (stochastic or classification methods, neural nets, etc.). Such techniques automatically extract prosodic regularities from speech corpora, such as fundamental frequency (Ross, 1995; Dusterhoff & Black, 1997), segment duration (van Santen, 1994; Shih & Ao, 1997), or pauses (Barbosa & Bailly, 1996). They can also be used to collect information about the speaker's characteristics and speaking styles (Abe, 1997).

One of the disadvantages of the automatic learning approach is that it requires preliminary manual labelling of the database corpus, a time-consuming task tedious enough to be a deterrent. There have been a few promising attempts to achieve automatic labelling (Taylor, 1993; Wightman & Campbell, 1994; Hirst, Di Cristo & Espesser, forthcoming; Campione, Flachaire, Hirst & Véronis, 1997; Campione, Hirst & Véronis, this volume) which, when applied to prosody in view of speech synthesis, seem to produce good results (de Tournemire, 1997; Véronis, Di Cristo, Courtois & Lagrue, 1997). The main shortcoming of the stochastic approach is that it masks the nature of the relationships between the levels of linguistic analysis and the acoustic properties of the signal (Beaugendre, 1996). Obviously, this gap is frustrating for the linguist insofar as it is not conducive to the integration of prosodic modelling into the linguistic environment, as an interdisciplinary approach would recommend. There are other more technical arguments against automatic learning systems. It is clear on the one hand that statistical methods reach a cutoff point above which performance no longer improves, regardless of the field of application, and on the other, that learning techniques as a whole are not very flexible or easy to parameterise. It is in fact quite difficult to modify a parameter (e.g. the speaker or the style) without recommencing the entire learning process with an appropriate corpus.

The quick overview just presented does not claim to be representative of the state of the art in speech synthesis. It is simply aimed at recalling a few crucial issues and indicating some of the anchor points that might contribute to understanding the context and theoretical choices behind our approach. We are currently developing two text-to-speech synthesis systems for French, one in which prosody generation relies on a stochastic technique (Véronis *et al.*, 1997) and one (the SYNTAIX system) where a rule-based system implementing our prosodic knowledge (Di Cristo, Di Cristo & Véronis, 1997). To the extent that the two systems use the same prosodic encoding method, the INTSINT alphabet, the former approach, although not expressly designed for this purpose, can be regarded as a heuristic tool that automatically learns the parameters explicitly generated by the linguistic model implemented in the latter approach. This point needs substantial development beyond the scope of the present chapter, which will only outline the theoretical basis and implementation stages of the prosodic model that supports the second approach. Part 1 gives an overall view of the system requirements, and presents the theoretical framework that guided us through the model development process. Part 2 describes some of the features of French prosody in general and of accent in particular, that underlie our approach and cast doubt on the traditional conception. Part 3 is devoted to presenting the overall architecture and modules of the SYNTAIX system. Finally, part 4 describes a perceptual test in which the prosodic quality of our system is compared to three competitors. We will show that prosodic quality of our system was judged by subjects very close to that of the best system, which is a well-established commercial product. This result is encouraging and demonstrates the validity of the principles adopted in our approach.

14.2 System Requirements and Theoretical Background

14.2.1 Some Prerequisites

The model of prosody described in this chapter represents the first layer of a more complete model that would take into account semantic and pragmatic factors. It is limited to accounting for the specific aspects of French prosody needed to develop a TTS system.

Four prerequisite conditions were set for this synthesis-oriented version of the model:

- The first prerequisite involves adapting the prosodic model to the current performance of the NLP module of the system, which is not yet capable of achieving a deep analysis of the syntactic structure or informational organization of arbitrary running text with an acceptable success rate. This limitation is relative, however, since shallow parsing seems to be sufficient for syntax (see infra), and the information structure of written text is subject to fewer constraints than dialogue.
- The second prerequisite is that the model must be capable of evolving as progress is made in text processing, and must therefore be based on explicit parameters. For example, the system must be capable of simulating style variations via a set of parameters, even if in the current state of the art the parameters implemented will correspond to rather formal reading. Satisfying this requirement is also necessary in order to address the monotony problem mentioned in the introduction.
- The third prerequisite is oriented towards optimization of development: the model must produce high-quality synthetic speech at a low cost. It is common to find efficient synthesis systems that use over a hundred prosodic rules. Our goal is to drastically reduce the number of rules without lowering quality, which might seem paradoxical at first. Our basic assumption is that the second and third prerequisites can be fulfilled if the prosodic module is driven by a model wherein an *abstract representation* of the rhythm and intonation is preserved until later stages of the prosodic derivation when the acoustic wave is generated. In this respect, our approach to French is comparable to Pierrehumbert's (1981) and Ladd's (1987) for English.
- Finally, the fourth prerequisite is related to our point of view that speech synthesis is not an end in itself. We therefore designed and intended our system to be used as a tool for testing the validity of the prosodic model we are developing.

14.2.2 Theoretical Background

14.2.2.1 The Concept of Prosodic Well-formedness

The quality and acceptability of synthetic speech depends for a large part on the prosodic "well-formedness" of utterances. Well-formedness is the outcome of various constraints, which can be classified into four categories: metrical, morphosyntactic, semantic-pragmatic, and related to alignment. An utterance will be prosodically well-formed if it abides by certain basic principles of eurhythmy, whose nature and operational values are determined by metrical phonology (Hayes, 1995). While the idea of congruence between prosody and syntax is not tenable, it is acknowledged that the latter imposes certain constraints on the former (Di Cristo, 1976; Rossi, 1993). The semantic-pragmatic constraints governing the organization of discourse are complex, and although there is an abundance of research on this topic, this type of constraint has not yet been exhaustively modeled. However, we might mention as examples the very interesting attempt by Grosz and Sidner (1986), and its application to the interpretation of intonation patterns in American English by Pierrehumbert and Hirschberg (1990) (see also Nakatani, 1997). Our criterion for this type of constraint is "the appropriateness of the utterance to its context", which takes into account characteristics related to both the text's coherence (Brown and Yule, 1983) and its information structure (Lambrecht, 1996). Finally, utterances will be prosodically well-formed - and this point is essential for successful synthesis - if the rules that associate the segmental and prosodic tiers are consistent with those governing the formation of prosodic patterns in the concerned language. The problem of alignment in English was recently studied by Pierrehumbert (forthcoming). For French, only partial data is available and most of the work has yet to be done.

It should be noted that the metrical, morphosyntactic, and semantic-pragmatic constraints governing the prosodic organization of utterances are often conflicting. Consequently, an integrative approach is necessary, that can account for the constraint hierarchy in effect at the different levels where linguistic and paralinguistic units are processed, thus explaining why some of the basic principles are violated. "Optimality Theory" (Prince & Smolensky, 1993) seems to offer some highly promising answers in this area. It is clear, though, that such a theory is not yet applicable to today's text-to-speech synthesis systems, if only due to the limited amount of information currently available to the prosodic module. The synthesis-oriented version of our model thus takes a low-profile approach, and places priority on *rhythmic organization*. However, the scope of this term as we define it (see below) enables us to incorporate various aspects of the text's information structure, and in doing so, to improve the naturalness of the synthesized speech. This entails partially neglecting syntax proper, which is not necessarily a disadvantage since it eliminates the need for complex syntactic parsing. Moreover, it has been shown elsewhere that utterances which meet certain conditions of eurhythmy may achieve greater prosodic well-formedness than others for which the prosody fits perfectly with the syntax (Dell, 1984; Choppy, Liénard & Teil, 1995). Similarly, according to Martin (1986), application of the eurhythmy criterion can produce a satisfactory prosodic structure even if it does not conform to the syntactic organization.

14.2.2.2 Levels of Prosodic Analysis and Representation

It is our contention that a prosodic theory should be able to clearly define the levels of analysis and representation that make up its interpretation framework (Hirst & Di Cristo, 1998; Hirst *et al.*, forthcoming).

Accordingly, for the purposes of describing and modeling intonation in a multilingual perspective, we propose to distinguish the following four levels of analysis, in increasing order of abstractness: (a) physical level, (b) phonetic level, (c) surface phonological level, and (d) deep phonological level. For the sake of brevity, we shall simply stress that the phonetic representation level is viewed as an interface between abstract cognitive representations and their concrete manifestations. This level is also regarded as an interface between production and perception constraints. The relationships between the various levels of analysis are subject to an interpretability condition stipulating that it must be possible to interpret each level at its adjacent levels.

To illustrate (Figure 14.1), a raw F_0 curve (a) can be interpreted at the adjacent higher level (the phonetic representation) in the form of a smooth, continuous curve determined by a sequence of target points connected by a monotonic interpolation function of the quadratic spline type (b). The F_0 curve is modeled automatically by an algorithm called MOMEL, proposed by Hirst and Espesser (1993) (see also a description in Campione *et al.*, this volume). The target points on the modeled curve are interpreted in turn as a sequence of abstract tonal segments which are (manually or automatically) labeled using the INTSINT system (International Transcription System for Intonation; Hirst & Di Cristo, 1998). The symbols in the INTSINT alphabet were chosen on the basis of the idea that F_0 targets are encoded either as absolute tones corresponding to the speaker's tonal registers: T(op), M(id), B(ottom), or as relative tones defined with respect to the preceding and following targets: H(igher), S(ame), L(ower), U(pstepped), D(ownstepped). Another useful distinction subdivides relative tones into non-iterated ones (H, S, and L) and iterated ones (U and D). A bracketed tonal segment sequence (output of F_0 target encoding) is the surface phonological representation of the intonation (c).

The underlying (or deep) representation level is specified in the model in terms of "templates" (L and H tones) associated with a hierarchical constituent structure (d). In the first version of the model (Hirst & Di Cristo, 1984), which is outlined here for illustrative purposes, there are two kinds of units in the constituent hierarchy: Intonation Units (IU) and Tonal Units (TU). In French, TUs in fact correspond to minor accent groups, i.e., prosodic groups comprised of an accented syllable plus one or more preceding syllables. A TU is the smallest unit used in describing the relationships between tonal segments and phonemic segments. In the example in Figure 14.1, the intonation of the utterance "*J'ai dit la veste*" (I said the jacket) (excerpt from a spontaneous speech corpus) is represented at the underlying level in the form of a single IU comprised of two TUs:

(*J'ai dit*) (*la veste*). Because it is a declarative utterance, the tone pattern of the IU is $_L[\ ]_L$. The prototypic TU pattern in French is $_L(\ )_H$. The deep phonological representation (d) is not pronounceable as such.

The surface structure (c) can be derived from the deep structure (d) by means of a set of rules (Figure 14.2), some of which are universal. One such rule is the downstep rule, which assumes that the detachment of an underlying L tone affects the subsequent H tone and thereby lowers it in the surface representation. Another example is the simplification rule, which amounts to applying the Obligatory Contour Principle or OCP (Leben, 1973). The tonal conversion rule consists simply of mapping the tone segments onto the registers. Unless otherwise specified, an M and a B are assigned to the edges of a final declarative IU.

The phonological representation must meet two criteria (Hirst *et al.*, forthcoming): it must supply (a) the information required to pronounce the utterance and (b) the information needed to interpret it semantically and syntactically. This conception is consistent with a broad definition of the interpretability condition, wherein prosodic phonology is viewed as an interface between the syntactic and semantic representation levels, and the phonetic and acoustic representation levels.

The set of rules presented above can account for the derivation of the basic intonation patterns of French from the underlying intonation structure of any utterance. Accordingly, the noun phrase "*Les enfants des amis du patron de son frère*" (The children of the friends of his brother's boss), whose deep representation can be denoted as in (01):

(01) $_L[\ _L$(les enfants)$_H\ _L$(des amis)$_H\ _L$(du patron)$_H\ _L$(de son frère)$_H\ ]_L$

could give rise to the following surface forms:

(02) M H L H L H D B
(unmarked assertion: single application of downstep rule in last TU)

(03) M H L H L H L H B
(marked assertion: local emphasis and inhibition of downstep rule)

(04) M H D D D T
(yes-no question: iterated application of downstep rule and final H converted into final T).

For additional illustrations and a more detailed presentation of the derivation procedures in French see Di Cristo and Hirst (1993a, 1996).

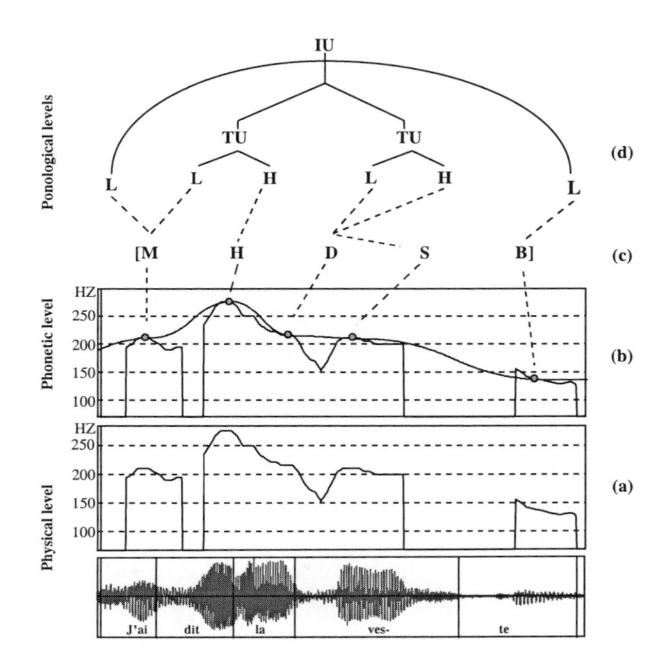

Figure 14.1. From bottom to top: (a) oscillogram of raw F_0 curve, (b) modelled curve, (c) representation of surface intonation structure, and (d) representation of deep intonation structure, for the French utterance: "J'ai dit la veste" (I said the jacket).

		J 'ai dit			*la veste*		
IU assignment	[						]
TU assignment		(	)	(		)	
IU T-segments	L						L
TU T-segments		L	H	L		H	
Downstep rule					D		
Liberalization	L	L	H		D		L
Simplification		L	H		D		L
Tonal conversion		M	H		D		B

Figure 14.2. Rules used to derive the phonological surface representation of the intonation of the French utterance "J'ai dit la veste" (I said the jacket).

14.2.2.3 Metrical Organisation and Rhythmic Organisation

For expository reasons, the previous section was entirely devoted to intonation. But accentuation is also an essential component of the model because, in line with a number of other authors, we consider it to precede the construction of intonation (Bruce, 1985; Rossi, 1985; Cruttenden, 1986; Bolinger, 1986). In this view, accented syllables, i.e., metrically strong ones, are the main anchoring points of the tonal segments of which intonation is comprised. Accordingly, in our approach, tonal segments are not attached to particular syllables but, as suggested by Hirst (1983), to metrical constituents like TUs and IUs.

We will also differentiate between meter and rhythm by making the distinction between "deep representation" and "surface representation" (Di Cristo, forthcoming). This distinction incorporates the opposition between "potential accent" and "actual accent", which has proved useful in a large number of studies. We contend that the constituents of meter can be represented at the deep level by metrical patterns or templates which form the core of the language's prosodic system. Metrical templates are "frozen" forms defined in terms of principles and parameters. For example, the parameter "location of head" in the smallest accent unit (the foot) is crucial and can be used to differentiate English and French on the basis of the left-headed vs. right-headed opposition (Hirst & Di Cristo, 1984, 1998).

Surface rhythmic structures give rise to *diversified patterns*, insofar as the rhythm of any utterance, as observation has shown, can be encoded and interpreted in several different ways (Guaïtella, 1991). This structural variability is mainly the result of the effects of semantic-pragmatic constraints which condition the prosodic organization of utterances. However, it is important in accounting for such "rhythmic patterns" to describe how these constraints interact with syntactic and metrical constraints proper, the latter of which are essentially the result of the need to achieve eurhythmy.

14.2.2.4 Functions of Accent, Prosodic Constituency, and Projection Devices

Our conception of rhythm has theoretical and methodological implications for establishing the *functional classification* of accentual prominences and for specifying the corresponding domain. One of the additional consequences of the priority we grant to an utterance's accentuation structure (see 1.2.3) is that its tonal and temporal organizations can be derived from the accentual representation by means of *projection devices*.

We have already proposed (Di Cristo, forthcoming) to define a functional typology of accentuation based on the following three categories: lexical accent, rhythmic accent, and semantic-pragmatic accent. Rhythmic accent refers to accents whose basic function is to satisfy eurhythmic constraints. Accents in the semantic-pragmatic category serve mainly to signal various forms of focalization and emphasis. It is clear, however, that in addition to their primary function, all accents contribute to the expression of rhythm as it is defined here.

The framework of our approach to accentuation is *metrical theory*, mainly conceived of as a device for calculating and representing levels of accentual prominence or metrical heads (Idsardi, 1992). However, in this framework we include the problem of prosodic constituency (or prosodic domains), with the understanding that the metrical organization and the prosodic constituency are part of *one and the same hierarchical representation* (Truckenbrodt, 1995). On this point, we adopt Halle and Vergnaud's (1987) view that representations of heads and prosodic domains are *conjugate representations*, which brings us to the definition of *government*. The property of being a head is the same as that of being a governing element in a constituent, and the property of belonging to a given domain is the same as that of being governed by a head. The implementation of the conjugate representation generates *bracketed grids*, a formalization more likely to serve the purposes of a metrical theory than one based on *strictly autosegmental* devices (for a complete presentation of the rules, principles, and parameters that dictate the grid bracketing process, see Idsardi, 1992; Halle & Idsardi, 1995; Hayes, 1995).

Projection is a key concept in conjugate representations, in that the assignment of a parenthesis to a line in the grid causes a head to be projected onto the line above it, and that heads projected onto a line trigger the bracketing process. In our approach, this concept is also crucial because it allows us to establish strict links between the three *structuring orders* of the phonological representation, i.e., metrical, tonal, and temporal. Accordingly, the metrical organization illustrated in the grid is used to project the tonal and temporal structures. The tonal organization is projected by associating tonal segments to metrical constituents as stipulated in section 1.2.2. The temporal organization is projected by quantifying the segment durations and then classifying them into four discrete categories labelled N(ormal), E(xpanded), X (eXtra-expanded), and R(educed). As we shall see later (section 3), quantification is performed on the heads and domains of the metrical representation. It can be applied either globally to the syllable, or locally to each of its constituents (onset, nucleus, and coda). A similar quantification process can also be used to *encode pauses* conceived of as discrete entities (Fant & Kruckenberg, 1996).

14.3 French Accentuation Revisited

Given the key role played by the metrical and rhythmic dimensions in our approach, it is necessary at this point to take a closer look at accentuation in French, often the object of conflicting theoretical views. French has sometimes been called a "*language without accent*", an idea which is no longer tenable in the light of recent research (for a discussion on this subject, see Di Cristo & Hirst, 1997; Di Cristo, forthcoming). Most authors now agree that the main characteristic of French accentuation is the presence of a *phrase-final accent* that falls on the last full syllable (without a schwa) of "accentable" words (content words). In fact, the accentual unit in French may be comprised of a single content word (*Mer**ci***, *Fran**çois***) or a content word accompanied by clitics (proclitics and enclitics). This accounts for *accent patterns* like "*Il les **vend***" (He sells them), "*Vends-**les***" (Sell them), "*Vends-**le***" (Sell it), and thus for the accentuation of function words which in theory are not accent-bearing. The final accent in this case has a demarcating value and indicates the boundary of a phonological constituent, called a *Prosodic Word* (PW) or *Phonological Phrase* (PP). Note that experimental research has shown that the realization of the final accent is necessarily accompanied by lengthening of the carrier syllable. This does not rule out the importance of F_0 variations, which play a crucial role in the identification of this type of accent (Astesano *et al.*, 1995). Phrase-final accent, also called *Primary Accent*, should not be confused with *Nuclear Accent*, which falls on the last accented syllable in the IU. However, given the characteristics of the French language, Primary Accent and Nuclear Accent may be the locus of *syncretism*, as in the utterance "*Il m'a é**crit***" (He wrote me), where the syllable "*crit*" is the carrier of both the phrase-final Primary Accent and the Nuclear Accent, in this case the marker of a conclusive IU. The "superposition" of these two accents triggers the realization of a specific boundary tone (e.g., a B(ottom) or a T(op), for conclusive and non-conclusive IUs, respectively) instead of the typical H tone found with Primary Accent, accompanied by substantial lengthening of the accented syllable (Behne, 1989; Fant *et al.*, 1991; Astesano *et al.,* 1995).

In addition to Primary Accent and Nuclear Accent, which fall on the right boundary of minor and major accent groups within the utterance, respectively, tradition also acknowledges the existence in French of an emphatic accent, called "*Accent d'Insistance*", which lands on the initial syllable of words and is mainly observed in speaking styles like public addresses, political discourse, and news broadcasting on radio or television (Séguinot, 1976; Düez, 1978; Lucci, 1983).

This traditional conception must be reexamined in the light of new findings indicating that French has evolved considerably and is still in an ongoing state of change. The two most noticeable manifestations of this evolution, so clearly demonstrated by Fónagy (1980), concern *the loss of strict word-final accent* and the *generalization of word-initial accent*, a tendency which is particularly pronounced today but is probably very old (Pensom, 1993). The combined occurrence of these two tendencies can lead to the realization of "*accentual bridges*" in which only the first syllable of the first word and the last syllable of the last word are accented, as in "*la **ma**jeure par**tie***" (the major part), "***po**litique étran**gèr**(e)*" (foreign politics), etc. These tendencies have led some authors to question the traditional conception of the Prosodic Word and the Phonological Phrase in French (Milner & Regnault, 1987; Lyche & Girard, 1995).

We hypothesize (Di Cristo & Hirst, 1997; Di Cristo, forthcoming) that initial accent is *rhythm*-motivated and as such, must not be confused with "Accent d'Insistance", whose function is to express emphasis. We shall see, however, that initial accent can also be rooted in semantic-pragmatic factors other than emphasis.

In our attempt to model French accentuation (Di Cristo, forthcoming), we propose to account for *accent pattern* generation on the basis of two fundamental principles: the *Accentual Bipolarization Principle* (BPP), which assigns an accent to the first and last syllables of a given unit (word, phrase, proposition, or even utterance), and the *Accentual Hierarchisation Principle* (AHP), which attributes a prominence level to the specified terminal element of a prosodic constituent, depending on the rank of the constituent in the grid. Figure 14.3 illustrates an application of these two principles and of the projection technique explained in the preceding section.

The bracketing of the first line of the grid (L1) is derived by applying the initial and final prominence rules described in Hirst and Di Cristo (1984). This bracketing step leads to the creation of minimal accentual units, which are *Metrical Feet*, and the projection of the heads of these prosodic constituents onto the next line up (L2). The grouping of these heads in turn triggers bracketing into *Prosodic Words*, whose heads are projected onto line L3. At the L3 level, the grouping process generates an *Intonation Unit* whose head is projected onto line L4. Obviously, in a representation of this type, the prominence level of each head is proportional to the height of the column in the grid.

```
                                                 x    L4
         x                x                      x)   L3
   x)    x)        x      x)         x           x)   L2
x  x)  x x)     x  x   x  x)      x  x       x   x)   L1
un fa bri quant de ma  té riaux   de cons   truc tion
```

Figure 14.3. Bracketed grid representation of the noun phrase "*un fabriquant de matériaux de construction*" (a maker of construction materials).

Strictly speaking, the representation in Figure 14.3 can be regarded as a *metrical representation* insofar as it highlights the potential basic accent patterns of French, irrespective of surface form constraints, and accounts for prosodic well-formedness at that level. This point of view is compatible with our hypothesis (Di Cristo, forthcoming) that the underlying representations (potential patterns) of content words are endowed with an initial prominence and a final prominence. An initial abstract prominence can be projected on the surface as a rhythmic accent (subject to eurhythmy-based phonotactics, see Pasdeloup, 1990); as a specific emphatic accent, the so-called "Accent d'Insistance", indicator of lexical or narrow emphasis; or as a pragmatic accent rooted in the information structure and focal scale. A final abstract prominence can give rise to the projection of a Primary Accent which marks off a Prosodic Word, a Nuclear Accent which marks off an IU, or a cumulative device with an emphatic sentential accent, indicator of a broader emphasis associated with the focus. These different types of accent are presented with illustrations in Di Cristo (forthcoming). Lack of space prevents us from going into this topic in detail here, so we shall limit our presentation to those aspects essential to understanding the content of this chapter.

The activation of the BPP and AHP principles and the inhibition of the BPP principle determine the probability with which the different accent patterns of French will be actualized at the surface structure level. Take for example a sequence like "*une jolie chanson*" (a pretty song), which, depending on the particular discourse setting and style, can generate the accent patterns and tonal patterns illustrated in examples (05) to (08). In these examples, the boundaries of the Metrical Feet (MF) and Prosodic Words (PW) are denoted by the symbols ( ) and < >, respectively.

(05)	< $_{L}$(une jolie chan***son***)$_{H}$ >	one MF, one PW (+ AHP, -BPP)
(06)	< $_{L}$(une jo***lie***)$_{H}$ $_{L}$(chan***son***)$_{H}$ >	two MFs, one PW (+AHP, -BPP)
(07)	< $_{L}$(une jo***lie***)$_{H}$ > < $_{L}$(chan***son***)$_{H}$ >	two MFs, two PWs (+AHP,-BPP)
(08)	< $_{L}$(une ***jo***)$_{H}$ $_{L}$(lie chan***son***)$_{H}$ >	two MFs, one PW (+AHP, +BPP)

Note that rhythmic pattern variability does not rule out potential ambiguities, as in:

(09)	(a) *des* ***per****sonnali****tés***	vs.	(b) *des* ***per****sonnes ali****tées***
	(some personalities	vs.	some bedridden persons)
(10)	(a) *des* ***a****noma****lies***	vs.	(b) *des* ***â****nes au Ma****li***
	(some anomalies	vs.	some donkeys in Mali)
(11)	(a) *un* ***lit*** *bien douil****let***	vs.	(b) *un* ***li****byen douil****let***
	(a very cozy bed	vs.	an oversensitive Libyan)

The ambiguity here is partly due to the "pressure" exerted by bipolarization, and partly due to the possibility of realizing all three of these examples as a single Prosodic Word. Nevertheless, French has the capability of removing the ambiguity in examples (09b), (10b), (11a), and (11b) by structuring the prosodic words in a way comparable to (07), which would generate the following accent patterns and groups: <(*des per****sonnes***)> <(*ali****tées***)>, <(*des* ***ânes***)> <*au Ma****li***)>, <(*un* ***lit***)> <(*bien douil****let***)>, and <(*un li****byen***)> <(*douil****let***)>.

BPP can take effect at different levels in the linguistic organization, ranging from the word (***décision***, ***a****vance****ment***, ***pri****o****rité***, etc.), to the phrase (***sou****liers* ***noirs***, ***can****tatrice* ***chauve***, ***po****litique étran****gère***, ***beau****jolais nou****veau***) and the proposition ("***Il*** *ne viendra* ***pas***"). In fact, *initial accent*, which is involved in the formation of BPP-based accentual bridges, seems to have two driving forces: a *rhythmic* one aimed at avoiding "clashes" and lapses" (Nespor and Vogel, 1989) and a *semantic-pragmatic* one which participates in the prosodic planning of discourse segments and the construction of the focal scale (Di Cristo, forthcoming). Accentual bridges with a semantic-pragmatic origin are generated by an *anticipated pitch accent* strategy comparable to that described for English by Shattuck-Huffnagel (1995). It would be more accurate in this case to speak of "pitch bridges", in that the discourse segment in question is bounded on the left and right by a pitch accent. Note that pitch bridge formation sometimes overrides the content word vs. function word opposition, insofar as a bridge-initiating function word can also receive a pitch accent, as in "***Il*** *ne viendra* ***pas***" (He won't come) or "***la*** *petite voi****ture***" (the small car). A final important point is that any accent, whether initial or final, can be assigned a mark of prominence for expressing emphasis, contrast, or at a more general level, the degree of speaker involvement in what he or she is saying. This last point will not be addressed in this chapter and the reader may wish to refer to Di Cristo (forthcoming).

In conclusion of this section, we will present two examples of phrasal bipolarization to illustrate some of the accent patterns used by our speech synthesis system.

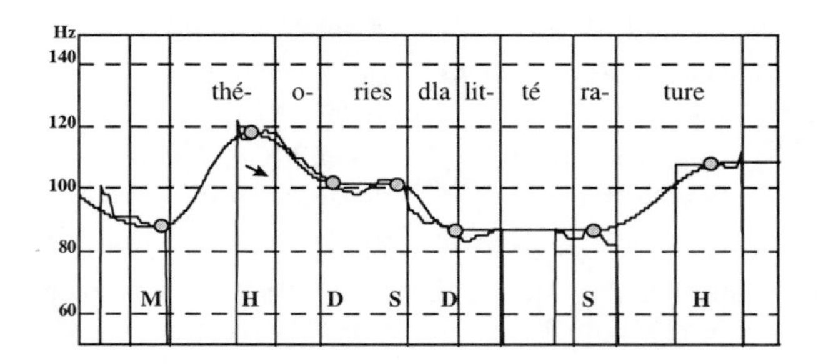

Figure 14.4. Modelled F_0 curve of the sequence "... *sur les théories de la littérature*" (on the theories of literature) taken from "*Il avait écrit un ouvrage sur les théories de la littérature qui ...*" (He had written a book on the theories of literature which ...). The vertical lines indicate segmentation into syllables. The target points are labelled using the INTSINT system.

The example in Figure 14.4 is particularly interesting since it illustrates how semantic-pragmatic and rhythmic constraints interact in order to ensure the prosodic well-formedness of the utterance. The accentual bridge is coextensive with the sequence "*théories de la littérature*" in which the syllables "*thé*" and "*tur(e)*" receive a pitch accent. However, the number of syllables separating these two accents is six, which might seem high for French, where this number usually varies between one and five (Fónagy, 1980; Léon, 1992; Jankowski, 1996). Two possible solutions arise, consisting either of keeping the bridge as it is by drastically increasing the articulation rate of the internal syllables, or of adding an intermediate, weaker accent in the form of a downstep accompanied by lengthening of the duration of the syllable "*rie*", as in Figure 14.4. The realization of the full pitch accent on the syllable "*tur(e)*" associated with an H tone would cause the bipolarization to be carried over to the lexical level and would thus destroy the cohesion of the bridge shown in figure 14.4. It is interesting to note that when the pitch bridge is in fact realized, the initial syllable generally exhibits an anticipated pitch peak and a falling contour, shown as arrows in figures 14.4 and 14.5, the latter of which illustrates the realization of two consecutive accentual bridges.

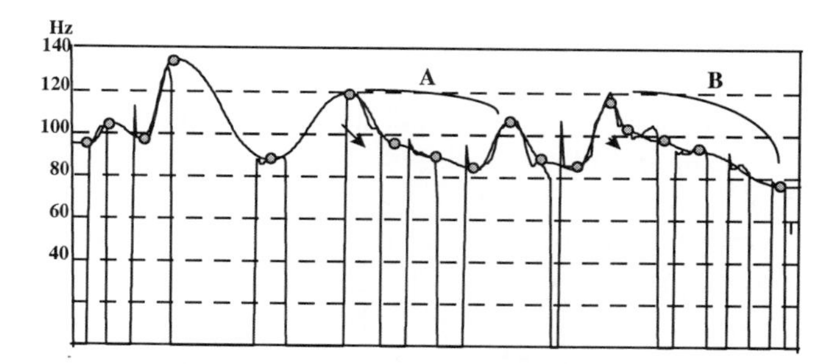

Figure 14.5. Modelled F_0 curve of the sequence "*Hier soir, avec ces sinistres industriels, nous avons longuement parlé de politique*" (Last night, with those industrial disasters, we spoke at length about politics). The accentual bridges "*sinistres industriels*" and "*longuement parlé de politique*" are labelled A and B, respectively.

In the end, then, it is the simultaneous exertion of both semantic-pragmatic constraints and rhythmic constraints that accounts for the great diversity of French accentuation patterns, and thus for the surface rhythmic motifs supported by our model. Taking this diversity into account in speech synthesis is indeed an effective way of introducing variability into the synthesized output to make it sound more natural.

14.4 Architecture and Implementation of System

The SYNTAIX system has a standard architecture consisting of two modules: a linguistic module comprising an NLP submodule, and a phonological (prosodic) submodule, followed by a phonetic (acoustic) module. The final output (consisting of triples <phoneme, duration, F_0>) is fed into the MBROLA synthesizer (Dutoit *et al.*, 1996).

14.4.1 The NLP Submodule.

The NLP submodule takes raw text as input, and performs tokenisation, sentence recognition, part-of-speech assignment, grapheme-to-phoneme conversion and phrase-level parsing. These procedures have for the most part been developed in the MULTEXT project (Véronis, Hirst, Espesser & Ide, 1994). We will not describe the first stages of processing, since they are very similar to what can be found in other systems, and will focus on the parsing module. This module performs only a very rough phrase-level analysis of the text, and does not attempt deep syntactic analysis, both for theoretical and practical reasons.

From a theoretical point of view, while many studies have shown that syntax is a major determining factor of the prosody of utterances, no one has ever been able to reduce that relationship to simple rules. It is now commonly accepted that prosody is organised in prosodic groups (also called prosodic domains) constituted by short, phrase-level segments of the text. Studies of the eye-voice span, i.e. the distance the eye is ahead of the voice when reading aloud, have shown that readers look ahead no further than two words (see Levin, 1979), and therefore cannot have at their disposal a complete parse of sentences when they speak, yet they are capable of producing correct prosodic structures. Various psycholinguistic studies indicate the importance and stability of small word groups in the sentence production and perception (e.g. Grosjean, 1980; Gee & Grosjean, 1983; Grosjean & Dommergues, 1983; Caelen-Haumont, forthcoming). In addition, it has been shown by several authors that a complete syntactic parsing of sentences was not necessary for producing acceptable prosody in speech synthesis (e.g. O'Shaughnessy, 1990; Monaghan, 1992 ; Quené & Kager, 1992; Ostendorf & Veilleux, 1994; Zellner, 1997).

From a practical point of view, the phrase-oriented relationship between prosody and syntax is fortunate, because no one, in today's state of the art, is capable of producing an automatic syntactic parsing of arbitrary texts with an acceptable success rate. Despite fifty years of efforts, no wide-coverage formal grammar of any language has ever been designed. In addition, even if such a grammar were available, human language presents too many syntactic ambiguities (such as prepositional phrase attachments) that can be solved only with semantic and pragmatic knowledge that is well beyond the reach of current technology. As a result, while there were early attempts to generate prosody based on complete syntactic parsing of sentences (e.g. Allen, Hunnincutt, Carlson & Granström, 1979; Allen, Hunnincutt & Klatt, 1987), research is nowadays moving toward the use of phrase-level parsing, also called *partial* or *shallow parsing*.

Many shallow parsing techniques have been described in the literature, which seem to cover the whole range of possibilities between simple part-of-speech tagging and complete parses (Liberman & Church, 1992; Ejerhed, 1988; Abney, 1991; Hindle, 1994; Karlsson, Voutilainen, Heikkilä & Anttila, 1995; etc. – see a survey in Abney, 1997). However, the majority of partial parsers is oriented towards the recognition of small groups of words that constitute the non-recursive kernels of major phrases. Depending on the authors, these groups are called *non-recursive phrases, core phrases* or *chunks*. Example (from Liberman & Church, 1992):

(12) *(A convicted murderer) (and rapist) (whose parole) (several years ago) (provoked public outrage) (has been charged) (with attacking) (a woman).*

The resulting groups are not necessarily consistent with the traditional syntactic trees, which are based on the notion of immediate constituents. In the example above, a traditional analysis would probably produce structures such as *(A (convicted (murderer and rapist)), (with (attacking a woman)),* etc. Yet these groups seem to be very close to the groups that psycholinguists found to be basic blocks of sentence production and perception, and closer to accentual/prosodic groups than the structures that result from traditional immediate constituent analysis, as can be seen in the above example.

The detection of phrase-level word groupings can be done with simple and robust heuristics. We use as a starting point one of the earliest and simplest techniques, the *chink and chunk* algorithm (the terms are from Ross & Tukey, 1975), used by Liberman and Church (1992) in the context of speech synthesis, and subsequently adapted to other languages (e.g. Bourigault, 1992 for French). In this algorithm, words are sorted into two categories: *chinks* and *chunks*. Chinks are words that are likely to start a new group (determiners, conjunctions, prepositions, etc.). Chunks are categories that are less likely to start a group (nouns, adjectives, etc.). Groups are then determined according to the following regular expression:

group = chink chunk**

The first step of our phrase-level module partitions sentences on the basis of punctuation (commas, dashes, colons, etc.), defining thus the IUs of utterances. Example:

(13) Cette théorie de la littérature, elle avait confirmé ses intuitions.

⇒ *[Cette théorie de la littérature]*
⇒ *[elle avait confirmé ses intuitions]*

Words are then grouped very roughly using a French adaptation of the *chink and chunk* algorithm. Example:

(14)*[Cette théorie de la littérature]*
⇒ *(Cette théorie)*
⇒ *(de la littérature)*

However, this crude strategy produces unbalanced groups: they are sometimes very small and sometimes very long. Liberman and Church observe the same phenomenon, and shift some word categories from chunks to chinks (e.g. tensed verbs). This enables them to produce the grouping (several years ago) (provoked public outrage) in the example cited, instead of a single chunk (several years ago provoked public outrage). However, this strategy leads to many difficulties and inconsistencies. We therefore preferred splitting the chunks into two groups, verbal chunks, containing the verb and its close satellites (pronouns, negation, adverbs) and nominal chunks, containing the noun and its close satellites (adjectives). This results in the following regular expression:

group = chink(V-chunk | N-chunk)+*

In addition, it seems important for the sake of naturalness that the word grouping heuristics not be deterministic, in order to reflect the great amount of variation observed among readers of the same text, and even for the same reader in different portions of the text. Therefore we used additional steps that can be triggered or not, thus producing a wide range of plausible groupings.

Some short groups (e.g. starting with a preposition) can then be merged with the preceding group. Example :

(15) *(Cette théorie) (de la littérature)*

⇒ *(Cette théorie de la littérature)*

Conversely, long groups can be split (e.g. after the last verb, before the last preposition, etc.). Example:

(16) *(et même avant cette nouvelle théorie)*

⇒ *(et même) (avant cette nouvelle théorie)*

Merging and splitting are triggered according to the syllabic structure of groups. Several studies have shown that some sequences are more acceptable than others (Dell, 1984; Martin, 1986; Pasdeloup, 1990), and that the most satisfying sequences are those in which the number of syllables is balanced across groups. However, balancing is not mandatory, and the same sentence can be finally grouped in many ways. In addition, profiles can be defined: uniformly balanced, mostly balanced with bursts of small groups, generally long groups with bursts of small groups, etc. For the time being, we use random strategies, but work is underway to relate grouping strategies to speech styles and emotions or intentions.

14.4.2 The Phonological (prosodic) Submodule

The phonological (prosodic) submodule takes the output of the NLP submodule as input, and builds an abstract representation of the rhythm and intonation of utterances (steps 1 to 4 in Figure 14.6).

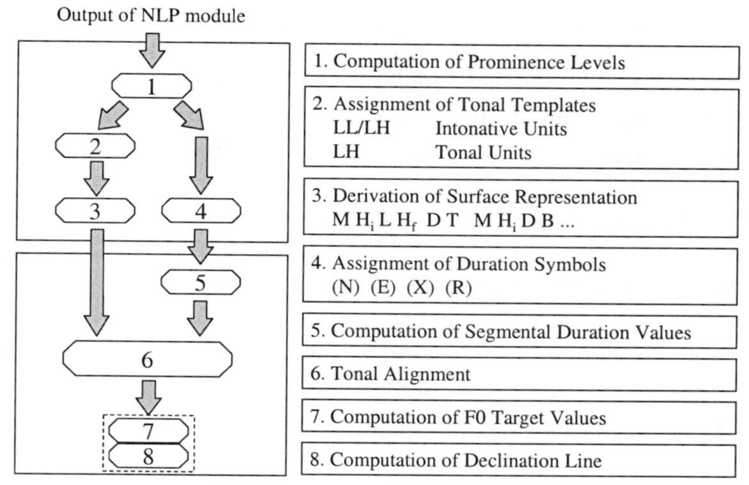

Figure 14.6. Prosodic submodules

The first stage in this module consists in computing the prominence levels and building the grid, according to the model developed in section 1.2.4. This step applies the BPP and AHP described in section 2. For instance, the following output of the NLP submodule:

(17) (Cette théorie) (de la littérature)

can produce the following accent patterns:

(18a) Cette ***thé***o***rie*** de la ***li***ttéra***ture***.

(18b) Cette ***thé***o***rie*** de la littéra***ture.***

(18c) Cette théo***rie*** de la ***li***ttéra***ture***.

The last example illustrated the inhibition of BPP. On the other hand, the following grouping:

(19) (Cette théorie de la littérature)

can be rhythmically interpreted as:

(19a) Cette ***thé***orie de la littéra***ture***.

(19b) ***Cette*** théorie de la littéra***ture***.

(19c) Cette ***thé***o***rie*** de la littéra***ture***.

(19d) ***Cette*** théo***rie*** de la littéra***ture***.

These last examples illustrate the production of *supra-lexical* accentual bridges. Examples (19c) and (19d) share the assignment of an internal prominence to the syllable "*rie*", which can be translated by a lengthening of that syllable in conjunction with a downstep (see Figure 14.4).

The tonal organization of the text is derived in two steps, by first projecting the deep tonal structure from the metrical grid, and then projecting the surface tonal structure from the representation of the deep tonal structure (Figure 14.7). the first operation amounts to associating tonal templates with the metrical constituents defined by the grid:

- L^L and L^H for major metrical constituents (UIs), which correspond to text segments comprised between punctuation marks such as commas, semi-colons and periods;
- L^H for minor metrical constituents (TUs).

The temporal organization is also projected from the metrical representation by the assignment of symbolic durations (step 4):

- N(ormal), to syllables which are not metrically marked, and to the heads of non-final metrical feet of prosodic words (initial accents);
- E(xpanded), to prosodic word heads;
- X (extra expanded), to UI heads;
- R(educed), to syllables embedded in a "perfect" pitch bridge (cf. 19a, 19b), excepted the first and last.

The second stage consists in the *derivation of the surface structure.* Simplification rules (OCP), downstep and tonal conversion are applied at this stage (see Figure 14.2). In our model, the downstep rule can be applied either punctually or iteratively. Punctual application is related to the formation of IUs and pitch bridges. A downstep is assigned to the *last TU of the IU.* The successive application of downstep, linearisation, simplification and tonal conversion rules will therefore produce the following transformations:

(a) terminal IU
(20) $_L(TU)_H\ IU]_L \Rightarrow$ D B

(b) non-terminal IU
(21) $_L(TU)_H\ IU]_H \quad \Rightarrow$ D T

Punctual assignment of the downstep rule is also applied within a supra-lexical pitch bridge, as shown in Figure 14.4.

The iterative application of the downstep rule concerns the total (Yes-No) and partial (Wh) questions (Di Cristo, forthcoming), and results in the derivation of surface structures such as:

(a) total question

(22) $_L[_L$(Tu as pris)$_H$ $_L$(le chapeau)$_H$ $_L$(de la femme)$_H$ $_L$(de jacques)$_H]_H$?

⇒ M H D D D T

(b) partial question

(23) $_L[_L$(Qui)$_H$ $_L$(a pris)$_H$ $_L$(le chapeau)$_H$ $_L$(de la femme)$_H$ $_L$(de Jacques)$_H]_L$?

⇒ M H D D D D B

These last examples show that the downstep rule applies to all the TUs, excluding the first. In addition, the final downstep related to the IU formation is maintained.

14.4.3 The Phonetic Submodule

The phonetic (acoustic) submodule interprets the output of the phonological (prosodic) submodule. It associates the prosodic and segmental levels, and converts the symbolic labels into acoustic data. It consists of steps 5 to 8 in Figure 14.6.

Segmental durations are computed relative to the mean duration of each phoneme, extracted from the recordings of the EUROM 1 corpus (Chan *et al.*, 1995). A multiplicative factor k is then applied according to the *temporal quantification* level (see 1.2.4): N(ormal), k =1; R(educed), k = 0.8; E(xpanded), k = 1.5; X(extra expanded), k = 2. Factor k can be applied either globally to the entire syllable, or separately to each constituent (onset, peak, coda). This second option enables taking into account recent results (Astesano *et al.*, 1995), in particular the fact that word-initial accents in French, whether rhythmic or emphatic, show significantly greater lengthening of the onset whereas word final accents, particularly before IU boundaries, show greater lengthening of the peak and coda with even greater lengthening of the peak before non-terminal boundaries.

The principles that govern the alignment of the surface tonal segments with the phonemic segments are the following:

- With the exception of tone D, all tones are by default aligned with the end of the rime of the final syllable of the unit concerned (TU, PW or IU).
- Tone D is globally assigned to the syllable in all cases of punctual or iterative downstep.
- H tones associated with an initial accent of an accentual bridge (indexed as Hi) are aligned with the beginning of the rime of the syllable (Figures 14.4 and 14.5 show examples of natural speech that justify this procedure).
- A tone L between two consecutive H tones can be treated with some flexibility. By experiment with our system, we noted that in a sequence such as "*la littérature*", the second L can be placed in either L_1 or L_2 (see 24) without other consequence than a variability beneficial to the synthesis. However, in the case of a perfect bridge, L_2 must be preferred, as shown with the group "*sinistres industriels*" of Figure 14.5.

(24)

F_0 target values for surface tonal segments are defined using the "intonation levels" proposed by Rossi and Chafcouloff (1972), which prove a reliable method to determine the speaker's registers and pitch range. The target points are interpolated using quadratic splines as described in Hirst and Espesser (1993) (see also Campione *et al.*, this volume).

Global "lowering effects" are roughly implemented for each of the IUs, according to a three-valued tonal range (see Figure 14.7):

- a tonal range TR_i, at the beginning of the IU;
- a tonal range TR_f, smaller than TR_i at the end of a non paragraph-final IU;
- a tonal range TR_p smaller than TR_f, at the end of a paragraph-final IU.

The declination line is defined by a function that tends asymptotically towards the bottom of the speaker's register (B).

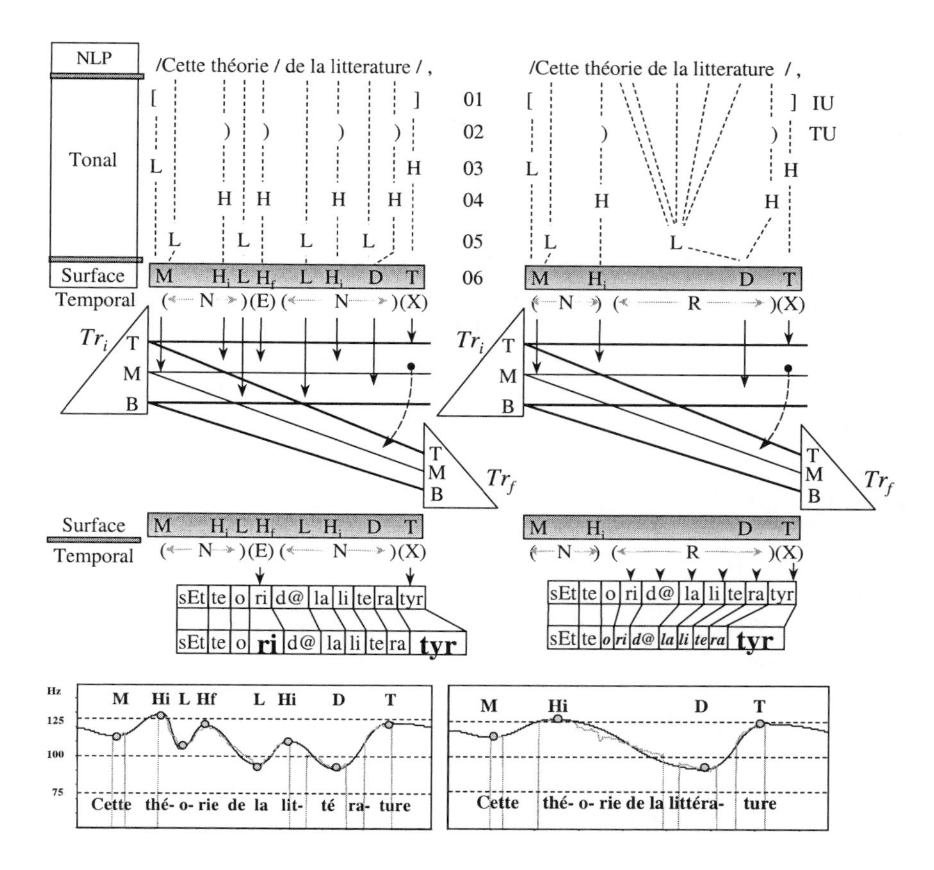

Figure 14.7. Synoptic view of the projection devices implemented for the generation of tonal and temporal organization in SYNTAIX. Target configurations issued from the derivational stages are represented in the lower part of the figure.

14.5 Evaluation

14.5.1 Methodology

Our system was compared to three other French TTS systems, whose evaluation versions are freely available on the World Wide Web. We will keep their identity hidden, and refer to them by letters. Two are commercial systems (X_1 and X_2), the other is a research system (X_3).

These various systems had a segmental component different from that of our system (S), and it was not possible to compare the outputs of the different systems directly, since it is well known that even expert subjects cannot completely dissociate segmental quality from suprasegmental quality. We therefore built an experiment in which we automatically segmented the output of each system in phonemes, and used the observed duration and F_0 values as input to the MBROLA synthesizer (Dutoit, Pagel, Pierret, Bataille & van der Vreken, 1996). The resulting material was then submitted to subjects who were asked to grade its prosodic quality.

14.5.2 Test Material

The test material was composed of ten passages of connected speech (five sentences each), drawn from the EUROM 1 corpus from the SAM project (Chan *et al.,* 1995). Each passage was submitted to the various TTS systems. A recording by a male speaker was also available from the SAM project.

The output of each system X_1,X_2, X_3 and the original recording were manually segmented using the MBROLIGN software (Malfrère & Dutoit, 1997) followed by correction by an expert. The F_0 for each passage was also extracted using the MOMEL algorithm (Hirst & Espesser, 1993; see also Campione *et al.,* this volume) followed by manual correction. This phase of material preparation resulted in a file composed of triples <phoneme, duration, F_0> for each passage. It was fed as input to the MBROLA synthesizer after a few errors in grapheme-phoneme conversions were manually corrected.

The various systems and original recording had different mean frequencies and speech rates. There was therefore the risk of a bias, since subjects could systematically associate a given combination of frequency and speech rate with a given quality. Normalizing all passages to the same frequency and speech rate by a multiplying factor did not seem fair, since it could be argued that each system was designed to work at a specific frequency and speech rate, and that modifying them was the source of potential bad results. We therefore decided to leave the frequency and speech rate unchanged for all of our competitors, and place *our* system in the worst situation by synthesizing the various passages using the parameters of our competitors. Each passage was therefore synthesized three times with different mean frequencies and speech rates, which were presented randomly to subjects. The original was also synthesized in three versions using the same strategy. Altogether, each frequency/speech rate combination was therefore associated to three different sources (one of our competitors, our system and the re-synthesized original), and it was felt that this variability was enough to prevent the above-mentioned bias.

14.5.3 Protocol

The passages were presented to 17 subjects, in two sessions separated by a short break. The subjects were Ph.D. students in phonetics and laboratory staff. All were native French speakers and none had known hearing deficiency. Each subject had to listen to 50 synthesized passages, i.e. 10 passages for each test set X_1, X_2, X_3, S and O. The passages corresponding to our system (S) and the original (O) were drawn randomly in a different way for each subject among the three versions with different frequencies and speech rates described above. The order of passages was also randomized and different for each subject.

The subjects were asked to grade the prosodic quality of passages on a scale ranging from 0 to 9. The test was run using the ASTEC test station developed in the laboratory (formerly EURAUD-ASTEC; see Pavlovic *et al.*, 1995): the scale was displayed as numbered boxes on the computer screen and subjects had to click on the desired value.

14.5.4 Results

The mean scores per system are summarized in Table 14.1.

Table 14.1. Evaluation results.

Set	X_3	X_2	S	X_1	O
Mean score	3.28	3.75	3.94	4.02	6.51

The test set derived from original parameters (O) receives the best score, as predictable. Our system (S) is ranked second among the systems, very close to the winner (commercial system X_1). Research system X_3 is last.

An analysis of variance showed that differences among systems are significant [$F(4, 612) = 137{,}04$, $p < 10^{-6}$). Post hoc comparison of means (LSD test) reveals that:

- the difference between the re-synthesized original (O) and all systems is highly significant (p < 10-6);
- the difference between our system (S) and the winner (X1) is not significant ($p = 0.58$);
- the difference between S and X2 is higher but not significant ($p = 0.22$);
- the difference between S and X3 is significant (p < 10-4).

14.5.5 Discussion

The SYNTAIX system achieves a good performance, since it comes very close to a well-established commercial system that was ranked first and better than two other systems including another commercial one. Since our system was voluntarily placed in the worst conditions (adapted to a mean frequency and speech rate that are not its optimal values), this result is particularly encouraging.

14.6 Conclusion

We have described in this chapter the theoretical background and the architecture of the SYNTAIX TTS system, with a focus on the prosodic component which is its most original part. In our approach, prosody generation is based on *conjugate representations* and *bracketed grids* proposed by Halle and Vergnaud (1987). The projection mechanism which is central for these authors is also used in our model. However, we use a different perspective in which the temporal and tonal organization is projected from the metrical grid. The metrical grid itself is constructed according to two principles: the *Accentual Bipolarization Principle* (BPP) and the *Accentual Hierarchisation Principle* (AHP), which specify the position and strength of accents, and their respective domains.

These principles can also explain the formation of accentual schemata (and therefore of rhythmic patterns), whose motivation can either be simply metrical or be based on semantic-pragmatic factors. In the current state of the art of text processing techniques, the principles cannot be fully exploited, especially with respect to the various forms of focus. However, the underlying prosodic model is open and could incorporate such information if it were available. For the time being, the model favors the use of a diversity of strategies for the encoding or rhythm which account for the great variability of French accentuation, and contribute to the naturalness of the resulting synthesized speech.

The evaluation of our system by means of perceptual tests is particularly encouraging since its prosodic quality was judged very close to that of a well-established commercial system, and better than two other systems. Given the small number of prosodic rules involved and the small amount of development effort, it seems therefore that the underlying model, described in this study, captures central aspects of French prosody.

Acknowledgements

The authors would like to thank Thierry Dutoit for the MBROLA synthesizer and Fabrice Malfrère for the MBROLIGN aligner, Martin Brousseau who developed the ASTEC test station and Robert Espesser who developed our signal editing tools. Thanks also to our reviewers Gérard Bailly and Alex Monaghan for their helpful comments. Finally, we are very grateful to Daniel Hirst for the numerous and productive discussions that we have had on various aspects of the model.

References

Abe, M. 1996. Speaking styles: statistical analysis and synthesis by a text-to-speech system. In van Santen *et al.* (eds.), 495-510.

Abney, S. 1991. Parsing by chunks. In Berwick, Abney and Tenny (eds.), 257-278.

Abney, S. 1997. Part-of-speech tagging and partial parsing. In Young and Bloothooft (eds.), 118-136.

Allen, J., S. Hunnincutt, R. Carlson and B. Granström. 1979. MITalk-79: The 1979 MIT text-to-speech system. *Proc. 97th ASA*, 507-510.

Allen, J., S. Hunnincutt and D. Klatt. 1987. *From Text to Speech: The MITalk System.* Cambridge University Press.

Astesano, C., A. Di Cristo and D.J. Hirst. 1995. Discourse-based empirical evidence for a multi-class accent system in French. *Proc. 13th ICPhS* (Stockholm, Sweden), vol. 4, 630-633.

Atkins, B. and A. Zampolli (eds.). 1994. *Computational Approaches to the Lexicon.* Oxford University Press.

Aubergé, V. 1992. Developing a structured lexicon for synthesis of prosody. In Bailly *et al.* (eds), 307-321.

Bailly, G. 1996. Pistes de recherches en synthèse de la parole. In Meloni (ed.), 109-121.

Bailly, G., C. Benoît and T. Sawallis (eds.). 1992. *Talking Machines: Theories, Models and Designs*. Amsterdam: Elsevier Science.

Barbosa, P. and G. Bailly. 1996. Generation of pauses within the z-score model. In van Santen *et al.* (eds.), 365-381.

Beaugendre, F. 1996. Modèles de l'intonation pour la synthèse. In Meloni (ed.), 97-107.

Behne, D.M. 1989. *Acoustic Effects of Focus and Sentence Position on Stress in English and French.* PhD dissertation, University of Wisconsin-Madison.

Benoît, C. 1995. Speech synthesis: present and future. In Bloothooft *et al.* (eds.), 119-123.

Berwick, R., S. Abney and C. Tenny (eds.). 1991. *Principle-based Parsing*. Dordrecht: Kluwer Academic Publishers.

Bolinger, D.L. 1986. *Intonation and its Parts*. Stanford University Press.

Bourigault, D. 1992. Surface grammatical analysis for the extraction of terminological noun phrases. *Proc. COLING* '92 (Nantes, France), vol. 3, 977-81.

Brown, G. and G. Yule. 1983. *Discourse Analysis*. Cambridge University Press.

Bruce, G. 1985. Structures and function of prosody. *Proc. French-Swedish Seminar on Speech* (Grenoble, France), 549-559.

Caelen-Haumont, G. Forthcoming. *Prosodie et Sens, une Approche Expérimentale*. Paris: Editions du CNRS, Collection Sciences du Langage.

Campbell, W.N. 1997. Synthesizing spontaneous speech. In Sagisaka *et al.* (eds.), 165-186.

Campione, E., E. Flachaire, D.J. Hirst and J. Véronis. 1997. Stylisation and symbolic coding of F_0. *Proc. ESCA Workshop on Intonation* (Athens, Greece), 71-74.

Campione, E., D.J. Hirst and J. Véronis. This volume. Automatic stylisation and modelling of French and Italian intonation.

Cartier, M., F. Emerard, D. Pascal, P. Combescure and L. Soubigou. 1992. Une méthode d'évaluation multicritère de sorties vocales. Application au test de 4 sustèmes de synthèse à partir du texte. *Proc. Journées d'Etude sur la Parole* (Bruxelles, Belgium), 117-122.

Chan, D., A. Fourcin, D. Gibbon, B. Granström, M. Hucvale, G. Kokkinakis, K. Kvale, L. Lamel, B. Lindberg, A. Mofreno, J. Mouropoulos, F. Senia, L. Trancoso, C. Veld and J. Zeiliger. 1995. EROM- A spoken language resource for the EU. *Proc. EUROSPEECH* '95 (Madrid, Spain), vol. 1, 867-870.

Choppy, C., J.S. Liénard and D. Teil. 1995. Un algorithme de prosodie automatique sans analyse syntaxique. *Actes des 6èmes Journées d'Etude sur la Parole* (Toulouse, France), 387-395.

Cohen, P., J. Morgan and M. Pollack (eds.). 1990. *Intentions in Communication*. Cambridge, Mass.: MIT Press.

Collier, R. and J.M.B. Terken. 1987. Intonation by rule in text-to-speech applications. *European Conference on Speech Technology* (Edinburgh, UK), 165-168.

Connell, B. and A. Arvaniti (eds.). 1995. *Phonology and Phonetic Evidence. Papers in Laboratory Phonology* IV. Cambridge University Press.

Cruttenden, A. 1986. *Intonation*. Cambridge University Press.

Cuttler, A and D.R. Ladd (eds.). 1983. *Prosody: Models and Measurements*. Berlin: Springer-Verlag.

Dechert, W. and M. Raupach (eds.). 1980. *Temporal variables in speech*. The Hague: Mouton.

Dell, F. 1984. L'accentuation dans les phrases en français. In Dell *et al.* (eds.), 65-122.

Dell, F., D.J. Hirst and J.R. Vergnaud. 1984. *Formes Sonore du Langage*. Paris: Hermann.

De Tournemire, S. 1997. Identification and automatic generation of prosodic contours for text-to-speech synthesis system in French. *Proc. EUROSPEECH* '97 (Rhodes, Greece).

Di Cristo, A. 1976. Indices prosodiques et structure constituante. *Cahiers de Linguistique, d'Orientalisme et de Slavistique* 7, 27-40.

Di Cristo, A. 1978. *De la Microprosodie à l'intonosyntaxe*. Thèse de doctorat d'Etat, Université de Provence (published 1985 by l'Université de Provence).

Di Cristo, A. 1998. Intonation in French. In Hirst and Di Cristo (eds.), 195-218.

Di Cristo, A. Forthcoming. Vers une modélisation de l'accentuation du français. *Journal of French Language Studies*.

Di Cristo, A. and D.J. Hirst. 1993a. Rythme syllabique, rythme mélodique et représentation hiérarchique de la prosodie du français. *Travaux de l'Institut de Phonétique d'Aix* 15, 9-24.

Di Cristo, A. and D.J. Hirst. 1993b. Prosodic regularities in the surface structure of French questions. *Proc. ESCA Workshop on Prosody* (Lund, Sweden), 268-271.

Di Cristo, A. and D.J. Hirst. 1996. Vers une typologie des unités intonatives du français. *Actes des 21èmes Journées d'Etudes sur la Parole* (Avignon, France), 219-22.

Di Cristo, A. and D.J. Hirst. 1997. L'accent non-emphatique en français: stratégies et paramètres. *Polyphonies à I. Fónagy*, 71-101. Paris: l'Harmattan.

Di Cristo, A., Ph. Di Cristo and J. Véronis. 1997. A metrical model of rhythm and intonation for French text-to-speech synthesis. *Proc. ESCA Workshop on Intonation* (Athens, Greece), 83-86.

Duez, D. 1978. *Essai sur la Prosodie du Discours Politique*. Thèse de Doctorat, Université de Paris III.

Dusterhoff, K. and A. Black. 1997. Generating F_0 Contours for Speech Synthesis Using the Tilt intonation Theory. *Proc. ESCA Workshop on Intonation* (Athens, Greece), 107-110.

Dutoit, T., V. Pagel, N. Pierret, F. Bataille and O. van der Vreken. 1996. The MBROLA Project: towards a set of high-quality speech synthesizers free of use for non-commercial purposes. *Proc. ICSLP* '96 (Philadelphia, USA), vol. 3, 1393-96.

Ejerhed, E. 1988. Finding clauses in unrestricted text by finitary and stochastic methods. *Proc. 2nd Conf. Applied Natural Language Processing* (Austin, USA), 219-27.

Emerard, F. 1977. *Synthèse par Diphones et Traitement de la Prosodie*. Thèse de Doctorat, Université de Grenoble.

Fant, G. and A. Kruckenberg. 1996. On the quantal nature of speech timing. *Proc. ICSLP* '96 (Philadelphia, USA), 2044-47.

Fant, G., A. Kruckenberg and L. Nord. 1991. Durational correlates of stress in Swedish, French and English. *Journal of Phonetics* 19, 351-365.

Fónagy, I. 1980. L'accent en français. *Studia Phonetica* 15, 123-33.

Furui, S. and M.M. Sandhi (eds.). 1991. *Advances in SSP*. The Bartlett Press Inc.

Furui, S. and M.M. Sondhi (eds.). 1992. *Advances in Speech Signal Processing*. New York: Dekker.

Gee, J.P. and F. Grosjean. 1983. Performance structures: A psycholinguistic and linguistic appraisal. *Cognitive Psychology* 15, 411-58.

Goldsmith, J. (ed.). 1995. *The Handbook of Phonological Theory*. Cambridge and Oxford: Blackwell.

Grosjean, F. 1980. Comparative studies of temporal variables in spoken and sign languages: A short review. In Dechert and Raupach (eds.), 307-312.

Grosjean, F. and J.Y. Dommergues. 1983. Les structures de performance en psycholinguistique. *L'Année Psychologique* 83, 513-536.

Grosz, B.J. and C. Sidner. 1986. Attention, intentions and the structure of discourse. *Computational Linguistics* 12, 175-204.

Guaïtella, I. Etude des relations entre geste et prosodie à travers leurs fonctions rythmique et symbolique. *Proc. 12th ICPhS* (Aix-en-Provence, France), vol. 3, 266-269.

Halle, M. and W. Isdardi. 1995. Stress and metrical structure. In Goldsmith (ed.), 403-43.

Halle, M. and J.R. Vergnaud. 1987. *An Essay on Stress*. Cambridge, Mass.: MIT Press.

Hayes, B. 1995. *Metrical Stress Theory*. The University of Chicago Press.

Higuchi, N., T. Hirai and Y. Sagisaka. 1996. Effect of speaking style on parameters of fundamental frequency contours. In van Santen *et al.* (eds.), 417-28.

Hindle, D. 1994. A parser for text corpora. In Atkins and Zampolli (eds.), 103-151.

Hirst, D.J. 1983. Structures and categories in prosodic representations. In Cuttler and Ladd (eds.), 93-109.

Hirst, D.J. and A. Di Cristo. 1984. French intonation: a parametric approach. *Die Neueren Sprachen* 83, 554-569.

Hirst, D.J. and A. Di Cristo, A. 1998. A survey of intonation systems. In Hirst and Di Cristo (eds.), 1-44.

Hirst, D.J. and A. Di Cristo (eds.). 1998. *Intonation Systems*. Cambridge University Press.

Hirst, D.J., A. Di Cristo and R. Espesser. Forthcoming. Levels of description and levels of representation in the analysis of intonation. In Horne (ed.).

Hirst, D.J., and R. Espesser. 1993. Automatic Modelling of Fundamental Frequency using a quadratic spline function. *Travaux de l'Institut de Phonétique d'Aix-en-Provence* 15, 75-85.

Horne, M. (ed.). Forhtcoming. *Prosody: Theory and Experiment*. Dordrecht: Kluwer Academic Publishers.

Idsardi, J. 1992. *The Computation of Prosody*. PhD dissertation, MIT.

Jankowski, L. 1996. *Le Marquage Prosodique des Mots*. Mémoire de Maîtrise, Université de Provence.

Karlsson, F., A. Voutilainen, J. Heikkilä and A. Anttila (eds.). 1995. *Constraint Grammars*. Berlin and New York: Mouton de Gruyter.

Keller, E. and B. Zellner (eds.). 1997. *Les Défis Actuels en Synthèse de la Parole. Etudes de Lettres* 3. Université de Lausanne.

Kohler, K. 1997. Modelling prosody in spontaneous speech. In Sagisaka *et al.* (eds.), 187-210.

Ladd, D.R. 1987. A model of intonational phonology for use in speech synthesis by rule. *European Conference on Speech Technology* (Edinburgh, UK), 21-24.

Lambrecht, K. 1996. *Information Structure and Sentence Form.* Cambridge University Press.

Leben, W.R. 1973. *Suprasegmental Phonology.* PhD dissertation, MIT (published 1989, New York: Garland).

Léon, P. 1992. *Phonétisme et Prononciation du Français.* Paris: Nathan Université.

Levin, H. 1979. *The Eye-voice Span.* Cambridge, Mass.: MIT Press.

Liberman, M.Y. and K. Church. 1992. Text analysis and word pronunciation in text-to-speech synthesis. In Furui and Sondhi (eds.), 791-831.

Lucci, V. 1983. *Etude Phonétique du Français Contemporain à Travers la Variation Situationnelle.* Publications de l'Université de Grenoble.

Lyche, C. and F. Girard. 1995. Le mot retrouvé. *Lingua* 95, 205-221.

Malfrère, F. and T. Dutoit. 1997. High quality speech synthesis for phonetic speech segmentation. *Proc. EUROSPEECH* '97 (Rhodes, Greece), 2631-34.

Martin Ph. 1986. Structure prosodique et rythmique pour la synthèse. *Actes des 15èmes Journées d'Etude sur la Parole* (Aix-en-Provence, France), 89-91.

Meloni, H. (ed.). 1996. Fondements et Perspectives en Traitement Automatique de la Parole. *AUPELF-UREF.*

Milner, J.C. and F. Regnault. 1987. *Dire le Vers.* Paris: Seul.

Monaghan, A.I.C. 1992. Heuristic strategies for higher-level analysis of unrestricted text. In Bailly *et al.* (eds.), 143-161.

Morlec, Y., V. Aubergé and G. Bailly. 1995. Synthesis and evaluation of intonation with a superposition model. *Proc. EUROSPEECH* '95 (Madrid, Spain), vol. 3, 2043-46.

Moulines, E. and F. Charpentier. 1990. Pitch-synchronous waveform processing techniques for text-to-speech synthesis using diphone. *Speech Communication* 9, 453-467.

Nakatani, C.H. 1997. Integrating prosodic and discourse modelling. In Sagisaka *et al.* (eds.), 67-80.

Nespor, M. and I. Vogel. 1986. *Prosodic Phonology.* Dordrecht: Foris.

Nespor, M. and I. Vogel. 1989. On clashes and lapses. *Phonology* 6, 69-116.

O'Shaughnessy, D. 1990. Relationships between syntax and prosody for speech synthesis. *Proc. ESCA Tutorial on Speech Synthesis* (Autrans, France), 39-42.

Ostendorf, M.F. and N.M. Veilleux. 1994. A hierarchical stochastic model for automatic prediction of prosodic boundary location. *Computational Linguistics* 20, 27-54.

Pardo, J.M., M. Martinez, A. Quilis and E. Muñoz. 1987. Improving text-to-speech conversion in Spanish: linguistic analysis and prosody. *European Conference on Speech Technology* (Edinburgh, UK), vol. 2, 173-76.

Pasdeloup, V. 1990. *Modèle de Règles Rythmiques du Français Appliqué à la Synthèse de la Parole*. Thèse de doctorat, Université de Provence.

Pavlovic, C., M. Brousseau, D. Howells, D. Miller, V. Hazan, A. Faulkner and A. Fourcin. 1995. Analytic assessment and training in speech and hearing using a poly-lingual workstation, EURAUD. In Placencia Porrero and Puig de la Bellacasa (eds.), 332-35.

Pensom, R. 1993. Accent and metre in French. *Journal of French Language Studies* 3, 19-37.

Pierrehumbert, J.B. 1981. Synthesizing intonation. *J. Acoust. Soc. Am.* 70, 985-995.

Pierrehumbert, J.B. Forthcoming. Tonal elements and their alignment. In Horne (ed.).

Pierrehumbert, J.B. and J. Hirschberg. 1990. The Meaning of the intonational contours in the interpretation of discourse. In Cohen, Morgan and Pollack (eds.), 271-311.

Pisoni, D.B., B.G. Greene and J.S. Logan. 1989. An overview of ten years of research on the perception of synthetic speech. *Proc. ESCA workshop on Speech Input/Output Assessment and Speech Databases*, 111-114.

Placencia Porrero, I. and R. Puig de la Bellacasa (eds.). 1991. *The European Context for Assistive Technology*. Amsterdam: IOS Press.

Pols, L.C.W. 1991. Quality assessment of text-to-speech synthesis by rule. In Furui and Sandhi (eds.), 387-416.

Prince, A. and P. Smolensky. 1993. *Optimality Theory: Constraint Interactions in Generative Grammar* (ms. Rutgers University, at New Brunswick and University of Colorado at Boulder).

Quené, H. and R. Kager. 1992. The derivation of prosody for text-to-speech from prosodic sentence structure. *Computer Speech and Language* 6, 77-98.

Ross, K. 1995. *Modelling Intonation for Speech Synthesis*. PhD dissertation, University of Boston.

Ross, I.C. and J.W. Tukey. 1975. Introduction to these Volumes. In Index to Statistics and Probability. The R&D Press, Los Altos (California), iv-x.

Rossi, M. 1977. L'intonation et la troisième articulation. *Bull. Soc. Ling.* Paris, LXII, 1, 55-68.

Rossi, M. 1985. L'intonation et l'organisation de l'énoncé. *Phonetica* 42, 135-153.

Rossi, M. 1993. A model for predicting the prosody of spontaneous speech (PPSS model). *Speech Communication* 13, 87-107.

Rossi, M. and M. Chafcouloff. 1972. Les niveaux intonatifs. *Travaux de l'Institut de Phonétique d'Aix* 1, 167-176.

Sagisaka, Y., N. Campbell and N. Higuchi (eds.). 1997. *Computing Prosody*. New York : Springer-Verlag.

Séguinot, A. 1976. L'accent d'insistance en français standard. In Carton *et al.* (eds.), 1-91.

Selkirk, E.O. 1984. *Phonology and Syntax: The Relation between Sound and Structure*. Cambridge, Mass.: The MIT Press.

Shattuck-Hufnagel, S. 1995.The importance of phonological transcription in empirical approaches to "stress shift" versus "early accent". In Connell and Arvaniti (eds.), 128-40.

Shih, C. and B. Ao. 1996. Duration study for the Bell Laboratories Mandarin text-to-speech system. In van Santen *et al.* (eds.), 383-399.

Silverman, K., M.E. Beckman, J. Pitrelli, M.F. Ostendorf, C.W. Wightman, P.J. Price, J.P. Pierrehumbert and J. Hirschberg. 1992. ToBI: a standard for labelling English prosody. *Proc. ICSLP* '92 (Banff, Canada), vol. 2, 867-870.

Sorin, C. and F. Emerard. 1996. Domaines d'application et évaluation de la synthèse de la parole à partir du texte. In Meloni (ed.), 123-131.

Sorin, C., D. Larreur and R. Llorca. 1987. A rhythm-based prosodic parser for text-to-speech systems in French. *Proc. 11th ICPhS* (Tallin, Estonia), vol. 1, 125-28.

Taylor, P. 1993. Automatic recognition of intonation from F_0 contours using the Rise/Fall/Connection model. *Proc. EUROSPEECH* '93 (Berlin, Germany), 2, 789-792.

Truckenbrodt, H. 1995. *Phonological Phrases: Their Relation to Syntax, Focus and Prominence*. PhD dissertation, MIT.

Vaissière, J. 1971. *Contribution à la Synthèse par Règles du Français*. Thèse de Doctorat, Université de Grenoble.

van Santen, J.P.H. 1993. Perceptual experiments for diagnostic testing of text-to-speech systems. *Computer Speech and Language* 7, 49-100.

van Santen, J.P.H. 1994. Assignment of segmental duration in text-to-speech synthesis. *Computer Speech and Language* 8, 95-128.

van Santen, J.P.H., R. Sproat, J. Olive and J. Hirschberg. 1996. *Progress in Speech Synthesis*. NewYork: Springer-Verlag.

Véronis, J., Ph. Di Cristo, F. Courtois and B. Lagrue. 1997. A stochastic model of intonation for text-to-speech synthesis. *Proc. EUROSPEECH* '97 (Rhodes, Greece), vol. 5, 2643-2646.

Véronis, J., D.J. Hirst, R. Espesser and N. Ide. 1994. NL and speech in the MULTEXT project. *Proc. AAAI '94 Workshop on Integration of Natural Language and Speech* (Seattle, USA), 72-78.

Wightman, C.W. and W.N. Campbell. 1994. *Automatic Labelling of Prosodic Structure*. Technical Report TR-IT-0061, ATR Interpreting Telecommunications Laboratories, Kyoto, Japan.

Young, S. and G. Bloothooft (eds.). 1997. *Corpus-Based Methods in Language and Speech Processing*. Dordrecht: Kluwer Academic Publishers.

Zellner, B. 1997. La fluidité en synthèse de la parole. In Keller and Zellner (eds.), 47-78.

15

Prosodic Parameters of French in a Speech Recognition System

KATARINA BARTKOVA

15.1 Introduction

The importance of integrating prosody into different speech processing systems is nowadays widely acknowledged. Prosodic parameter modelling was first carried out for text-to-speech synthesis, since this speech processing technique could not work without prosody (Emerard, 1977; Klatt, 1979). It appeared to speech researchers that prosody would also be helpful in speech recognition (Carbonell, Haton, Lonchamp & Pierrel, 1982; Waibel, 1987; Ljolje & Fallside, 1987; Wang & Hirschberg, 1992). However, the way to use prosodic parameters in a speech recognition system is less straightforward than in text-to-speech synthesis. A good predicting model in speech recognition has to forecast different varieties of speaking styles while in speech synthesis one correct speaking style prediction, appropriate to a given application, is sufficient.

15.1.1 Prosodic Parameters in Speech Recognition

Most of the current speech recognition systems are based on HMM modelling. The acoustic parameters, which are used in these systems, are related to spectral information of the speech signal. The good performance obtained in speech recognition has provided an interesting challenge for experts to integrate prosodic knowledge into these systems. In addition, software researchers are more and more aware of the usefulness of the prosody in speech recognition.

One of the first studies on the use of fundamental frequency (F_0) in speech recognition systems to help segment the speech signals into meaningful units was carried out by Vaissière at the CNET. The F_0

A. Botinis (ed.), Intonation, 357-382.

information was used in the KEAL speech recognition rule-based system (Mercier, Bigorgne, Miclet, Le Guennec & Querré, 1983).

A robust way to use phone duration in speech recognition is to impose a minimal segment duration (Bush & Kopec, 1987), often handled simply by adding or duplicating states which cannot be skipped in the dynamic programming procedure (Gupta, Lenning & Mermelstein, 1992; Soong, 1998). In segment based recognisers, the duration of the segment has been used as an additional parameter (André-Obrecht, 1990). It is acknowledged that the main characteristic of duration is its elasticity. Attempts were thus made to cope with this elasticity by normalising the phone duration using utterance length or by using variable phone duration according to speech rate (Suaudeau & André-Obrecht, 1993; Gong & Treurnier, 1993). This approach makes it possible to reduce the variability of the speech rate due both to the speaker's habit of speaking (some speakers talk faster than others (McNeilage & De Clerk, 1968)) and to the length of the word (phones in short words are longer than those in long words (Fónagy & Magdics, 1960; Lindblom & Rapp, 1973)). A recent study (Pols, Wang & ten Bosch, 1996) reconsidered the duration issue, by carrying out a thorough analysis of the phone duration in English using hand-segmented data with a view to integrating the results into a speech recognition system. In French, Langlais (1995) attempted to integrate micro-prosodic information into a speech recognition system. Papers published by Ostendorf, Wightman and Veilleux (1993) and Ostendorf and Ross (1997) presented prosodic event labelling and score calculation for prosodic parameters used in speech recognition. Predicting prosodic models often involves the segmentation of speech signals into prosodic units. Their aim is to detect syntactic or lexical boundaries (Ostendorf, Price, Bear & Wightman, 1990; Carbonell & Laprie, 1993; Wang & Hirschberg, 1992). Since the output of these models is rarely connected to a speech recognition system, their efficiency in terms of speech recognition performance therefore remains unknown.

So far, few studies have integrated predicting models into the recognition system to find out to what degree these models can improve recognition performance. Models described in (Bartkova & Jouvet, 1995) were tested on French isolated word and connected word corpora to re-score the N-best solutions provided by HMM decoding. Although simple, this approach made it possible to evaluate to what extent non-spectral parameters could improve speech recognition performance. Wang (1997) studied context-dependent and context-independent English phone duration modelling in the post-processing mechanism to re-score the first-phase recognition hypotheses. Chung and Seneff (1997) use a knowledge-based duration model to constrain a speech recognition system. The efficiency of the predicting model was measured in a keyword detection task where a significant improvement of the recognition system performance was achieved.

15.1.2 Framework

The present study provides an insight into prosodic parameter modelling experiments conducted in the CNET-Lannion laboratory. The aim of parameter modelling was to improve HMM-based automatic speech recognition systems in vocal services accessible to the general public via the telephone (Sorin, Jouvet, Gagnoulet, Dubois, Sadek & Toularhoat, 1995). Prosodic parameter models are studied in keyword spotting applications and in isolated word or expression (several words) recognition tasks.

Prosodic parameter constraints are implemented in a post-processing procedure since they are calculated for linguistic units (phonemes, syllables...) while HMM decoding (first run) is a frame synchronous procedure. Using prosodic information in HMM decoding would be time consuming and computationally expensive as many segment boundaries must be taken into account to achieve optimal decoding (Bush & Kopec, 1987; Suaudeau & André-Obrecht, 1993). Moreover, in frame synchronous decoding, predicting models and normalisation procedures cannot be handled efficiently since only a part of utterances (past frames) is known. However in the post-processing stage, all the phonemes, which build up a word or a sentence, are already known (decoded by HMM) (Lokbani, Jouvet & Monné, 1993). Rule based models can therefore be easily applied.

The difficulties in using prosodic parameters in speech recognition based on keyword spotting are twofold. Firstly, prosodic parameter detection is not always correct, and could corrupt both the modelling and the use of prosodic parameters in speech recognition. Secondly, keyword detection drastically limits the possibility of interpreting the manner in which prosodic parameter values contribute to the structuring of speech. In the keyword recognition approach, the speech signal, which does not correspond to a keyword, is captured by "garbage" models trained to reject non-keywords (Rose & Paul, 1990). For this reason, no information is available regarding rejected speech, and only a very limited prediction about syntactic structuring can be made.

It should be kept in mind that prosodic parameters (phone duration, F_0 movement or phone energy) contain little information with respect to the spectral representation of phones. When Mel frequency coefficients are used in HMM modelling, spectral information for each frame is represented by several coefficients (8 in our case). Most of the time, these coefficients are augmented by their first and second order temporal derivatives, which double or triple the number of acoustic coefficients used. In contrast, prosodic parameters typically represent only one coefficient per phonetic segment. In addition, two different phonemes, with very different spectral representations, can have perfectly identical prosodic parameters (duration, F_0 or energy). Therefore one cannot expect to recover all HMM recognition errors by using one single coefficient (or even three, when they are applied together). However, one can reasonably expect to improve the performance of the automatic recognition system, since, in some cases, prosodic features are distinctive while the spectral representation is not. For example, when the burst of a [t] is recognised as an [s], its short duration can penalise the recognition of [t] (which has a much longer duration).

15.2 Database and Recognition System Overview

15.2.1 Application Field

Public applications of automatic speech recognition include Interactive Vocal Services (IVS), accessible through the telephone network. These IVSs have to be robust and speaker-independent in order to recognise anybody's speech. One way to make a system more robust is to restrict the number of words to be recognised, since it is easier to recognise a smaller rather than a larger number of words. Most IVSs process the input speech to identify one or several keywords or key-expressions (words or expressions known by the service). The speech signal portions, which are not recognised as keywords, are captured by garbage models trained to eliminate out-of-vocabulary words as well as different kinds of noise in the users' environment. Unfortunately, garbage models also reject some of the keyword utterances.

Errors committed by the system have various impacts on service information delivery. For instance, a false alarm (when an out-of-vocabulary word or a noise token is recognised as a vocabulary word) or a keyword substitution (when a vocabulary word is recognised as another vocabulary word) is less tolerated by the user than a keyword rejection. In fact, false alarms and keyword substitutions will deliver unwanted information to the user, while a keyword rejection simply forces the user to repeat that keyword (Sorin *et al.,* 1995). A trade-off must always be found between false alarm errors and false rejection errors since a too high false rejection rate is not acceptable.

Current recognition systems are becoming increasingly robust and the substitution rate is relatively low, especially for small vocabularies.

However, the rate of false acceptance still remains a serious issue (Morin, 1991). Therefore the efficiency of the prosodic parameters in reducing error rate has been evaluated within the framework of re-scoring the N-best hypothesis and within the framework the out-of-vocabulary token rejection procedure. In the rejection procedure, the answer delivered by HMM is rejected when the prosodic parameter information yields a score below a previously chosen threshold. Unfortunately, the score computed for to a correctly recognised word could also be below this threshold, thus causing a false rejection of this word. An excessively high threshold will lead to a high keyword rejection, which would not be tolerated by IVS users. The efficiency of prosodic parameter modelling is therefore evaluated by the trade-off between the keyword rejection (false rejection) rate and the non-keyword acceptance (false alarm) rate.

15.2.2 Speech Database

The training and testing of prosodic parameters was carried out using different speech corpora recorded through the telephone network under "laboratory" and "field conditions". When a recording is made for a laboratory corpus, the speaker repeats the speech material or reads it from a paper. During such a recording, speech intensity is controlled and some information is also collected about the speaker, such as age, sex and geographic location (dialect). Field data corpora are collected from calls to vocal services. Here, the speaker is not limited as to what to say or how to say it.

The laboratory corpora, used in this study, contain digits (between 0 and 9) and numbers (between 00 and 99) in French, as well as 36 French words and expressions (words used in information requests). About 800 speakers were recorded for every "laboratory" corpora used here.

The field corpora used in the following experiments were collected from several thousand calls to three operational vocal services. The first two vocal services ("Baladins_1" and "Baladins_2") recognise 26 keywords and expressions and deliver information about the cinema program in the Trégor area (North West Brittany, France). Both services are based on keyword recognition: the first service ("Baladins_1") is able to recognise only one keyword in the speech signal. Conversely, the second service ("Baladins_2") can detect several keywords allowing users to utter unrestricted sentences. The third service is the "Automatic Dialling Service" in operation in the CNET-Lannion laboratory. The vocabulary of the "CNET Automatic Dialling Service" contains more than 1,300 surnames and more than 500 first names. This service permits a call to be made by simply pronouncing the name of the correspondent.

The corpora (field and laboratory) were hand transcribed in order to indicate keywords, noise tokens and out-of-vocabulary words (words that are not part of the vocabulary to be recognised). They were split into 2 subsets, dedicated to training and testing of the prosodic parameters respectively.

15.2.3 Speech Recognition System

The recognition system used for this study is the CNET system PHIL90. The HMM acoustic modelling units are allophones (context-dependent modelling of phonemes) (Jouvet, Bartkova & Monné, 1991). HMM Gaussian density modelling was carried out using 8 Mel cepstral coefficients and their first and second order time derivatives. Garbage models were trained using out-of-vocabulary words and noise tokens. For all but one of the corpora, the training of the Gaussian densities was carried out on the same vocabulary (vocabulary dependent training). For the "Automatic Dialling" corpus the evaluation was carried out with a vocabulary independent training, that is with Gaussian densities trained on a different vocabulary.

15.2.4 Post-processing Using Prosodic Parameters

An HMM-based decoder provides the best or the N-best hypotheses using spectral information (Mel frequency cepstral coefficients). Post-processing is then carried out using prosodic parameters. Figure 15.1. shows how the different calculation and decision stages were implemented in the recognition system. The aim of post-processing is either to accept or to reject the best HMM hypothesis, or to re-score and reorder the N-best hypotheses. In the latter case, the best reordered hypothesis give the recognised solution (answer).

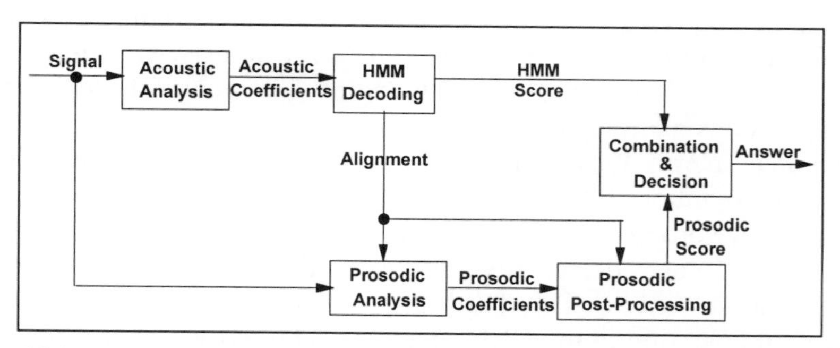

Figure 15.1. Implementation of the calculation and decision modules in the speech recognition system.

Three prosodic parameters are evaluated in this study: phone duration, phone energy and F_0 pattern associated with keywords. The choice of these parameters is motivated by the fact that neither phone duration nor F_0 movement are used in HMM modelling. Their introduction into the post-processing procedure could therefore provide supplementary information for speech recognition. Frame energy is included in the acoustic vector parameters in HMM modelling. However, predicting phone energy according to relevant phonetic conditions is a new approach which may improve speech recognition. These three prosodic parameters allow a prosodic score to be computed that is subsequently used to confirm or to penalise the spectral HMM score of the speech signal. The prosodic score is calculated according to a twofold modelling of the prosodic events: the first score is calculated from a model of prosodic events belonging to words with correct alignment (words correctly recognised by the HMM) and the second score from a model of prosodic events belonging to incorrect alignments (substitution errors or false alarms made by the HMM decoder). The following assumption is made: if the modelled data reflecting the correct alignments are sufficiently different from the data observed on incorrect alignments, then an HMM recognition error could be detected. The prosodic score is obtained by calculating the sum of the logarithm of the likelihood ratio associated with each prosodic parameter (ratio between correct and incorrect event modelling) for the whole keyword or expression.

In the following sections, experiments carried out with the 3 prosodic parameters will be presented. Phone duration is discussed more thoroughly than the other parameters, since it gives the highest recognition performance improvement.

15.3 Phone Duration

In automatic speech recognition, phone duration is difficult to use, since the recognition system often yields a phonetically incorrect segmentation (Cosi, Flalaviga & Omologo, 1991; Boeffard, Cherbonnel, Emerard & White, 1993). A comparison carried out in our laboratory between hand-segmented and HMM-segmented data revealed that the weaknesses of HMM segmentation are mainly related to the statistical distribution of the contexts in which a phone occurs. When a phone occurs in many different contexts in the training set, HMM segmentation is generally precise and phonetically correct. However, if the context representation (in the training set) is poorly balanced, the HMM training is not sufficiently constrained, and the boundaries obtained by HMM segmentation can be phonetically incorrect. A previous study (Bartkova & Jouvet, 1995) showed that duration models trained from hand-segmented data lead to worse recognition results than duration models trained from automatically-segmented data. This confirms the following hypothesis: even if an HMM segmentation is not correct, it may be assumed that the phonemic segmentation is performed in a similar way throughout the whole corpus (the same boundary shifts are observed in the same contexts). This implies that the difference between two phone duration values is smaller when both of them are obtained using the same segmentation technique. As a matter of fact, the models trained on automatically segmented data reflect the system segmentation drifts while the models trained on hand-segmented data do not.

Two phone duration models were evaluated in this study: word-independent modelling based on explicit phonetic rules, and word-dependent modelling using implicit phonetic rules.

15.3.1 Rule-based Modelling

Explicit phonetic rules, often described in the literature for text-to-speech synthesis (Klatt, 1979; O'Shaughnessy, 1984; Bartkova & Sorin, 1987; Campbell and Isard, 1991) are used to set up the duration predicting model. They are related to the acoustic characteristics of the phones (Delattre, 1966), their left and right contexts (Di Cristo, 1978), the length (in syllables) of the word in which the they occurs and the place of their host syllable in the word (Fletcher, 1991). In our approach, vowels are clustered into three main groups to distinguish between oral and nasal French vowels (oral vowel are shorter than nasal ones) and the schwa vowel. The use of the schwa group is motivated by the fact that in French the schwa may not be pronounced, or may appear even if it is not present in the standard transcription: for example, it can be pronounced at the end of each word ending with a consonant, especially when it is followed by a pause. This latency feature influences the duration of the schwa, which may be much shorter than the other vowels. Unlike the vowel group, the consonant group is less homogeneous, therefore 7 sub-groups are created so that the consonants integrated into each group have the same "duration characteristics".

It was observed that despite the automatic segmentation yielded by HMM and the possibility of errors during alignment, the mean duration values obtained for the different phoneme groups are phonetically correct; the possible segmentation errors are compensated for by the large amount of data.

One of the peculiarities of the French language is the phonological lengthening of the last syllable of a prosodic group, which is considered to be the stressed syllable (lexical stress does not exist in French (Fónagy, 1980)). The final syllable of a prosodic group is always much longer than the same syllable in a non-final position. A distinction is therefore made by the model between final and non-final positions.

A longer duration, observed in some studies, in the first syllable of the prosodic group (Vaissière, 1977) is not sufficiently significant to be pertinent for modelling. Moreover, this syllable length characteristic appears to be speaker dependent: some speakers do the opposite, they shorten the first syllable instead of lengthening it.

Two procedures were implemented to handle rule-based phone duration modelling: the first was based on Gaussian density modelling and the second on discrete density modelling.

15.3.1.1 Gaussian Density Modelling

This approach is carried out in two stages. First, the duration of every phone of the alignment is predicted using their mean values and standard deviations according to relevant phonetic rules. In order to take the speech rate variation into account (some speakers talk faster than others), two coefficients minimising the prediction error (difference between predicted duration of the phones and actual duration of the segments for a given alignment) are calculated, one for the consonants and the other for the vowels. These two speech rate coefficients aim to adjust the phone duration prediction according to the speech rate of a given word. In the second stage, the prediction error is modelled for every phone using Gaussian densities. One set of densities is associated with the correct events (correct alignments) and a second one with incorrect events (resulting from substitutions and false alarms). The density parameters are estimated on the training database. The goal of the prosodic score is to decide which set of models (correct event models vs. incorrect event models) best reflects the duration values of the word being processed.

Figure 15.2 shows standard deviation values (of the difference between predicted and actual phone duration) obtained for phones in correct and in incorrect alignments for the corpus of French numbers. The standard deviation values are systematically lower for correct alignments than for incorrect ones. This means that the duration values of the phones along an incorrect alignment are not consistent with the predicted values, hence the assumption that this parameter enables incorrect alignments to be rejected.

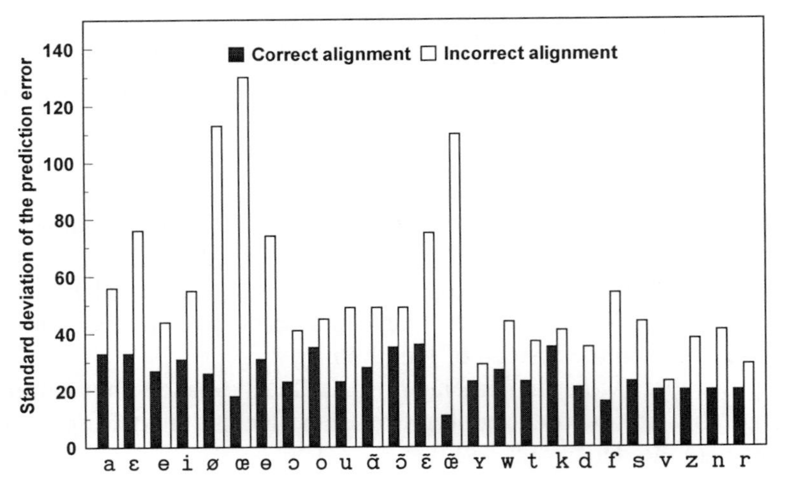

Figure 15.2. Standard deviation of the prediction errors for correct and incorrect alignments on French numbers.

Unfortunately, this approach is not relevant for alignments containing only one consonant and one vowel. For such words the prediction error is always 0 since the adjustment carried out by the two speech rate coefficients minimises the prediction error. In order to overcome this drawback, speech rate coefficient values are also modelled. It is believed that the speech rate coefficient values obtained in correct alignments are sufficiently different from those obtained for incorrect alignments. It should therefore be possible to use them as a pertinent parameter in the case of short words. The diagram in Figure 15.3. illustrates the histogram values of the speech rate coefficient for the vowels calculated for the corpus of French digits. It shows a clear difference in coefficient values between the correctly aligned words and the incorrectly aligned words.

It was observed (Lokbani *et al.,* 1993) that by maintaining several (N-best) word candidates the probability that the correct one could be found among them was significantly increased (the optimal number of candidates is about 5 for small vocabularies). The rule-based duration model is evaluated using laboratory corpora for re-scoring the 5-best HMM solutions. The prosodic score is recombined with the HMM score. A new ranking of the 5 candidates is performed, taking the combined HMM and prosodic scores as a basis, and the top ranking solution then provides the answer. Depending on the different corpora, the reduction of the recognition error rate ranges from 5% to 8% when only the duration score was used and from 8% to 17% when other post-processing parameters were added to the duration score (Bartkova & Jouvet, 1995). Although the gain is always positive, the number of recovered errors was not high, since the HMM substitution error rate was low (from 0.5% to 3.1% for the corpora studied here).

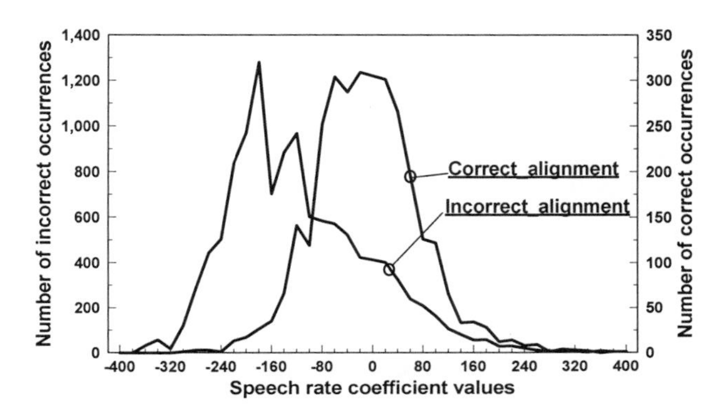

Figure 15.3. Speech rate coefficient values measured on correctly and incorrectly aligned French digits.

15.3.1.2 Discrete Density Modelling

The second duration-based approach, is carried out using discrete duration modelling. Here again two sets of discrete densities are used to model the normalised duration of the phones: one set for correct alignments and the other for incorrect ones. The phone duration is normalised in this case according to word duration. Discrete densities are defined for classes of phones according to the phonetic rules specified above and according to the word duration in syllables. The densities are estimated by computing histograms over correct and incorrect alignments. This approach is evaluated on field data from the "CNET Automatic Dialling" corpus. The discrete duration model is used to improve HMM rejection of out-of-vocabulary data such as non-keywords or noise tokens. It also helps to reduce the substitution rate by transforming a recognition error into a false rejection (from the users' point of view a rejection is considered as a minor error).

Figure 15.4 represents examples of prosodic duration score histograms calculated for two error categories: errors due to substitutions and errors due to false alarms on noise tokens. According to the computed histograms, it appears that it is easier to differentiate a vocabulary word from a noise token than to differentiate it from a substitution error. The difference between the score values obtained for the correctly recognised words and those obtained for the set of noise tokens is clear. However, the difference between the score values obtained for correctly recognised vocabulary words and those obtained for keyword substitutions is less obvious.

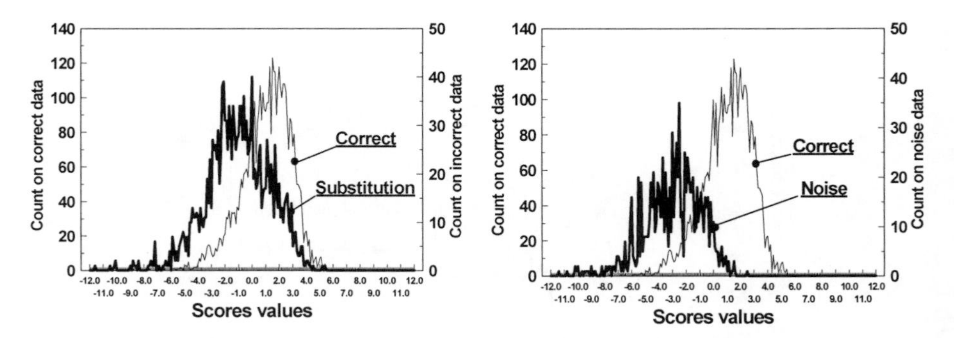

Figure 15.4. Histograms of duration scores: comparison between correctly recognised words and substitution errors (left) and false alarms on noise tokens (right), for the "Automatic Dialling Corpus".

The prosodic score is compared to a previously chosen threshold, which is varied during the test. When the keyword score is below the threshold, the sentence is rejected, otherwise it is considered as being correct.

The discrete duration model is particularly efficient for the "CNET Automatic Dialling" corpus. By applying slightly different rejection threshold values on the duration post-processing score, depending on the length of the word, numerous false alarms are eliminated from out-of-vocabulary data. Figure 15.5 represents the application of the duration post-processing starting from three rejection threshold values used with HMM modelling. It illustrates the reduction in the false alarms on noise tokens and out-of-vocabulary words as a function of the false rejection rate. The HMM rejection threshold «threshold_1» leads to the lowest rejection rate (but the highest false alarm rate), whereas «threshold_3» leads to a high rejection of noise tokens and out-of-vocabulary words (i.e. small false alarm rates) but at the same time it also leads to a much higher false rejection rate. It is to be noted that, after applying the post-processing procedure, the substitution errors and the correctly recognised utterances, rejected by the duration post-processing, are counted as false rejections. Hence the false rejection rate is the sum of the HMM false rejection rate and the duration model false rejection rate.

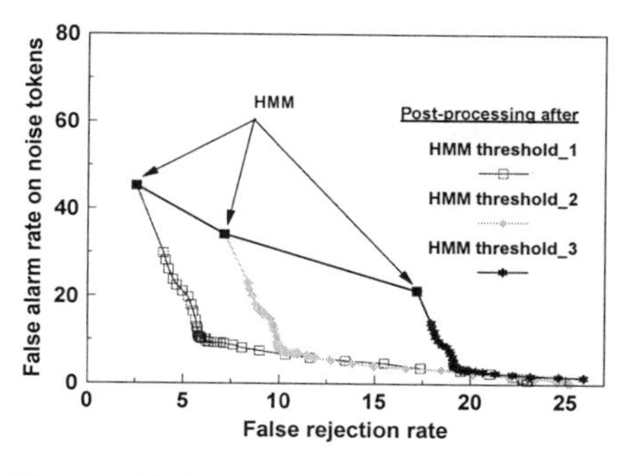

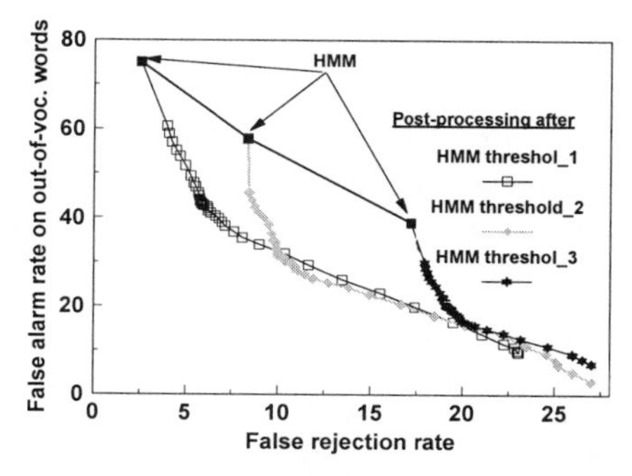

Figure 15.5. Rejection rate of the noise tokens and out-of-vocabulary words when applying the discrete phone duration model on the "Automatic Dialling Corpus". Duration post-processing is carried out, starting from 3 different trade-offs between false alarm rate and false rejection rate resulting from 3 choices of the HMM rejection threshold.

The recognition performance improvement achieved by duration modelling is very positive on "CNET Automatic Dialling" corpus. For example, for a 10% to 20% false rejection rate (which can be considered as acceptable for this corpus) about 80% of the false alarms for noise tokens and about 50% of the false alarms for out-of-vocabulary words are rejected. The reduction of the substitution error rate is not as spectacular as that of false acceptances on noise tokens or out-of-vocabulary words: only a 20% reduction in the substitution error rate is achieved under the previously described conditions.

15.3.2 Word-dependent Modelling

In this approach no explicit rules were used since modelling was carried out for each phone according to its specific position in each given word. All phone duration variations were therefore captured in a phonetically consistent way. Two models, associated with an identical phone in the same word, occurring in the same context but in different positions, were not merged together: they were maintained as different models. In the word-dependent model, discrete phone duration modelling was carried out. As for the other approaches, two sets of models were estimated, one on correct alignments and the other on incorrect alignments.

Figure 15.6 illustrates the reduction of false alarms when duration is used in the post-processing procedure on "Baladins_1" field data (the energy feature will be discussed in section 15.5.). The duration rejection performance was compared with results obtained by HMM alone as the value of the HMM rejection threshold varied. As Figure 15.6 illustrates, noise token rejection is significantly improved when duration prediction is used. For a false rejection rate of about 10%, HMM by itself accepts 6.3% of noise tokens. After the duration post-processing, only 1.4% of noise tokens are still accepted (yielding a 77% reduction in the false alarm rate, see arrow in Figure 15.6). Unfortunately, in this corpus, the same approach does not improve out-of-vocabulary word rejection.

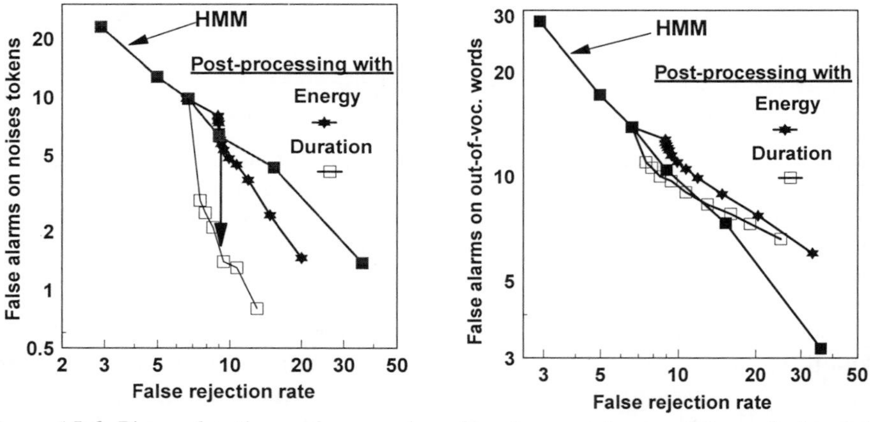

Figure 15.6. Phone duration and energy-based post processing used for reducing false alarm rate on noise tokens and on out-of-vocabulary words on "Baladins_1" corpus.

To summarise, the results achieved by a duration model are either similar to those of HMM by itself (out-of-vocabulary word rejection), or significantly better (noise token rejection). This approach is therefore considered as sufficiently efficient to be used in the post-processing procedure to double-check the first HMM hypothesis.

Although duration modelling is better adapted to the rejection of non-vocabulary tokens, it was also tested on "French numbers" laboratory data too. Since the only laboratory data errors stem from the substitution of vocabulary words, the duration model was tested on the rejection of these errors. Figure 15.7 shows that the duration model can also efficiently reject substitution errors. For a false rejection rate of 2.2% an error reduction of 32% is achieved with duration-based post-processing (the energy feature will be discussed in section 15.5). Assuming that a false rejection rate of 5% is acceptable for this particular application, then the error rate can be reduced by about 50% (1.5% instead of 3.1% using only HMM decoding).

In the approach presented here, a keyword occurrence having an unexpected F_0 pattern, that is a pattern that was not encountered in the training database, should have a poor prosodic score. However, if a recognised word is aligned at the end of a final or non-final prosodic group, its F_0 pattern can hardly be contested, even in the case of a recognition error.

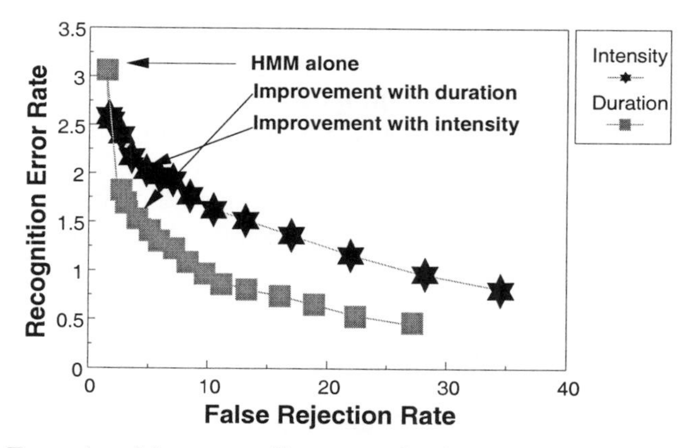

Figure 15.7. Example of incorrect alignment of a keyword, having an untypical F_0 pattern.

15.3.3 Discussion

One of the difficulties encountered in phone duration modelling is that the duration yielded by automatic segmentation is not always phonetically correct. Thus, in some cases, a discrepancy may exist between model-predicted duration and measured duration.

In general, the duration parameter provides better results in noise token rejection tasks than in word token (false alarms on out-of-vocabulary words and substitution errors) rejection tasks. This worse performance can be explained by the fact that a spectrally incorrect word candidate can be perfectly correct as far as its phone duration is concerned.

The word-dependent model predicts duration for every phone of each word; since all word related contexts are taken into account for each particular phone, HMM segmentation defaults are overcome. The drawback of a word dependent approach is the constraint induced by vocabulary size: such a method can only be applied to small vocabularies (about 100 words) and cannot predict phone duration values for new words (in flexible tasks). However, when the vocabulary to be recognised is not very large, a word-dependent model can be more appropriate. Moreover, no explicit phonetic rules are necessary in such a model, since the phonetic events are implicitly captured with great precision.

The advantage of a word-independent model is that it can predict the duration of any new word to be recognised. It is therefore well adapted to flexible recognition tasks where the recognition vocabulary is independent of the training vocabulary. A word-independent model is also required when the vocabulary to be recognised increases. The drawback of this model is that it cannot deal with all the different phone contexts, therefore poor HMM segmentation is not always compensated for.

The rule-based discrete duration model yields very good results on the "Automatic CNET Dialling" corpus, especially in rejecting false alarms on noise tokens. However, as the results for substitution error rejection are less conclusive, new approaches are currently being studied in order to combine the duration parameter with other parameters more appropriate to this specific rejection task.

15.4 Fundamental Frequency Pattern

The main characteristic of the fundamental frequency (F_0) is how it evolves with time. The F_0 movement structures the utterance at a syntactic, semantic and pragmatic level (Selkirk, 1986; Nęspor & Vogel, 1986). It can help to break up the speech chain into smaller parts, i.e. prosodic groups (Lehiste, 1970). The F_0 pattern characterises the whole prosodic group while the duration and the energy characterise the speech chain at the syllable or phone level.

Anybody working in speech signal processing will acknowledge that automatic detection of the fundamental frequency of the speech signal is not always satisfactory for some voices or under some noisy conditions (Rouat, Liu & Morissette, 1997). Even if the F_0 is correctly determined, its automatic use in keyword spotting task is still difficult since this recognition approach limits the hypothesis which can be made about the utterance prosodic structures.

15.4.1 F_0 Movement Modelling

Since in French lexical stress does not exist, the F_0 pattern essentially reflects the utterance prosodic structure. F_0 use in isolated word corpora is therefore irrelevant in French. For this reason, F_0 modelling is evaluated on a continuous speech corpus. The post-processing is applied only on the hypotheses provided by the keyword-spotting recogniser. The continuous speech corpus "Baladins_2" was collected from the cinema's voice activated program inquiry service. A hand analysis of the corpus, collected from this service, revealed that system keywords were generally perceived by users as important words (they also became user keywords). Keywords often occur in a stressed position, on prosodic boundaries, having a characteristic upward F_0 movement in a non-final utterance position.

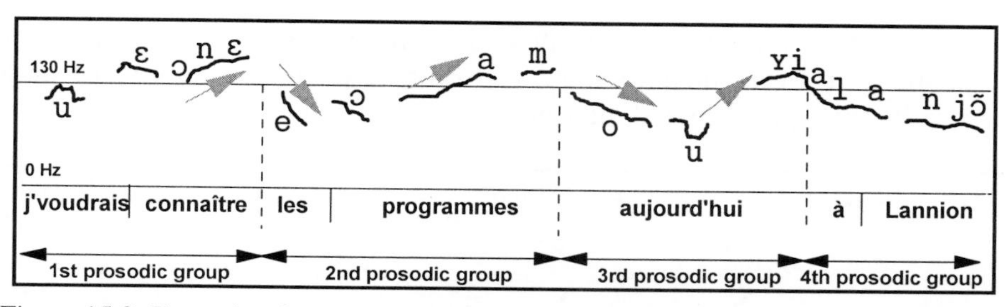

Figure 15.8. Example of a correct alignment of 3 keywords ("programme" (program), "aujourd'hui" (today) and "Lannion" (a city name)) placed on prosodic boundaries.

In one of the requests addressed to the system by a user: "Je voudrais les programmes aujourd'hui à Lannion" (I'd like to know today's program in Lannion) the system keywords are "program", "Lannion" and "today". Figure 15.8 illustrates the F_0 pattern of this particular utterance; the keywords are in a stressed position, on prosodic boundaries situated at the end of the prosodic groups. Automatic F_0 slope modelling is carried out using the vowels, semi-vowels, liquids and nasals belonging to the two last syllables on correctly and incorrectly recognised words. The slope of the whole F_0 pattern, with which the keyword is aligned, is also taken into account. Thus, a keyword, aligned to a F_0 pattern which continue to rise after its last syllable, can be penalised. It is believed that if a recognised keyword, generally in a stressed position and expected with a characteristic upward F_0 movement, has an "untypical" F_0 pattern, it probably corresponds to a wrong answer and should be rejected.

In the approach presented here, a keyword occurrence having an unexpected F_0 pattern, that is a pattern that was not encountered in the training database, should have a poor prosodic score. However, if a recognised word is aligned at the end of a final or non-final prosodic group, its F_0 pattern can hardly be contested, even in the case of a recognition error.

Figure 15.9 shows how a recognition error can be penalised when F_0 pattern modelling is used. A keyword HMM model "aujourd'hui" (today) is incorrectly aligned with the speech signal containing "Bonjour" (Hello) and "j'voudrais" (I'd like to). The last syllable [dYi] of the keyword matches a relatively low F_0 value (arrow 1 on the Figure 15.9), and the F_0 continues to rise (arrow 2 on the Figure 15.9). The keyword is expected to be in a stressed position (on a prosodic group boundary) while it is not. In this particular case, the F_0 slope prediction helps to reject a bad keyword candidate.

F_0 pattern application was evaluated in the N-best re-scoring task but error recovery was insignificant. Unfortunately, new errors were also caused by this approach and the final error rate was not improved.

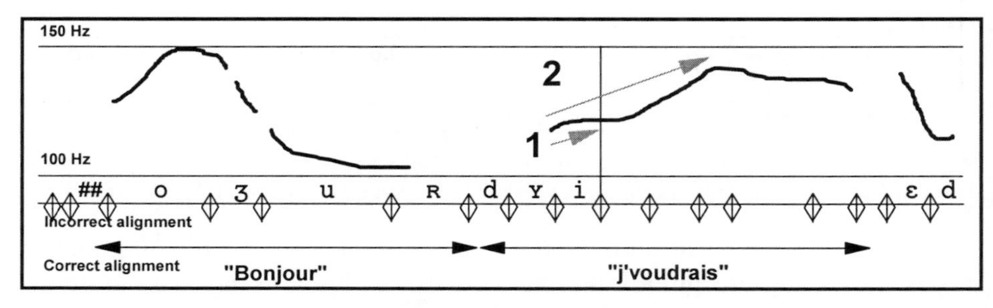

Figure 15.9. Example of incorrect alignment of a keyword, having an untypical F_0 pattern.

15.4.2 Discussion

The rather negative results obtained in the above experiments do not mean that F_0 has a minor role in speech recognition. On the contrary, it must be acknowledged that from a linguistic point of view, use of F_0 movement can be efficient in continuous speech recognition tasks, such as text dictation and speech understanding applications. For example, F_0 movement can help a language model to structure the different words into syntactic units; it can also detect the speaker's attitude to the meaning of the sentence by highlighting prosodic words, hence detecting the semantic focus. Another very useful application is to model the F_0 contour in utterance final position where a characteristic F_0 pattern can differentiate between a question and a statement (Batliner, Weiand, Kiesling & Nöth, 1993; Daly & Zue, 1995).

The use of F_0 in the keyword spotting approach is greatly limited, since F_0 patterns reflect the structuring of the meaning of the sentence and, in a keyword spotting approach, a part of the prosodic group is rejected, hence not recognised. This accounts for the difficulty of integrating F_0 information into such an approach.

Some recognition tests showed that even in a keyword recognition application, a F_0 related information (the phone voicing degree) could still be used. Several experiments were carried out using the voicing degree of the phones, leading to a significant improvement of the rejection of non-keywords (Bartkova & Jouvet, 1997).

15.5 Phone Energy

The third prosodic parameter, often considered as being of the least importance in phonetic studies and automatic speech processing (Di Cristo, 1978), is the phone energy. As far as phone energy use is concerned, one of the main issues is the normalisation of its value. As a matter of fact, it is the most fragile of the three prosodic parameters; it is subject to variations (one can speak softly or loudly without really altering the meaning of the sentence) and it is easily corrupted by a noisy environment (Sorin, 1989).

Although the energy of the speech signal is taken into account in HMM decoding (at the frame level), as it is one of the HMM modelling parameters, it might be interesting to predict phone energy by using a model based on phonetic rules. The phone energy information is evaluated to reject substitution errors and false alarms on out-of-vocabulary words and noise tokens. Since the energy of the speech signal may vary greatly, normalisation is necessary. Several normalisation approaches are tested: normalisation by the energy of the whole word, by the mean energy of all the vowels in the word, by the highest energy of the vowels in the word or by the energy measured in the middle part (three middle frames) of the vowels in the word. The drawback of vowel energy normalisation is that in monosyllabic words the normalised vowel energy is equal to 0 dB. In such words, consonants remain the only elements to provide information about correct or incorrect energy values.

In phone energy modelling, two approaches are tested: the first is based on word-dependency and the second one on phonetic rules. Both approaches are carried out using discrete modelling. The discrete models are estimated by computing histograms over the normalised phone energy measured on correct and incorrect alignments. The two approaches are tested in post-processing rejection procedures on laboratory data (French numbers) and on two field corpora ("Baladins_1" and "CNET Automatic Dialling corpus"). The test results discussed here are obtained after normalising the phone energy by the mean vowel energy.

15.5.1 Word-dependent Modelling

In word-dependent modelling, no explicit phonetic rules are used since the phonetically relevant influences are already captured by the model.

Figure 15.6 illustrates the reduction in the false alarm rate when energy is used in the post-processing procedure on "Baladins_1" field data. The results from the rejection procedures are compared with those obtained by HMM used alone when the rejection threshold value varies. It appears that sound energy post-processing also reduces HMM false acceptance of noise tokens, however the reduction is smaller than that provided by the duration parameter.

Although phone energy is less efficient than phone duration, the phone energy model yields encouraging results both on field (Figure 15.6) and laboratory data (Figure 15.7).

15.5.2 Rule-based Modelling

In the rule-based modelling the phones are clustered according to their relevant characteristics related to their inherent energy (4 vowel groups and 7 consonant groups are created) and their position in the sentence (Di Cristo, 1978; Barkova, Haffner & Larreur, 1993). For each phoneme group a distinction is made between voiced and voiceless left and right contexts.

Figure 15.10 compares phone duration and phone energy post-processing results for two different tests carried out with two rejection threshold values. These thresholds are chosen in order to achieve a 10% false rejection rate in the first test (left part of Figure 15.10) and a 20% false rejection rate in the second test (right part of Figure 15.10). Although the duration model works better in every test, the most significant differences between the two parameters are found in the noise token rejection task.

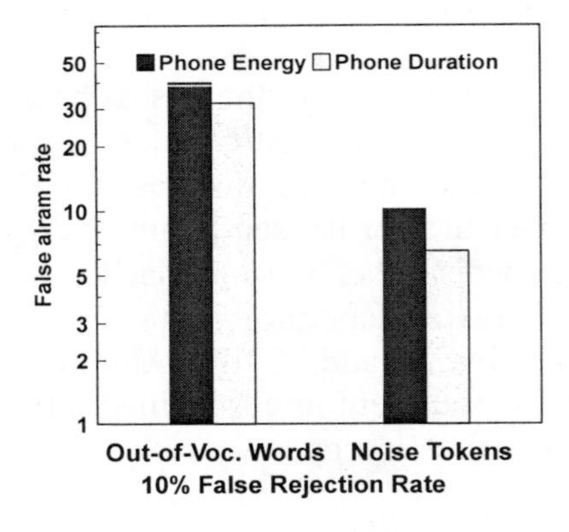

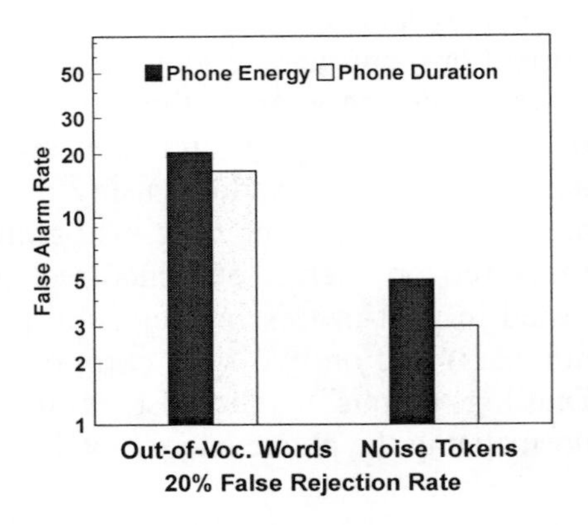

Figure 15.10. Comparison between results obtained with phone energy and phone duration post-processing on the "CNET Automatic Dialling" corpus for 10% (left) and 20% (right) false rejection rates.

15.5.3 Discussion

In the experiments carried out in this paper, phone energy proved to be less efficient than phone duration for rejecting out-of-vocabulary data. This is probably due to its great variability and also to the fact that speech energy values (at the frame level) are integrated into HMM modelling in two forms: firstly, as an absolute value and secondly as first and second derivative parameters. It is perhaps for this reason that the use of energy in the post-processing procedure, even under different conditions, is not as efficient as the use of duration, which is not explicitly taken into account in HMM modelling.

15.6 Summary and Conclusion

Nowadays, many general public speech recognition applications work over the telephone network and rely on limited vocabulary keyword recognition. The present study showed the possibilities and the limitations encountered using prosodic parameters in speech recognition systems for keyword spotting tasks. One of the major drawbacks of using prosodic parameters in such tasks is that it is not always possible to access syntactic information since only system keywords are recognised and the rest of the speech signal is rejected.

Three prosodic parameters were evaluated in this study: phone duration, phone energy and F_0 patterns associated to keywords. The parameter efficiency was tested in two tasks. The first task was the N-best keyword re-scoring task. The second task was the out-of-vocabulary data rejection since one of the major problems of recognition systems working in real applications remains the too high acceptance of out-of-vocabulary tokens.

The duration prediction was firstly evaluated in a re-scoring task where the N-best solutions yielded by HMM had to be re-arranged, by adding an extra duration score to the HMM score. The result achieved by this duration post-processing was only slightly positive. Secondly, the duration application had to reject out-of-vocabulary tokens (noises and out-of-vocabulary words) thus improving the system performance. In this task, the duration parameter proved to be very efficient since there was a 80% reduction in false alarms on noise tokens and a 50% reduction in false alarms on out-of-vocabulary words on the “CNET Vocal Dialling” corpus for a 10% to 20% false rejection rate.

The F_0 pattern models were evaluated in the N-best solution re-scoring task. This parameter does not improve the HMM recognition performance. Although the F_0 pattern is difficult to use in a keyword spotting task, its important role in continuous speech recognition cannot be neglected. In French speech recognition, F_0 is essential in helping to detect word or sentence boundaries.

Results obtained by phone energy are positive and very encouraging though not as good as those obtained using sound duration prediction.

Phone energy is extremely variable, hence further studies are necessary to test new normalisation procedures. The smaller improvement achieved by phone energy modelling is probably due to the fact that frame energy is already one of the HMM parameters.

Researchers working in speech technology are more and more interested in applying linguistic knowledge to constrain speech recognition systems. It has become clear that statistic modelling alone cannot completely improve system performance. The keyword-spotting task is quite an unnatural way of using prosodic information since the part of the speech that does not contain the predefined vocabulary words cannot be recognised (is captured by garbage models). Nevertheless, this study showed that although prosodic parameters were used in a simple manner, they can successfully improve recognition systems.

Results presented here proved that all the parameters are not equally useful. The sound duration parameter is more efficient than phone energy or F_0 patterns of the keywords. However the aim of this paper was to report the different experiments which have been carried out in the CNET-Lannion laboratory and not to describe only the best parameter.

It has been currently found in our laboratory that the efficiency of the different parameters can depend on the corpus. Therefore, one of the important issues that still has to be dealt with, is how to combine the various prosodic parameters in order to obtain the best post-processing procedure. A preliminary analysis of recognition errors suggests that such a combination has probably to be phoneme and position dependent, which means that rather than using theses parameters systematically, they should be applied only in cases where they are efficient.

In the present paper, the results obtained on the HMM error recovery, after re-scoring the N-best hypothesis, were less conclusive than those obtained on the out-of-vocabulary data rejection. However, the error recovery is still a challenging issue, especially when words are difficult to differentiate by means of spectral information only such as the French numbers 80 [katrəvɛ̃] and 81 [katrəvɛ̃ɛ̃]. In fact, when there is no pause between the last two vowels in number 81, the only feature which distinguishes these two numbers (80 and 81) is the vowel length (much longer in 81 than in 80). Therefore, in this case and in similar cases, the use of the duration parameter would increase recognition efficiency.

Acknowledgements

I wish to thank Björn Granström, Gérard Bailly, Jean Véronis, Mario Rossi and Denis Jouvet, whose helpful comments have improved the presentation of this paper.

References

André-Obrecht, R. 1990. Reconnaissance automatique de parole à partir de segments acoustiques et de modèles de Markov cachés. *18ième Journées d'Etudes sur la Parole* (Montreal, Canada), 212-216.

Bartkova, K., P. Haffner and D. Larreur. 1993. Intensity prediction for speech synthesis in French. *Proc. ESCA Workshop on Prosody* (Lund, Sweden), 280-283.

Bartkova, K. and D. Jouvet. 1995. Using segmental duration prediction for rescoring the N-best solution in speech recognition. *Proc. 13th ICPhS* (Stockholm, Sweden), vol. 4, 248-51.

Bartkova, K. and D. Jouvet. 1997. Usefulness of phonetic parameters in a rejection procedure of an HMM based speech recognition system. *Proc. EUROSPEECH* '97 (Rhodes, Greece), 267-70.

Bartkova, K. and C. Sorin. 1987. A model of segmental duration for speech synthesis in French. *Speech Communication* 6, 245-260.

Batliner, A., C. Weiand, A. Kiesling and E. Nöth. 1993. Why sentence modality in spontaneous speech is more difficult to classify and why this fact is not too bad for prosody. *Proc. ESCA Workshop on Prosody* (Lund, Sweden), 112-115.

Boeffard, O., B. Cherbonnel, F. Emerard and S. White. 1993. Automatic segmentation and quality evaluation of speech unit inventories for concatenation-based, multilingual PSOLA text-to-speech systems. *Proc. EUROSPEECH* '93 (Berlin, Germany), 1449-52.

Botte, M.C., G. Canévet, L. Demany and C. Sorin. 1989. *Psychoacoustique et Perception Auditive*. INSERM/SFA/CNET.

Bush, M.A. and G.E. Kopec. 1987. Network-based connected digit recognition. *IEEE Transactions on Acoustics, Speech and Signal Processing,* ASSP 35, 1401-13.

Campbell W.N. and S.D. Isard. 1991. Segment duration in a syllable frame. *Journal of Phonetics* 19, 37-47.

Carbonell, N., J.P. Haton, F. Lonchamp and J.M. Pierrel. 1982. Indices prosodiques pour l'analyse syntaxico-sémantique dans Myrtille II. *Actes du Séminaire Prosodie et Reconnaissance* (Aix-en-Provence, France), 59-61.

Carbonell N. and Y. Laprie. 1993. Automatic detection of prosodic cues for segmenting continuous speech into surpralexical units. *Proc. ESCA Workshop on Prosody* (Lund, Sweden), 184-87.

Chung G. and S. Seneff. 1997. Hierarchical duration modelling for speech recognition using the Angie framework. *Proc. EUROSPEECH* '97 (Rhodes, Greece), 1476-78.

Cosi, P., D. Flalaviga and M. Omologo. 1991. A preliminary statistical evaluation of manual and automatic segmentation discrepancies. *Proc. EUROSPEECH* '91 (Genova, Italy), 693-96.

Daly, N. and V. Zue. 1995. Acoustic, perceptual and linguistic analyses of intonation contours in human /machine dialogues. *Proc. ICSLP* '95, 497-500.

Delattre, P. 1966. *Studies in French and Comparative Phonetics*. London: Mouton.

Di Cristo, A. 1978. *De la Mircroprosodie à l'Intonosyntaxe*. Thèse d'Etat, Université de Provence.

Emerard, F. 1977. Les diphones et le traitement de la prosodie dans la synthèse de la parole. *Bulletin de l'Institut Phonétique de Grenoble*, vol. VI, 103-47.

Fletcher J. 1991. Rhythm and final lengthening in French. *Journal of Phonetics* 19, 193-212.

Fónagy, I. 1980. L'accent français, accent probabilitaire (dynamique d'un changement prosodique). *Studia Phonetica* 15, Montreal.

Fónagy, I. and K. Magdics. 1960. Speech of utterance in phrases of different lengths. *Language and Speech* 3, 179-192.

Gong, Y. and C.W. Treurnier. 1993. Duration of phones as function of utterance length and its use in automatic speech recognition. *Proc. EUROSPEECH* '93 (Berlin, Germany), 315-18.

Gupta, V., M. Lenning and P. Mermelstein. 1992. Use of minimum duration and energy contour for phonemes to improve large vocabulary isolated-word recognition. *Computer Speech and Language* 6, 331-344.

Jouvet, D., K. Bartkova and J. Monné. 1991. On the modelisation of allophones in an HMM based speech recognition system. *Proc. EUROSPEECH* '91 (Genova, Italy), 923-26.

Klatt, D.H. 1979. Synthesis by rule of segmental durations in English sentences. In Lindblom and Öhman (eds.), 287-300.

Langlais P. 1995. *Traitement de la Prosodie en Reconnaissance Atomatique de la Parole*. Thèse de l'Université d'Avignon.

Lehiste, I. 1970. *Suprasegmentals*. Cambridge, Mass.: MIT Press.

Lindblom, B. and S.E.G. Öhman (eds.). 1979. *Frontiers of Speech Communication Research*. New York: Academic Press.

Lindblom, B. and K. Rapp. 1973. Some temporal regularities of spoken Swedish. *Symposium on Auditory Analysis and Perception of Speech* (Leningrad, USSR), 21-23.

Ljolje, A. and F. Fallside. 1987. Modelling of speech using primarily prosodic parameters. *Computer Speech and Language* 2, 185-204.

Lokbani, M.N. , D. Jouvet and J. Monné. 1993. Segmental post-processing of the N-best solutions in a speech recognition system. *Proc. EUROSPEECH* '93 (Berlin, Germany), 811-14.

McNeilage, P. and J.L. De Clerk. 1968. Cinefluorographic study of speaking rate. *Proc. 76th Meeting of the Acoustic Society of America* (Cleveland, USA), 19-22.

Mercier, G., D. Bigorgne, L. Miclet, L. Le Guennec and M. Querré. 1990. Recognition of speaker-dependent continuous speech with KEAL. In Waibel and Lee (eds.), 225-234.

Morin D. 1991. Influence of field data in HMM training for a vocal server. *Proc. EUROSPEECH* '91 (Genova, Italy), 735-38.

Nespor, M. and I. Vogel. 1986. *Prosodic Phonology*. Dordrecht: Foris

O'Shaughnessy, D. 1984. A multispeaker analysis of durations in read French paragraphs. *J. Acoust. Soc. Am.* 76, 1664-72.

Ostendorf, M.F. and K. Ross. 1997. A Multi-level model for intonation labels. In Sagisaka *et al.* (eds.), 291-308.

Ostendorf, M.F., P.J. Price, J. Bear and C.W. Wightman. 1990. The use of relative duration in syntactic disambiguation. *Proc. DARPA Speech and Natural Language Workshop* (Hidden Valley, USA).

Ostendorf, M.F., C.W. Wightman and N.M. Veilleux. 1993. Parse scoring with prosodic information : an analysis/synthesis approach. *Computer Speech and Language* 7, 193-210.

Pols, L.C.W., X. Wang and L.F.M. ten Bosch. 1996. Modelling of phone duration (using the TIMIT database) and its potential benefit for ASR. *Speech Communication* 19, 161-176.

Rose, R.C. and D.B. Paul. 1990. A Hidden Markov model based keyword recognition system. *Proc. IEEE International Conference on Acoustic, Speech and Speech Processing* (Albuquerque, USA), 129-132.

Rouat, J., Ch.Y. Liu and D. Morissette. 1997. A pitch determination and voiced/unvoiced decision algorithm for noisy speech. *Speech Communication* 2, 191-207.

Sagisaka, Y., N. Campbell and N. Higuchi (eds.). 1997. *Computing Prosody*. New York: Springer-Verlag.

Selkirk, E.O. 1986. On derived domains in sentence prosody. *Phonology Yearbook* 3, 371-405.

Soong, F.K. 1998. A phonetically labelled acoustic segment (PLAS) approach to speech analysis-synthesis. *Proc. IEEE International Conference on Acoustics, Speech and Signal Processing* (Glasgow, UK), 584-587.

Sorin, C. 1989. Perception de la parole continue. In Botte *et al.* (eds.), 123-139.

Sorin, C., D. Jouvet, C. Gagnoulet, D., Dubois, D., Sadek and M. Toularhoat. 1995. Operational and experimental French telecommunication services using CNET speech recognition and text-to-speech synthesis. *Speech Communication* 17, 273-286.

Suaudeau, N. and R. André-Obrecht. 1993. Sound duration modelling and time-variable speaking rate in a speech recognition system. *Proc. EUROSPEECH* '93 (Berlin, Germany), 307-310.

Vaissière, J. 1977. Premiers essais d'utilisation de la durée pour la segmentation en mots dans un système de reconnaissance. *8ème Journée d'Etudes sur la Parole* (Aix-en-Provence, France), 345-352.

Waibel, A. 1987. *Prosody and Speech Recognition*. London: Pitman.

Waibel, A. and K.F. Lee (eds.). 1990. *Readings in Speech Recognition*. San Mateo, CA: Morgan Kaufmann Publishers.

Wang, M.Q. and J. Hirschberg. 1992. Automatic classification of intonation phrase boundaries. *Speech Computer and Language* 6, 175-96.

Wang, X. 1997. *Incorporating knowledge on segmental duration in HMM-based continuous speech recognition*. IFOTT, Amsterdam, The Netherlands.

Index of Names

K

L

M

R

S

T

U

Subject Index

O

P

Q

R

U

V

W

Y

Text, Speech and Language Technology

1. H. Bunt and M. Tomita (eds.): *Recent Advances in Parsing Technology.* 1996 ISBN 0-7923-4152-X
2. S. Young and G. Bloothooft (eds.): *Corpus-Based Methods in Language and Speech Processing.* 1997 ISBN 0-7923-4463-4
3. T. Dutoit: *An Introduction to Text-to-Speech Synthesis.* 1997 ISBN 0-7923-4498-7
4. L. Lebart, A. Salem and L. Berry: *Exploring Textual Data.* 1998 ISBN 0-7923-4840-0
5. J. Carson-Berndsen, *Time Map Phonology.* 1998 ISBN 0-7923-4883-4
6. P. Saint-Dizier (ed.): *Predicative Forms in Natural Language and in Lexical Knowledge Bases.* 1999 ISBN 0-7923-5499-0
7. T. Strzalkowski (ed.): *Natural Language Information Retrieval.* 1999 ISBN 0-7923-5685-3
8. J. Harrington and S. Cassiday: *Techniques in Speech Acoustics.* 1999 ISBN 0-7923-5731-0
9. H. van Halteren (ed.): *Syntactic Wordclass Tagging.* 1999 ISBN 0-7923-5896-1
10. E. Viegas (ed.): *Breadth and Depth of Semantic Lexicons.* 1999 ISBN 0-7923-6039-7
11. S. Armstrong, K. Church, P. Isabelle, S. Nanzi, E. Tzoukermann and D. Yarowsky (eds.): *Natural Language Processing Using Very Large Corpora.* 1999 ISBN 0-7923-6055-9
12. F. Van Eynde and D. Gibbon (eds.): *Lexicon Development for Speech and Language Processing.* 2000 ISBN 0-7923-6368-X; Pb: 07923-6369-8
13. J. Véronis (ed.): *Parallel Text Processing.* Alignment and Use of Translation Corpora. 2000 ISBN 0-7923-6546-1
14. M. Horne (ed.): *Prosody: Theory and Experiment.* Studies Presented to Gösta Bruce. 2000 ISBN 0-7923-6579-8
15. A. Botinis (ed.): *Intonation.* Analysis, Modelling and Technology. 2000 ISBN 0-7923-6605-0

KLUWER ACADEMIC PUBLISHERS – DORDRECHT / BOSTON / LONDON